OPPORTUNITIES IN BIOLOGY

Committee on Research Opportunities in Biology
Board on Biology
Commission on Life Sciences
National Research Council

NATIONAL ACADEMY PRESS
Washington, D.C. 1989

National Academy Press • **2101 Constitution Avenue, N.W.** • **Washington, DC 20418**

NOTICE: The project that is the subject of this report was approved by the Governing Board of the National Research Council, whose members are drawn from the councils of the National Academy of Sciences, the National Academy of Engineering, and the Institute of Medicine. The members of the committee responsible for the report were chosen for their special competences and with regard for appropriate balance.

This report has been reviewed by a group other than the authors according to procedures approved by a Report Review Committee consisting of members of the National Academy of Sciences, the National Academy of Engineering, and the Institute of Medicine.

The National Academy of Sciences is a private, nonprofit, self-perpetuating society of distinguished scholars engaged in scientific and engineering research, dedicated to the furtherance of science and technology and to their use for the general welfare. Upon the authority of the charter granted to it by the Congress in 1863, the Academy has a mandate that requires it to advise the federal government on scientific and technical matters. Dr. Frank Press is president of the National Academy of Sciences.

The National Academy of Engineering was established in 1964, under the charter of the National Academy of Sciences, as a parallel organization of outstanding engineers. It is autonomous in its administration and in the selection of its members, sharing with the National Academy of Sciences the responsibility for advising the federal government. The National Academy of Engineering also sponsors engineering programs aimed at meeting national needs, encourages education and research, and recognizes the superior achievements of engineers. Dr. Robert M. White is president of the National Academy of Engineering.

The Institute of Medicine was established in 1970 by the National Academy of Sciences to secure the services of eminent members of appropriate professions in the examination of policy matters pertaining to the health of the public. The Institute acts under the responsibility given to the National Academy of Sciences by its congressional charter to be an adviser to the federal government and, upon its own initiative, to identify issues of medical care, research, and education. Dr. Samuel O. Thier is president of the Institute of Medicine.

The National Research Council was organized by the National Academy of Sciences in 1916 to associate the broad community of science and technology with the Academy's purposes of furthering knowledge and advising the federal government. Functioning in accordance with general policies determined by the Academy, the Council has become the principal operating agency of both the National Academy of Sciences and the National Academy of Engineering in providing services to the government, the public, and scientific and engineering communities. The Council is administered jointly by both Academies and the Institute of Medicine. Dr. Frank Press and Dr. Robert M. White are chairman and vice chairman, respectively, of the National Research Council.

This Board on Biology report is based upon work supported by the U.S. Department of Agriculture (Grant No. 59-32U4-6-52); Department of Navy, Office of Naval Research (Grant No. N00014-86-G-1007); the National Institutes of Health and the National Science Foundation (Contract No. DCB-8515516); and the National Research Council Fund.

Any opinions, findings, conclusions, or recommendations expressed in this publication are those of the authors and do not necessarily reflect the views of the U.S. Department of Agriculture, Department of Navy, the National Institutes of Health, or the National Science Foundation. The United States Government has a royalty-free license throughout the world on all copyrightable material contained in this publication.

Library of Congress Cataloging-in-Publication Data

Opportunites in biology/Committee on Research Opportunities in
   Biology, Board on Biology, Commission on Life Sciences, National
   Research Council.
       p. cm.
       Includes bibliographic references.
       ISBN 0-309-03927-4
       1. Biology—Research.   I. National Research Council (U.S.).
   Committee on Research Opportunities in Biology.
   QH315.O66   1989   89-13098
   574.072—dc20   CIP

Printed in the United States of America

Cover photograph: Photomicrograph courtesy of D. Lansing Taylor.

COMMITTEE ON RESEARCH OPPORTUNITIES IN BIOLOGY

Panel on Cell Organization

THOMAS POLLARD (*Co-chairman*), Johns Hopkins Medical School, Baltimore, Maryland
BRUCE ALBERTS (*Co-chairman*), University of California, San Francisco
THOMAS DEUEL, Jewish Hospital at Washington University Medical Center, St. Louis, Missouri
LELAND HARTWELL, University of Washington, Seattle
PHILLIP SHARP, Massachusetts Institute of Technology, Cambridge

Panel on Ecology and Ecosystems

SIMON LEVIN (*Chairman*), Cornell University, Ithaca, New York
JAMES EHLERINGER, University of Utah, Salt Lake City
THOMAS EISNER, Cornell University, Ithaca, New York
STEPHEN HUBBELL, University of Iowa, Iowa City
HOLGER JANNASCH, Woods Hole Oceanographic Institution, Woods Hole, Massachusetts
PETER RAVEN, Missouri Botanical Garden, St. Louis
THOMAS SCHOENER, University of California, Davis
PETER VITOUSEK, Stanford University, Stanford, California

Panel on Evolution and Diversity

DANIEL HARTL (*Co-chairman*), Washington University School of Medicine, St. Louis, Missouri
MICHAEL CLEGG (*Co-chairman*), University of California, Riverside
J. WILLIAM SCHOPF, University of California, Los Angeles
DOUGLAS FUTUYMA, State University of New York at Stony Brook
DAVID RAUP, University of Chicago, Illinois
EDWARD WILSON, Harvard University, Cambridge, Massachusetts
CARL WOESE, University of Illinois, Urbana

Panel on Genome Organization and Expression

GERALD FINK (*Co-chairman*), Whitehead Institute and Massachusetts Institute of Technology, Cambridge
JOSEPH GALL (*Co-chairman*), Carnegie Institution of Washington, Baltimore, Maryland
PETER QUAIL, Plant Gene Expression Center, Albany, California
MELVIN SIMON, California Institute of Technology, Pasadena
JOAN STEITZ, Yale University School of Medicine, New Haven, Connecticut
RAYMOND WHITE, Howard Hughes Medical Institute and University of Utah Medical School, Salt Lake City

Panel on Growth and Development

MARY-LOU PARDUE (*Co-chairman*), Massachusetts Institute of Technology, Cambridge
MARC KIRSCHNER (*Co-chairman*), University of California Medical School, San Francisco
SUSAN BRYANT, University of California, Irvine
COREY GOODMAN, University of California, Berkeley
PHILIPPA MARRACK, National Jewish Hospital, Denver, Colorado
DAVOR SOLTER, Wistar Institute, Philadelphia, Pennsylvania

Panel on Immune System, Pathogens, and Host Defenses

WILLIAM PAUL (*Co-chairman*), National Institutes of Health, Bethesda, Maryland
BERNARD FIELDS (*Co-chairman*), Harvard Medical School, Boston, Massachusetts
ZANVIL COHN, Rockefeller University, New York, New York
STANLEY FALKOW, Stanford University, Stanford, California
MALCOLM GEFTER, Massachusetts Institute of Technology, Cambridge
DAVID SACHS, National Institutes of Health, Bethesda, Maryland
MATTHEW SCHARFF, Albert Einstein College of Medicine, New York, New York
THOMAS WALDMANN, National Institutes of Health, Bethesda, Maryland

Panel on Integrative Approaches to Organism Function and Disease

RUSSELL ROSS (*Chairman*), University of Washington, Seattle
HENRY BOURNE, University of California, San Francisco
MICHAEL CZECH, University of Massachusetts Medical School, Worcester
FRED FOX, University of California, Los Angeles
BERTIL HILLE, University of Washington School of Medicine, Seattle
PHILIP NEEDLEMAN, Washington University, St. Louis, Missouri
ALEXANDER NICHOLS, University of California, Berkeley

Panel on Molecular Structure and Function

CHARLES CANTOR (*Co-chairman*), Columbia University College of Physicians and Surgeons, New York, New York
FREDERICK RICHARDS (*Co-chairman*), Guilford, Connecticut
MARK PTASHNE, Harvard University, Cambridge, Massachusetts
LUBERT STRYER, Stanford Medical School, Stanford, California
NIGEL UNWIN, Stanford University, Stanford, California
DONALD WILEY, Harvard University, Cambridge, Massachusetts

COMMISSION ON LIFE SCIENCES

Acknowledgments

The committee would like to acknowledge the thoughtful work of the contributors and reviewers:

Edward Adelberg
Bruce Baker
Edwin Beachey
Roger Beachy
Hans Bode
Lawrence Bogorad
Marianne Bonner-Fraser
Dan Brower
Bob Buchanan
Guy Bush
Judith Campbell
Thomas Cech
Robert Chanock
Bruce Chesebro
Tom Cline
C. Robert Cloninger
R. John Collier
Robert J. Collier
John Collins
Jonathon Cooke
George Cross
Kathryn Crossin
Deborah Delmer

William Earnshaw
Robert Fraley
Michael Freeling
Larry Gerace
Norton Gilula
Timothy Goldsmith
Leslie Gottlieb
Antonio Gotto
J. Frederick Grassle
Paul Green
Douglas Hanahan
Richard Harlan
George Haughn
Ari Helenius
Samuel Hellman
John Hildebrand
H. Robert Horvitz
Rudolf Jaenisch
Larry Kahn
Arthur Kelman
Eric Knudsen
Thomas Kornberg
Stephen Krane

Michael Levine
Jane Lubchenco
Harry MacWilliams
Thomas Maniatis
Lynn Margulis
Paul Marks
David Martin
Gail Martin
Victor McKusick
Douglas Melton
Harold Mooney
J. Anthony Movshon
Howard Nash
June Nasrallah
Daniel Nathans
Stephen O'Brien
William Ogren
Michael Oldstone
Gordon Orians
David Page
David Pisetsky
Peter Quail
Calvin Quate

Richard Root
Michael Rosenzweig
John Roth
Erkki Ruoslahti
Frank Ruddle
Jay Savage
Luis Sequeira
David Schlessinger
James Schwartz

Matthew Scott
Michael Sheetz
Emil Skamene
Allan Spradling
Malcolm Steinberg
William Sugden
James Tavares
D. Lansing Taylor

Howard Temin
Samuel Thier
Robert Trelstad
David Van Essen
Graham Walker
David Wake
Allan Wilson
Thomas Woolsey

Preface

In 1970 the National Research Council published *Biology and the Future of Man*. This report, edited by Philip Handler, then president of the National Academy of Sciences, summarized the state of biology at that time. Now, almost 20 years later, the National Research Council, with the publication of this report, reevaluates research opportunities in biology and attempts to convey the current excitement in the field of biology. Our report differs from *Biology and the Future of Man* because of the enormous advances that have occurred in biology over the past two decades. The field has, in fact, changed to the point that no single individual can hope to grasp all of the new activities and opportunities.

To address this daunting task, a committee of distinguished scientists began in late 1985 to determine the major research areas that exist in the field of biology and then to discuss how advances in each of these areas can be maximized and how and where possible interactions among biologists of various subdisciplines and biologists and other scientists can be facilitated to lead to new interdisciplinary insights and approaches. This committee of 20 individuals soon realized that such a goal would require the assistance of many other experts. Toward this end the committee organized 11 panels, each with at least one individual from the committee. Each panel was asked to produce a report of about 50 pages stressing the current and future opportunities for exciting research in their area of expertise and also to stress the interconnections of biological scientists with others from different areas of biology and from other disciplines. Even the panels found that they needed assistance; thus the input of additional individuals was solicited for the report.

The steering committee edited these reports, in some cases combining the efforts of several groups to produce a single chapter, in other instances letting panel topics stand as chapters in the final report. The committee report then underwent an extensive review process. The reviewers' comments were extremely useful to the committee in polishing and producing the final draft of the report.

This report reflects the panoply of interesting and exciting topics that constitute biology. Some areas are barely covered and still others not at all. This is by necessity: There is just too much information today in biology, a fact illustrated by the length of high school and college introductory textbooks that can characteristically exceed 1,000 pages. In selecting areas to be included we chose those that represent major themes in biological research. In many cases such selections were difficult, since most of biology today is expanding as new techniques and ideas are applied to old fields or to new areas.

We have produced this volume for a large audience: biologists; policymakers both in government, universities, and in industry; and other scientists from a variety of disciplines who may interact with biologists. We hope that each of these groups will learn from this book. For all of you who read this report, our goal is to leave you with some understanding and appreciation of the diversity of problems and opportunities that await the biologist. Some of these opportunities will provide us with a better understanding of the basic workings of life, and others will have immediate application in our lives through medicine, agriculture, or environmental management.

In any such effort it is important to acknowledge the work of the many individuals, in addition to the committee members, who contributed to this effort. Those who served on the panels and contributed material to the panels are listed in the section following the committee list. My thanks to all of them. I also thank the staff of the National Research Council for their efforts. Frances Walton cheerfully provided the administrative support necessary for the committee to meet and function. Kathy Marshall spent long hours skillfully preparing the many drafts of the manuscript. Caitilin Gordon provided expert editorial assistance. Walt Rosen and David Policansky provided panel support early in the project and continued to contribute as the project progressed. In particular, I acknowledge the dedicated staff support of John Burris and Cliff Gabriel. John was the project director throughout the effort, organizing the committee and helping it through the early drafts. Cliff shepherded the effort through its later stages, assuming oversight of the manuscript in later drafts and the review process, a labor of great effort and dedication. To them, and all others who worked so hard to produce this study, I offer my most sincere appreciation.

PETER H. RAVEN, *Chairman*
Committee on Research
Opportunities in Biology

Contents

OPPORTUNITIES IN
BIOLOGY

Executive Summary

The opportunities for exciting advances in the biological sciences, many of them capitalizing on our newfound knowledge of the structure of biological macromolecules, have never been greater than they are at present. Starting with the establishment of the structure of DNA in 1953 and continuing especially with the demonstration 20 years later that genes could be modified and moved precisely from one organism to another, the flow of biological discovery has swelled from a trickle into a torrent. We base much of what we regard as our civilization—including agriculture, forestry, and medicine—directly on our ability to manipulate the characteristics of plants, animals, and microorganisms. Thus, these discoveries have profound implications for our welfare. They teach us to utilize the productive capacity of the global ecosystem on a sustainable basis.

Many of the recent advances in biology have been driven by new methods. Among the most significant have been those involving the uses of recombinant DNA, monoclonal antibodies, and microchemical instrumentation. Today, given a small amount of protein, we can clone the corresponding gene through the combined use of protein sequencing, DNA synthesis, and recombinant DNA techniques. From the readily determined nucleotide sequence of the cloned gene, the entire amino acid sequence of the protein can be derived. Alternatively, if part of a gene sequence is known, a peptide fragment can be produced that in turn can be used to generate antibodies that will detect the protein produced by the gene. Moreover, if the protein sequence of a particular gene product is known, a corresponding gene can be constructed for optimum expression in bacteria, yeast, or mammalian cells—which makes even the minor proteins of a cell available in virtually unlimited amounts.

As a result of the applications of these techniques, of powerful computers, and of imaginatively constructed new instruments, all fields of biology are being

1

revitalized. The methods necessary for a complete understanding of living systems at the molecular level now seem to be at hand. We have already learned a great deal about how the structures of proteins determine their functions in living cells, and we shall learn a great deal more during the next few years, provided that we continue to support these efforts adequately. Discoveries of structure-function relationships are of fundamental importance for understanding all of the interactions within biological systems, from the determination of our ability to resist infection, to our understanding of the way a fertilized egg develops into an adult human, to the very ways we dream, imagine, and reason. Biology holds the key to understanding these and all other processes fundamental to life.

Not only the fates of research institutions and universities, but increasingly the fates of nations will be decided by their abilities to use the facts and principles of biology with wisdom and ingenuity. Biotechnology will provide the basis of many of the advances of the future and will lay the foundations for the accumulation of wealth at many levels; will the United States maintain its current lead in this area over Japan and the European community? Who will make the most effective use of monoclonal antibodies and develop the chemical engineering basis for producing adequate supplies of the new products that are being developed so rapidly? The answers to questions such as these are of fundamental importance, but they will also have much to do with determining the contours of the world of the twenty-first century.

Over the next few years, great advances should be seen in health care with the development of powerful new therapeutic drugs and improved methods of diagnosis. Some of the the world's greatest health problems will reach critical levels in the next two decades. The disease most discussed is acquired immune deficiency syndrome (AIDS), but other problems may be of equal importance, such as atherosclerosis, Alzheimer's disease, and many forms of cancer. All of these problems, however, should be increasingly controlled through improved diagnosis and treatment. Many diseases are far better understood at a functional and molecular level than they were a few years ago. Recombinant DNA techniques will help greatly in the development of improved vaccines and of specific DNA probes that can be used for carrier detection. In addition, recombinant DNA techniques will facilitate the production of powerful cell-regulating molecules such as growth factors. It is hoped that these compounds will play an important role in the therapy of many intractable diseases. Furthermore, as we understand more about the complete sequence of nucleotides in the genomes of organisms, including humans, our power to deal with such problems will increase dramatically.

For improvements in plant-based agriculture, research on photosynthesis, nitrogen fixation, growth and development, plant-pathogen interactions, and the ways plants cope with environmental stress is of fundamental importance. As these processes become better understood in molecular detail, significant advances in agricultural productivity can be made. For example, anything that would allow plants to grow well in conditions of limited water or light or in environments colder or hotter than normal would be significant; yet we currently understand too little about the molecular mechanisms of plant survival under adverse conditions.

It is ironic that a time so filled with great opportunity should also be a time when a major fraction of the diversity of life on earth, perhaps a quarter over the next few decades, is in danger of extinction. Each species that disappears takes with it an elaborate and unique pattern of gene expression that evolved over many millions of years, about which we know little. Acting prudently for the future demands that we greatly accelerate the pace of taxonomic inventory and that we chart the poorly understood contours of life on earth, especially in the tropics, while we still have the chance. Our continued failure to do so will cause us to lose the opportunity to understand the baseline conditions for many biological phenomena, including numerous biological strategies that might have been applied for human benefit.

The broadening of ecological principles and their application to the characterization and protection of the world's biological diversity will remain important issues for the foreseeable future. For example, efforts are now being mounted to slow the loss of biological diversity and the destruction of tropical forests through improved land management, better application of the principles of conservation biology, and the formation of seed banks. It is crucial that the United States provide leadership and support in this area. The changes in the atmosphere that have become evident in the 1980s (involving significant increases in carbon dioxide, methane, and other greenhouse gases; the depletion of the ozone layer in the stratosphere; and air pollution, such as that associated with acid precipitation, on a regional scale) all highlight the intensity with which we are changing the earth on which we all depend for survival. Findings from biology and the other sciences must be used to provide stability for our children and grandchildren.

As so often in human affairs, a time of maximum opportunity is also a time of maximum challenge. In preparing this report, we have attempted to provide a set of guideposts to the great advances in biology that are occurring today and to point out specific directions that appear to be worthy of special effort. We have also made some general recommendations about how the enterprise of biology might be strengthened. Finally, we believe that our narrative demonstrates the importance of vigorous funding for biological research.

THE NEW BIOLOGY

During the past two decades, biological research has been transformed from a collection of single-discipline endeavors to an interactive science in which traditional disciplines are being bridged. Biologists have been aided by the development of new, powerful techniques and instruments, including recombinant DNA techniques, production of monoclonal antibodies, and microchemical techniques such as those used in synthesizing or sequencing macromolecules. Flow cytometry, microscopy, magnetic resonance spectroscopy, and computer applications in data collection and analysis have also been improved and have produced synergistic interactions that have shortened the time between fundamental research and its application. To encourage the development of new, improved techniques and instrumentation for biology, the barriers separating biology, chemistry, physics, and engineering must be breached by a new generation of well-trained scientists and engineers.

STRUCTURAL BIOLOGY

All biological functions depend ultimately on events that occur at the molecular level. These events are controlled by macromolecules, which can be viewed as machines designed to perform specific tasks. These macromolecules include proteins, nucleic acids, carbohydrates, lipids, and complexes among them. The ultimate goal of research in this area is to predict the structure, function, and behavior of these macromolecules from their chemical formulas.

The study of three-dimensional structure at the atomic level emphasizes x-ray diffraction techniques, which produce revealing images of these molecules. Among the structures determined thus far are those of nucleic acids, antibodies, and enzymes. In addition, these techniques have been used to determine the structure of complex molecular assemblies such as virus particles and the photosynthetic reaction center of *Rhodopseudomonas viridis*. It is likely that with the extension of x-ray analytical methods, the structures of even larger complex molecular assemblies such as ribosomes will be determined over the next few decades.

A key element in determining how proteins function is understanding how they fold into their three-dimensional structures. This folding problem now seems ripe for major advances, largely because new experimental tools and expanding data bases of protein sequences have become available. We can now design and construct new macromolecules through the use of recombinant DNA technology and chemical synthesis methods. These proteins will provide useful new pharmaceuticals and experimental models of protein function.

GENES AND CELLS

The combined application of many molecular techniques is leading to rapid and impressive advances in our understanding of the cell, the fundamental unit of which all living organisms are built. The ways in which cellular components interact to propagate, store, and express an organism's genetic information are steadily becoming clearer. Questions that have concerned biologists for decades, such as the mechanisms of DNA replication, recombination, repair, and gene expression, are being elucidated at a level of detail that was unimaginable earlier.

Biologists are discovering the molecular basis for the complex interactive matrix of chemical reactions and transport systems inside the cell. Protein synthesis occurs in the cytoplasm and is followed by protein traffic to the correct intracellular location, which involves the regulation of protein synthesis, targeting, and sorting. The mitochondrion, a cellular organelle possibly derived from ancient symbiotic bacteria, produces most of the adenosine triphosphate (ATP) for the cell, and ATP is in turn the direct source of most of the energy required for cell functions. Studies of the interaction between the nuclear and mitochondrial genomes and of the tertiary structure of the oxidative phosphorylation apparatus are of fundamental importance in appreciating the ways cells function as living units.

An understanding of the basis of cellular motility is also central to our understanding of cells. Three types of protein polymers contribute to the cytoskeleton and interact with force-producing enzymes to cause the motion of cells and organelles. Investigators have isolated and started to characterize the major molecular components of this system. This work has already contributed many new insights into the molecular basis for cellular organization and dynamics, but much remains to be learned.

An area of cell research of special relevance to medicine is that centering on cell-to-cell communication mechanisms. Growth factors, hormones, and their receptors, coupled with signal-transduction mechanisms and second-messenger activity (cyclic adenosine monophosphate, inositol trisphosphate-diaclygylcerol, calcium ions and probably others) within the cell create a communication network that allows the cell to react to its surrounding environment. Many disease conditions, including cancer and certain types of mental illness, are believed to result directly from abnormal cellular regulation. Therefore, research in cellular structure, function, and communication should provide a deeper understanding of human disease and possibly result in improved treatments or even cures.

Some diseases, such as atherosclerosis—the disease that causes heart attacks and strokes—represent a complex interaction of several different types of cells and of several factors elaborated by these cells that can lead to their abnormal multiplication. The development of monoclonal antibodies to unambiguously identify cell types and the application of tools of molecular biology to determine the nature of the growth factors and modifiers that regulate these cells are crucial for the understanding of many diseases of cell proliferation.

Plant cells have many features in common with animal cells, but also differ from animal cells in many ways. For example, plant cells have plastids, large vacuoles, and rigid cell walls, and they are connected to one another by strands of cytoplasm called plasmodesmata. Plant cells also differ from animal cells in their metabolism. Research on many of these unique plant characteristics is progressing at a remarkable rate, with profound implications for agriculture.

DEVELOPMENT

Advances in understanding how biological molecules work are leading to understanding how an organism develops from a single cell to a complex individual. The action of individual master control genes in directing and regulating the process of development is being dissected and is becoming better understood as a result. The differentiation of eggs and sperm, the chemical reactions that take place during and after fertilization, and the special regulation of the developmental control of gene expression have all been recent subjects of fruitful investigations.

Gene expression has increasingly been shown to be highly regulated with respect to tissue, time, and position. Research is providing a clear view of how this information is utilized at the molecular level. One exciting area of research involves master control genes, such as the homeotic genes in *Drosophila*, which control the body plan of organisms. Similar genes are also being found in mammals. Mutation in these genes can drastically affect the morphology of an organism. To further the study of gene expression on basic developmental mechanisms, better techniques must be devised to selectively delete the action of a given gene at a specific time and place in the developmental process.

We are now beginning to unravel the mechanisms that control cell movement and cell adhesion. It is these mechanisms that determine cell positions in an organism and the morphogenetic processes that shape tissues. The ability of cells to move correctly during the events of morphogenesis is controlled in part by their ability to adhere selectively to one another and to their extracellular environment. Various cell-adhesion molecules have been discovered, and some of them and their genes have been characterized in molecular detail. Research on these molecules, as well as on the extracellular matrix, are providing fascinating insights into cellular movement, the selectivity of cell association, and their roles in morphogenesis.

Species differ not only in their cell types, but also in the pattern in which the cell types are arranged. Pattern formation operates during development to ensure the correct spatial arrangement of cells, tissues, and organs. Even though pattern formation lies at the heart of developmental biology, until recently it was relatively unstudied compared with other problems in development. Factors thought to be important in pattern formation are cell lineage, external molecules such as morphogens, and local cell-cell interactions. The extent to which these and possibly other factors play a role in pattern formation is now actively being investigated.

Development does not stop with the formation of an adult organism. Developmental processes regulate the number of cells in different tissues of the adult and the life-spans of adult organisms and individual cells. We anticipate that studies of cell interactions in the adult will lead to a greater understanding and control of such processes.

Special circumstances exist that are unique to the developmental process in plants. For example, plants have indeterminate growth, no motile cells, and no defined germ line. Thus, even though plants and animals share many other developmental features, special insights are required into the investigation of plant development. In each of the areas of development mentioned—experimental control of gene expression, the molecular analysis of morphogenesis and pattern formation, the elucidation of the cellular interactions that regulate growth, development, and senescence—great opportunities exist for continued investigation and discovery.

THE NERVOUS SYSTEM AND BEHAVIOR

One of the most challenging and complex of all biological frontiers—a field of enormous potential for future research—is an understanding of the ways in which the nerve cells of the brain direct behavior. Nerve cells are the processing and signaling units of the brain; they communicate by electrical and chemical means. Contemporary research on nerve cell signaling focuses on the variety of neurotransmitters and neuromodulators, their receptors, and transduction into electrical events or second-messenger activity. A technique called patch-clamp measurement has enabled neuroscientists to study the individual physiological events that occur in nerve cells and that underlie the transmission and processing of information. Recombinant DNA techniques have made it possible to identify and sequence some neurotransmitter and neuromodulator genes and their receptors.

We are beginning to understand the mechanisms underlying simple forms of learning and the short-term memory to which they give rise. For example, short-term memory involves second-messenger systems similar to those used for other cellular processes, and long-term memory is likely to involve alterations in gene expression. These links between research on learning and molecular and cellular biology will provide a greater understanding of the mechanisms of learning.

The development of the nervous system can now be described in broad outline. Research on neural induction, neuronal proliferation, migration, cell aggregation, cytodifferentiation, axonal outgrowth, and nerve cell death and process elimination is providing clues to the early assessment of the developing nervous system. Although many important questions still need to be answered, biologists are now armed with the techniques to answer them.

One fascinating area of research involves the mechanism of the transport of materials within the neuron: slow and rapid axonal transport. New techniques of microinjection, quantitative fluorescence video microscopy, and labeling of samples with antibody probes for light and electron microscopy are providing insights into such questions as: What materials are transported at each rate? What is the nature of the transport mechanism? What role does axonal transport play during the growth of neural processes and during normal neuronal function?

Our understanding of how the brain controls the movement of the body is undergoing dramatic changes. For more than a century, the primary goal of those interested in the control of movement was to map the areas of the brain concerned with movement. We are now beginning to investigate how it all works: How does the brain decide when we move? How does the brain select targets? What is the speed, accuracy, and force of particular movements? Again, new methods make possible collaborative efforts that increase the yields of scientific studies. Important methods used in research on motor control include brain imaging techniques, robotics, and artificial intelligence.

Fundamental understanding of the neurobiology of cognition will have important practical applications. For example, research has begun to guide efforts to produce specific remediation techniques in developmental dyslexia. Here too, methodological developments provide unprecedented opportunities for exploring the human brain, and they are yielding important findings relating cognition to neurobiology in such areas as sensory perception, learning and memory, attention, language, and the psychobiology of development.

Central to the study of the nervous system is the desire to understand the abnormalities of behavior produced by various neurological and psychological disorders. The study of the nervous system will continue to provide the scientific and therapeutic underpinnings for neurology and psychiatry—for example, the application of molecular genetic approaches to diagnose neurological diseases and the application of modern biochemical and imaging techniques to diagnose and treat psychiatric disorders. Even the few examples cited here make it clear that the study of the nervous system will be a subject of central importance in biology for decades into the future.

THE IMMUNE SYSTEM AND INFECTIOUS DISEASES

The immune system is a potent defense developed by vertebrates to deal with the challenge presented by pathogenic microbes, malignant cells, and foreign macromolecules. Perhaps the most remarkable aspect of the immune system is the specificity of antibodies and of T-cell receptors for the antigens (foreign molecules) with which they must cope. Elegant studies have revealed the complicated mechanism of genetic rearrangement that is partly responsible for the specificity in antibody and receptor molecules. The processes through which the individual immunoglobulin and T-cell-receptor genes are activated and through which the rearrangements are controlled are currently a subject of intense investigation. One of the important tasks that lies ahead is the identification of membrane molecules through which immunoglobulin molecules signal the activation of the inositol phospholipid metabolic pathways, which play a major role in the intracellular signaling process. An additional task will be the detailed description of the molecular events that occur as a result of the activation of this signaling pathway.

Structural studies of the key molecules in the immune system are now yielding the details of major structures involved in immune regulation, deepening our understanding of the molecular mechanisms of the regulation of the immune response, the complement system, and other critical elements of the immune system. Ultimately, the results of these studies will lead to practical benefits, including the more efficient use of organ transplants by limiting rejection, the therapeutic use of molecules (such as lymphokines) that regulate the immune response, and the possible amelioration of the consequences of infection by the human immunodeficiency virus. Similarly, advances are being made in our understanding of the molecular mechanisms of microbial pathogenicity, the extension of which will create a solid foundation for disease prevention and treatment.

EVOLUTION, SYSTEMATICS, AND ECOLOGY

The mechanisms of evolution provide the key for understanding the marvelous diversity of life on earth. In the 1980s, the applications of molecular analyses have provided significant new insights in this area, as in all other biological disciplines. For example, more than 15 million nucleotides of DNA sequence have now been determined, revealing the molecular construction of genes and proteins and the detailed evolutionary relationships they have to one another. The application of these and related molecular and genetic techniques has made it possible to analyze the genetic differences between species with much greater precision than was formerly possible, and thus to better understand these differences and the ways in which they have arisen in the past. The study of the evolution of genome organization and composition is just beginning, but major long-range opportunities are available through the application of new techniques for cloning and mapping large DNA molecules.

The field of systematics charts the diversity of life on earth, but its capabilities are severely challenged in this age of rapid extinction. Approximately 1.4 million species of plants, animals, and microorganisms have been described to date, but we do not know even to an order of magnitude the number of species of organisms that exist on earth. Thus, it is estimated that between several million and perhaps as many as 30 million or more await discovery. This is not only an academic matter, but one with enormous potential consequences. Because we base our civilization to a large extent on our ability to manipulate the properties of organisms to produce food, shelter, clothing, and many other commodities, a carefully planned effort to identify additional useful ones will prove richly rewarding. The rapid pace of destruction and loss, especially in the tropics, makes it necessary for us to accelerate our efforts to understand the nature of evolution of species and communities.

The field of paleontology has been rejuvenated by the introduction of powerful modeling techniques that have enriched our understanding of extinction. Theories have been advanced about the possible role of asteroid collisions in major extinction events such as that which ended the Cretaceous period 65 million years ago, and we have gained a new appreciation of the ways major evolutionary lines originate. Meanwhile, the introduction of electron microscopic techniques, an improved understanding of which geological formations to examine, and special types of chemical analysis have pushed our knowledge of the origin of life back to at least 3.5 billion years ago, within a billion years of the origin of the solar system itself.

The structure and functioning of communities and ecosystems represent the most complex levels of biological integration. Drawing theories and practical methodologies from all of the other parts of biology, ecology provides feedback to them by illuminating the adaptive significance of characteristics of all kinds. As the theoretical basis of the field improves, its major concepts are being pressed into service to assist the human race (which has doubled its numbers since 1954) to manage the global ecosystem in a sustainable manner.

Over the past several decades, ecosystem studies have proceeded from simple descriptions of the amount of energy or materials in an area to measurement of the rates and regulation of these flows. The development of new approaches to ecosystem studies, such as the watershed-ecosystems approach and the establishment of parameters for realistic models of air and water circulation, offers an opportunity for a much broader array of studies. Present attempts to establish predictive models for global change, utilizing remote sensing and similar techniques must be continued.

Mathematical methods are being used to provide estimates of the relative importance of different factors at ecosystem, community, and population levels, so as to enhance predictive capabilities. At the same time, new refined technologies make it possible to analyze very small quantities of chemicals in ecosystems, thus helping us to understand communication within and between species. Generally, the field of behavioral ecology provides a bridge to neurobiology, which is clarifying the neural substrates of behavior. Future investigations in this area will improve understanding of such critical areas as communication, foraging behavior, sexual behavior, kinship, learning, and the roles of the sexes in an ecological and evolutionary context. These results have both theoretical and practical importance, as for the control of pests and diseases in man-made ecosystems, such as cultivated fields.

Ecological principles have been employed in formation of the new discipline of conservation biology, in which modern studies of systematics, evolution, and ecology are being applied to the problem of species and community survival. They also provide the basis for understanding biological invasions of pests such as weeds and the gypsy moth. Ecological principles will also help us better utilize genetically altered organisms that provide such outstanding opportunities for enhancing agricultural productivity and sustainability.

PLANT BIOLOGY AND AGRICULTURE

The advent of the biological revolution in which we are now engaged has provided an extraordinary opportunity to broaden our approach to plant agriculture. The modern discipline of molecular biology and its associated recombinant DNA technology are simultaneously driving rapid advances on two frontiers—the basic and the applied. The concepts and tools that have emerged from this discipline have rapidly accelerated the pace at which fundamental knowledge about plants is expanding. These tools are being applied to real-world, practical problems, with the result that genetically engineered plants have already been produced that resist herbicides, certain viral infections, and some insects. The rapidity with which this success has been achieved illustrates not only that the manipulation of plant genetic material to our advantage in attacking practical agricultural problems is feasible, but that the rate at which we have arrived at this point far exceeds the expectations of those who embarked on this enterprise a few years ago. It is equally clear, however, that these successes represent the very smallest of beginnings.

The interactions between plants and their pathogens are also biologically intricate and of fundamental scientific and commercial interest. The study of these interactions has been enhanced by new techniques in cell culture, chemical analysis, and molecular biology. Significant breakthroughs can be expected in the molecular basis of host-pathogen interactions, the mechanisms controlling the expression of virulence and resistance genes, and the molecular details of the transfer and integration of T-DNA (DNA from the bacterium *Agrobacterium tumefaciens*) into the plant chromosome. Central to much of this research is our current ability to study mechanisms of communication between organisms and between cells. Transmembrane signaling, second-messenger activity, and long-distance communication will be active areas of research during the next decade. In addition, expanding the applications of computer technology to epidemiological studies will allow better predictions of plant disease epidemics and thus the more rational application of control procedures.

INFRASTRUCTURE OF BIOLOGY RESEARCH AND RECOMMENDATIONS

In order to accomplish much of the research described in this report, an increased effort will need to be mounted to adequately support the infrastructure of biology research. Many components contribute to this infrastructure, and these must be strengthened in order to ensure the health of biology over the next decade. Among these components are training, employment, equipment and facilities, and funding. In addition, the role of large data bases and repositories needs special consideration, as do the relative merits of developing large research centers compared with additional support for individual investigators.

Training

• Despite impressive advances and great opportunities in biology, we are rapidly approaching a crisis in training of biology researchers. Current levels of support appear inadequate in light of the shortages of trained personnel predicted for the late 1990s.

• Shortages of trained technical personnel in biology are now occurring at the bachelor's and master's level. Attempts should be made to enhance university training programs at these levels, especially for biotechnology-related areas (biochemistry, cell biology, microbiology, immunology, molecular genetics, and bioprocess engineering).

• Shortages of Ph.D.s in biotechnology-related areas are anticipated in the late 1990s. Therefore, appropriate educational programs should be initiated and supported immediately.

• The recent employment and educational advances made by women in the life sciences must be fostered and encouraged to provide an attractive research and career environment.

• Every attempt should be made to encourage complete representation of minorities in the biological sciences. This will require in turn that greater attention be paid to the precollege education of minorities so that equal training opportunities will exist in college.

• As biological research becomes more sophisticated, the need increases to develop interdisciplinary and flexible training programs for students, postdoctoral fellows, and established scientists.

Equipment and Facilities

• Because of the ever-increasing need for and expense of laboratory equipment, funds to provide for specific pieces of equipment should be available. This is especially true when requested equipment is to be placed in a shared facility.

• The development of instrumentation to be applied to a variety of biological problems should be accelerated.

• Centers can provide a valuable approach to research, but the operation of a center should not interfere with the funding or creativity of the individual investigator.

Funding

• Agencies should increase their programs that provide long-term and start-up funding and should look with favor on innovative projects by qualified investigators that propose new, creative research directions.

Information Science and Collections

• An assessment of the information-handling requirements for biology should be made. Special emphasis should be given to the training of biologists in the information sciences and to the maintenance and enhancement of large-scale data bases.

• A unified approach needs to be adopted in organizing and maintaining collections of preserved and living specimens and other biological materials. This will require increased funding and attention.

International Cooperation

• The United States has long encouraged and benefited from international cooperation in biological research. As other countries increasingly emerge as valuable sources of quality research, this policy of cooperation should be strengthened.

1

The New Biology

Brightly colored cells can be followed as they move, their structural changes can be monitored, and even their internal chemistry is visible with advanced imaging techniques. Such a wide variety of features is now readily observed with specific fluorescent probes and a fluorescence microscope and with computer-imaging methods. This new ability is illustrated on the cover of this report, where cells migrating into a wound are shown. The biochemical, physical, and data-processing techniques that allow such pictures represent just some of the many new research tools that have revolutionized biology and allowed us to observe and begin to understand what is happening inside living organisms and their cells.

This report attempts to elucidate the present state of biology and to predict some of this science's goals and research opportunities. Such an attempt is filled with uncertainties, for biological research is dynamic and scientific breakthroughs unpredictable. Our primary objectives are to show the currently exciting activity in biology and to explore foreseeable research opportunities; we fully appreciate the explosive nature of biological research and the near impossibility of anticipating the many directions that may assume central importance in the immediate future.

We are afforded some help in anticipating the future by our recognition of the unity of biology. Valuable information on intractable research systems can be gained by drawing direct parallels from more tractable research systems. For example, although substantial differences exist among yeasts, plants, and animals, their mechanisms of gene expression at the molecular level are strikingly similar. As a result, each new bit of information obtained by studying yeasts and other model organisms has a good chance of being useful for understanding many other experimental systems. The principles of physics, chemistry, and biology that apply to living systems, be they entire ecosystems, unicellular organisms, or individual organelles, operate at the core of biological function. The main role for biologists today is to elucidate these basic principles and determine how they are modified to produce the diverse world we inhabit.

Divisions Between Traditional Disciplines Are Being Removed

Over the past two decades, contemporary biology research has been launched into an era in which many new biological principles can be determined and studied experimentally. Quantum leaps in understanding have been made in such areas as enzyme catalysis, molecular recognition, transmembrane signaling, genetics, and organismal relationships. With these advances, disciplines such as molecular, cellular, and developmental biology have become impossible to separate, so that the borders between traditional fields of study have become largely artificial. Similarly, the organismal sciences, such as ecology, are now closely linked to physiology, behavior, and molecular biology. These developments reflect a synergism among disciplines that is unprecedented in the history of biological research. An example is the discovery and characterization of an important new class of antibiotics from the skin of the clawed African frog, *Xenopus laevis*. Here, the observation that these frogs resist infection led rapidly to the isolation of the protein antibiotic, the sequencing of this protein and its complementary DNA (cDNA), and now to the possibility of controlling infections by using the antibiotic.

Spectacular advances are being made in determining the molecular structure of proteins and nucleic acids by physical methods such as x-ray crystallography and nuclear magnetic resonance spectroscopy. Here too, techniques of molecular biology are being used, allowing scientists to analyze the effects of designed changes in the amino acid sequence of enzymes on structure and catalytic function. In the future, the results from these studies will have direct application to the development of improved agriculture and health care by guiding the production of modified preexisting enzymes as well as the production of other new molecules that can be used as therapeutic drugs.

Advances made in determining ways in which organisms interact have revolutionized the study of ecology. For example, the pheromones, or chemical signals, of insects and mammals have been shown to play crucial roles in the organisms' biology. Pheromones influence the attraction of mates, the recognition of specific mating groups, and the synchronization of biological cycles. Recently, pheromones have been exploited to alter the behavior of pest organisms in agroecosystems to bring about their control without the use of synthetic chemical insecticides.

A recent application of the techniques of molecular biology has made it possible to monitor the fate of microorganisms released into the environment and has thus contributed to our understanding of ecological principles. Molecular markers produced by recombinant DNA techniques are being developed to facili-

---

FROG SKIN ANTIBIOTICS

The operation has been performed hundreds of times—the removal of the ovaries of the African clawed frog, *Xenopus laevis*. The incision is sutured and the frog is returned to an aquarium, which characteristically teems with a diverse array of microorganisms. In 1987, however, a biologist observing that these frogs seldom experience any postoperative infection, even though they receive no postoperative care, began to search for the reason for their resistance.

A standard antibiotic assay, which involved observing the bactericidal effects of substances on the common colon bacterium, *Escherichia coli*, was used to determine the location in the frog of the hypothesized antimicrobial agent. Antimicrobial activity was demonstrated and shown to be highest in skin extracts. Now that the location of the antibiotic was determined, its characterization could begin. Standard biochemical purification procedures were used to isolate the antibiotic. It was eventually shown to be a new class of proteins, subsequently named magainins, from the Hebrew word magain, meaning shield.

The amino acid sequence of the magainins was determined. This sequence was then used to synthesize short DNA probes that were used to screen a library of complementary DNA clones constructed from messenger RNA isolated from the frog's skin. This experiment was of interest since the isolated complementary DNA clones could be used to determine the mechanism of magainin production. Specific complementary DNA clones were isolated and their nucleic acid sequence determined and analyzed.

Thus, in a single year, a scientist progressed from an original observation of a biological phenomenon to the isolation and characterization of what might be a very important antibiotic. Rapid progress of this sort would not have been possible without the coupling of organismal biology with microbiology, biochemistry, and molecular biology.

---

tate the determination of organismal spread and survival. Such findings have both theoretical and practical applications.

Great advances should be seen in health care, with the development of powerful new therapeutic drugs and improved methods of diagnosis. Some of the world's greatest health problems will reach critical levels in the next two decades. The disease most discussed is acquired immune deficiency syndrome (AIDS), but other problems may be of equal importance, such as atherosclerosis, Alzheimer's disease, and many forms of cancer. We can expect significant advances through improved diagnosis and treatment. Recombinant DNA techniques will help greatly in the development of improved vaccines and in the development of specific DNA probes that can be used for carrier detection. In addition, these techniques make possible the commercial production of powerful cell-regulating molecules such as growth factors. It is hoped that these compounds will have an important role to play in the therapy of many intractable diseases.

The genomes of several different types of organisms, including human beings, should be mapped and sequenced by early in the next century. Advances currently being made in the automation of DNA handling and sequencing should greatly facilitate this effort. The enormous amount of sequence information then available will provide an invaluable tool for the medical sciences, as well as for basic biology. Obstacles in program funding and organization are being dealt with and must be overcome before such a project can be carried out efficiently.

The characterization and protection of the world's biological diversity will be an important issue during the next two decades. The destruction of the tropical forests, resulting in the loss of species habitat, is largely responsible for the accelerated rate of species extinction. Efforts are now being mounted to slow this extinction rate through improved land management and through such methods as the application of the principles of conservation biology and the formation of seed banks and other collections of living organisms. In addition, biologists are trying to understand more completely the interactions of the organisms that live in rainforest ecosystems. International cooperation is vital for these efforts to succeed. It is crucial that the United States provide leadership and support in this area.

Agriculture should experience many significant changes in the next two decades. There is an urgent need to make American agriculture more efficient through research. Good prospects exist to realize benefits from biotechnology in animal and crop agriculture. The production of transgenic agricultural organisms with desirable simple agronomic traits, such as herbicide resistance, has already been accomplished. Scientists are making considerable progress in the production of transgenic organisms with more complex traits, such as drought tolerance. Obstacles to the transfer of complex traits should be overcome with increased basic research on gene expression and regulation.

This report identifies many of the research opportunities that currently exist in biology and attempts to predict some of those that may emerge in the near future. This task is difficult, since biological research is progressing so rapidly. The report takes the reader through the disciplines of biology from molecular structure and function to evolution and diversity, while stressing the interdisciplinary approach taken by most biologists today. After a discussion of basic research, the report then addresses the applications of biological research in medicine, agriculture, and other areas. The report concludes with recommendations on the infrastructure needs of future biological research.

2

New Technologies and Instrumentation

Many of the recent advances in biology have been driven by the development of new technologies and instrumentation, such as recombinant DNA techniques, monoclonal antibody techniques, and microchemical instrumentation. Each of these technologies has opened new opportunities to explore both fundamental and applied biological problems. Moreover, these technologies have proven to be synergistic—each operating in conjunction with the others to amplify their potentials.

RECOMBINANT DNA TECHNIQUES

Recombinant DNA Techniques Permit Us to Isolate a Single Gene from the Tens of Thousands Encoded in a Complex Genome

After a gene has been isolated by recombinant DNA techniques, studies of its structure, regulation, and function can begin. These techniques depend on the molecular complementarity of DNA molecules, which are the backbones of chromosomes and the dictionary of the genetic code, and on the two categories of enzymes that can manipulate DNA molecules.

The DNA molecule is composed of two strands, each made up of a linear chain of four different building blocks: guanine (G), cytosine (C), adenosine (A), and thymine (T) (Figure 2-1). The DNA chains exhibit molecular complementarity in that the Gs on one strand always pair with the Cs on the other; likewise, the As always pair with the Ts. This molecular complementarity means that in a mixture of unpaired DNA fragments, one fragment will always be able to find its complement by virtue of the precise pairing of their nucleotide bases. For this

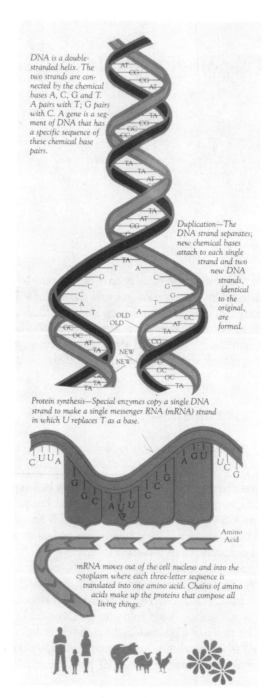

FIGURE 2-1   DNA. [Figure courtesy of the Monsanto Company]

reason, a small fragment of a gene can be used to find the complete gene in a complex mixture of DNA fragments.

A unique set of three contiguous DNA bases specifies one particular amino acid subunit. Given the DNA sequence of a gene, we can predict the order of amino acids in the protein it encodes. Conversely, given the amino acid sequence of a protein, one can, with certain ambiguities, predict the DNA sequence of the gene encoding it. The ability to use the genetic code dictionary to go from genes to proteins or proteins to genes is of fundamental importance to certain recombinant DNA strategies.

Two categories of enzymes have played a critical role in the development of recombinant DNA techniques—DNA-cutting enzymes, or restriction endonucleases, and DNA-joining enzymes, or ligases. The DNA restriction enzymes cut double-stranded DNA at precise short DNA sequences (Figure 2-2). Thus, they provide a means of taking a large DNA molecule and cutting it into uniformly determined smaller fragments. The DNA ligases permit any two DNA fragments to be joined together. Accordingly, the essence of the recombinant DNA techniques is the ability to take a DNA fragment containing any particular gene of interest, say α-interferon, and join it into an appropriate vector sequence, such as a plasmid, to create a hybrid or recombinant DNA molecule. Such a recombinant sequence can be inserted into bacteria, yeast, or mammalian cells and amplify itself by a factor of 30 to 1,000. In this way, many copies of the α-interferon gene can be produced for study. In addition, if the appropriate regulatory machinery is available in the vector, the gene can be expressed, so that large quantities of the corresponding protein are produced for biological study or application.

Transformation of Higher Organisms

Biologists Can Specifically Insert a Functioning Gene into the Genome of Complex Organisms

The revolution in molecular genetics has led us into an understanding of how genes are functionally constructed and has allowed unprecedented access to specific genes and the protein products they encode. Transformation is the process by which DNA molecules that have been isolated (usually by recombinant DNA techniques) are introduced into the cell in a way that allows a gene's perpetuation from generation to generation. Much of what we have learned about genes has been through experimental approaches in transformation that transfer those genes from one place to another—either from cell to cell, organism to organism, or cell to test tube, and then from test tube back into cells and organisms. This has allowed us to dissect the DNA of a gene and separate it into specific components; in particular, it has allowed the DNA of a particular gene to be separated into its protein coding information, as specified by the genetic code,

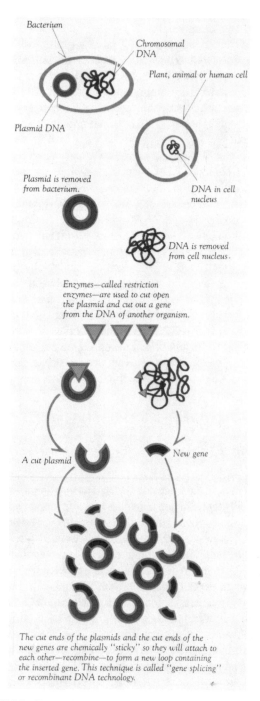

FIGURE 2-2   Recombinant DNA Technology. [Figure courtesy of Monsanto]

and the control elements that participate in the decision about when, where, and under what circumstances a specific gene expresses its encoded information.

This decision about when and where a given gene should "play back" its information is of critical importance when one considers that individual plants and animals often contain billions of cells, all of which must act and interact so that the harmony of a healthy, living organism can be preserved. If a gene is expressed in the wrong place, or fails to express in the right place, disastrous problems follow. For this reason, understanding the nature of gene regulation is central to basic research in all aspects of biology.

Both the nature of genes and the mechanisms associated with specific diseases can be studied effectively by introducing (or transferring) genes into the chromosomes of an individual animal or plant. When transferred to such plant or germ cells of an animal, the new genetic traits are passed on to successive generations, where they reside in every cell of the offspring. Individuals produced by such alterations are called transgenic organisms.

Making a Transgenic Animal

A Transgenic Animal Is Produced Initially by a Combination of Microsurgery and Embryological Techniques

Much of the experimental work on transgenic animals has been carried out with laboratory mice. Mice have been selected for such experiments because they have the advantages of being small and easy to maintain, having short generations (6 to 8 weeks on the average), and producing relatively large numbers of progeny (usually about eight).

To produce a transgenic mouse, one first removes fertilized eggs from a female mouse about 12 hours after copulation. After being cultured for a few hours in an incubator, each fertilized egg is injected with a solution of pure DNA through a fine glass capillary needle and monitored through a high-powered microscope. The DNA is injected into one of the two pronuclei, one of which contains the sets of chromosomes originating from the mother and the other from the father. The injected embryos are then reimplanted into the oviducts of a female mouse and allowed to develop. Typically, if one injects 100 fertilized eggs, about 50 will survive the injection process and perhaps 10 will develop into living mice.

The mice born are analyzed at 2 to 3 weeks of age to determine which, if any, have incorporated the injected DNA and are therefore transgenic. On average, 2 of every 10 mice born will be transgenic. Several copies of the injected gene will have become integrated into a chromosome of a transgenic mouse, generally in one location, as shown by the Mendelian transmission of the gene to half of its progeny. The transgenic mice that develop from these injected embryos are each potentially the founder of a unique family: Even if the same gene is injected into multiple embryos, the different transgenic mice will almost certainly have the gene in a different chromosomal location. Often, the different mice will even show distinct responses to the presence of the transgene. Thus each initial transgenic mouse is used to found a lineage (or family) by mating it to a normal mouse and maintaining its unique genetic properties by selection.

Transgenic Mice Have Been Used for a Variety of Experiments

Transgenic mice have been used to perform detailed analyses of the consequences of the presence of particular genes on the organism in which they occur. For example, transgenic mice were produced that carried hybrid genes designed to overproduce human growth hormone. The mice that inherit these growth-hormone genes grow unusually large—about twice the normal body weight. Among the characteristics associated with large size are lethargy, a shortened life-span, poor reproductive performance of males, and sterility in females. Such complications probably reflect the fact that the size of an organism results from a variety of constraints and compromises, which have been reached in a long evolutionary process. In another line of transgenic mice, human growth hormone is produced under the control of a different regulatory element. The mice of this second line are not sterile, but they are still "big," suggesting that in the long run we may be able to design animals with a desired characteristic but without the accompanying undesirable side effects.

Another application using transgenic mice has been the dissection of the regulatory segments of genes expressed in different cell types of the body. In human beings and other mammals, for example, different hemoglobin genes come into play sequentially in embryos, fetuses, and young individuals; the switching mechanisms involved in this sequence have been studied in transgenic mice. The difference in gene action in different parts of the body has also been studied. For example, the spatial control elements of the genes for serum albumin and the digestive enzyme elastase, which specify expression in hepatocytes or pancreatic acinar cells, respectively, are being localized by the application of these methods. The results of such experiments will contribute to our eventual understanding of the mechanisms of tissue-specific control of gene expression.

Transgenic mice are also being used in gene therapy experiments that seek to cure genetic abnormalities. In one such example, a mutant strain of mice called "little," which have too little growth hormone, have been "cured" by transferring a growth hormone gene into their chromosomes. The transgenic "little" mice, treated in this way, become nearly normal in size. In similar experiments, a gene encoding a transplantation antigen that had been deleted from a mouse chromosome was restored to inbred strains of mice. Similarly, mice deficient in gonadotropin-releasing hormone, which is synthesized in the hypothalamus, remain sexually immature and hence are infertile. This defect has been cured genetically by transferring an intact copy of the gene that encodes the hormone protein into the embryos of mice that are predisposed to develop this inherited abnormality. Such mice become sexually mature and fertile, and therefore "cured."

Such experiments can show that specific defects are associated with particular genes. The defects can then be cured by the introduction of normal alleles of these genes. In addition, transgenic mice afford excellent tools for learning about physiology and endocrinology because individuals that produce particular hormones in unusual quantities can be produced. In such experiments, the genes that are responsible for the production of the hormones can be placed in different regulatory environments, and they can be active at new sites of synthesis. Examining all of these variables allows a more precise study of mechanisms to be carried out.

Creating Transgenic Plants

The Creation of Transformed Plants Has Been One of the Most Exciting Developments in Modern Biology

During the past 3 years, systems for the transformation and regeneration of plants have emerged, and the insertion of novel traits has opened a new domain for the study of gene regulation and expression. In addition, the insertion of foreign genes into plants has also provided a powerful tool for modifying plants genetically for applied purposes as well.

The successful development of genetic transformation systems in plants has depended on the use of the soil bacterium *Agrobacterium tumefaciens*, which is capable of infecting plants through wounds to result in crown-gall tumors in many dicotyledonous plants. The tumor-inducing organism carries a large circular plasmid (the Ti plasmid), which contains a piece of DNA (T-DNA) that can be inserted into a plant chromosome. This T-DNA contains genes that encode enzymes for phytohormone synthesis, and the overproduction of these hormones results in tumor formation. Additionally, other genes carried on the Ti plasmid facilitate transfer of the T-DNA. It has been possible to delete most of the genes within the T-DNA, thereby disarming its tumor-inducing genes while preserving its capacity to be transferred. By "filling-in" this modified T-DNA with genes of choice, a powerful vector system for plant transformation has been created. The foreign genes inserted in the T-DNA are often incorporated into the host cell genome; they may subsequently be transmitted in the course of the plant's normal reproduction.

The actual transformation system has been greatly simplified so that the process of injection, transformation, and plant regeneration can be carried out easily, producing transformed plants relatively frequently. The system involves the incubation of a genetically modified *Agrobacterium*—that is, one carrying a gene of choice—with either leaf disks or other pieces of tissue capable of forming

TRANSGENIC ANIMALS AS DISEASE MODELS

An important application of the transgenic experimental system is the creation of animal disease models. For example, certain human oncogenes, or cancer-causing genes, can be introduced into transgenic animals in which they induce specific tumors. This procedure provides a convenient animal system in which the genesis and progression of the tumor can be studied. The same model may also be used to develop new methods for diagnosing and treating the human version of the tumor. This approach is now being considered for experimental studies on acquired immunodeficiency syndrome (AIDS). In this disease, the viral agents seem to regulate host-cell genes. If this is an important causative mechanism of the disease, it might be possible to simulate AIDS in animals by transferring the viral gene into animal embryos. The construction of animal models for AIDS would represent a significant step in furthering our understanding of the disease process and in formulating an effective therapy and prophylaxis. For example, several laboratories are already pursuing the possibility of over-producing a receptor to AIDS virus in vivo. AIDS virus binds to a receptor on immune cells. The surrogate gene would produce a receptor that would act as a scavenger, removing AIDS virus so that the genuine immune cells would not be attacked.

shoots. After the tissue and the bacteria are cultured together for a short time, the bacteria are killed with an antibiotic, and the plant tissue is grown on a medium that allows for the selection and subsequent regeneration of transformed plant cells.

Future Prospects

The Potential for Using Transgenic Organisms to Make Discoveries over the Next 5 or 10 Years Is Vast

The techniques of producing transgenic organisms are fast becoming an important part of a wide variety of experimental approaches to questions in biology and medicine. These range from the study of gene regulation, to the development of the immune system, to tests of theories in endocrinology and physiology, to mechanisms of self-tolerance and autoimmune disease, and finally to the study of cancer and other important human diseases.

In the long run, it may prove possible to alter the characteristics of farm animals such as pigs, sheep, and cattle, perhaps to provide them with disease resistance or improved physiological responses. This seems a long-range prospect at present, and the central focus of research and results will likely remain on transgenic mice for now. In plants, partly because of the ease with which new genotypes can be created in large numbers, the practical applications are apt to come sooner.

On the basic side, these techniques have shown that relatively short DNA sequences, only a few hundred nucleotides long, are capable of providing highly specific regulation of gene expression. These results have made possible recent efforts to discover the proteins or nucleic acids that interact with specific DNA regulatory sequences in the gene complex. By such methods, the DNA sequences involved in the regulation of gene expression in specific tissues and developmental processes are being discovered. Other areas for application of these techniques include such processes as embryogenesis and morphogenesis, sexual transmission and inheritance of genes, and the development of new models for the study of diseases.

The new technologies in biology are being applied to research problems related to improvements in crop productivity. The commercial development of genetically transformed plants and animals has just begun to emerge as a viable application of these technologies. For example, the development of new transformation techniques for monocotyledonous plants—such as wheat, maize, and rice—coupled with appropriate regeneration technologies will provide results of the greatest economic importance.

In medicine the transformation of microbial cells with foreign genes has resulted in the commercial production of such valuable products as insulin, human growth hormone, interferons, and tissue plasminogen activator.

MONOCLONAL ANTIBODIES

Monoclonal Antibodies Can Be Used as Biological Probes for Specific Molecules

Another significant area of biological advance has been the development and application of monoclonal antibodies. Antibody molecules exhibit exquisite specificity for the foreign macromolecular patterns (antigenic determinants generally contained on viruses and bacteria) that initiate their synthesis. Antibodies are synthesized by one class of blood cells, the B lymphocytes, with each B cell having the capacity to synthesize just one type of antibody molecule. The typical immune response to a bacterial antigen is self-limiting and extremely heterogeneous, both because mature B cells have a short lifetime (a few days) and because the myriad different B cells that are turned on produce many different types of antibody molecules. The development of the monoclonal antibody technique has made it possible to produce virtually unlimited quantities of homogeneous antibody molecules to antigens of particular interest. The basic idea is to take a transformed (malignant) antibody-producing cell and mutate it so that it can no longer produce its own antibody (Figure 2-3). Then it is fused to a normal antibody-producing cell to generate a hybrid cell line, having the chromosomes of both parents. Such a cell line, which is potentially immortal, has the capacity to produce unlimited quantities of one particular type of antibody molecule. Thus, large quantities of homogeneous antibodies of any particular specificity can be produced.

An additional powerful tool for obtaining antibodies of particular specificities arises from our ability to synthesize peptide fragments of proteins through the use of a peptide synthesizer. By this procedure, a fragment of protein is synthesized, coupled to an appropriate carrier protein (a larger molecule), and used to immunize animals. Some fraction of the time, antibodies that can recognize the protein from which the peptide fragment was derived, as well as the peptide fragment itself, are generated. Thus, the antibody response can be directed precisely to a particular region in the protein molecule. In a sense, the peptide antibody approach allows us to fine-tune the specificity of the immune response. Monoclonal antibodies can then be generated from these immune responses.

The use of monoclonal antibodies has revolutionized many aspects of fundamental biology and clinical medicine. It is now possible to obtain monoclonal antibodies for rare and biologically significant or medically interesting molecules. These antibodies can be used to identify and purify these key molecules, and they have enormously facilitated many aspects of the study of development and molecular structure-function relationships. In addition, they have provided countless critical diagnostic reagents [for example, the antibody specific to HIV, the virus associated with acquired immune deficiency syndrome (AIDS)]; in the future, they will be used increasingly as therapeutic reagents. Antibodies have also played a critical role in permitting the genes that synthesize these protein products to be isolated by a blending of recombinant DNA and monoclonal antibody techniques.

MICROCHEMICAL TECHNIQUES

Microchemical Instrumentation Has Had a Powerful Impact on Modern Biology That Is Just Beginning to Be Felt

Many of the advances in modern biology are dictated not only by the development of new technologies but also by the development of instrumentation. For example, the instrumentation for the automated synthesis and sequencing of protein and of DNA has been developed over the past 20 years.

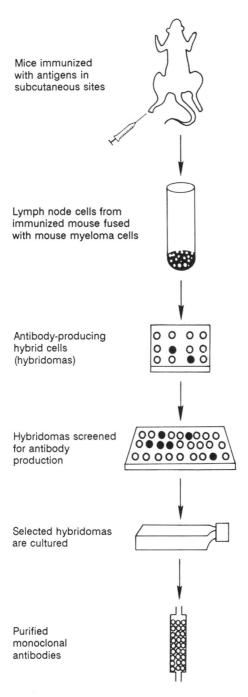

Mice immunized
with antigens in
subcutaneous sites

Lymph node cells from
immunized mouse fused
with mouse myeloma cells

Antibody-producing
hybrid cells
(hybridomas)

Hybridomas screened
for antibody
production

Selected hybridomas
are cultured

Purified
monoclonal
antibodies

FIGURE 2-3   Production of monoclonal antibodies. [Adapted from J. F. Kearney, in Fundamental Immunology, W. E. Paul, ed. (Raven, New York, 1984), p. 756]

Protein Sequencing. In 1967 the first protein sequencer, an automated device for determining the linear order of amino acid subunits in a polypeptide chain starting at one end (the amino terminus) was developed. In the ensuing 20 years, the amount of protein required for sequencing has dropped to 1/10,000 of the earlier amount, to the point that experienced protein chemists can sequence 10 picomoles of protein. The ability to sequence very small quantities of protein is important because even limited amounts of protein sequence data can facilitate cloning of the corresponding gene. Currently a variety of new approaches are being used for even more sensitive detection of amino acids, including fluorescent detection, which should permit sequencing of 10 to 100 femtomoles within the next few years.

The importance of this level of sensitivity is that it would permit the direct sequencing of most proteins separated by the most sensitive methodology available—two-dimensional gel electrophoresis. For example, this procedure separates in one dimension by size and in the second dimension by charge and is capable of separating 1,000 to 5,000 different proteins. When this sensitivity in protein sequencing is attained, many of the genes whose products have been visualized only as spots on a two-dimensional gel can be cloned.

Protein Synthesis. A method for protein synthesis has been developed in which the carboxyl terminal amino acid subunit is attached to a resin support and the polypeptide chain is synthesized by a repetitive chemistry that adds one subunit at a time to the growing chain. This approach has been automated, and current state-of-the-art peptide synthesis (and subsequent purification) can produce relatively homogeneous polypeptides as long as 60 residues. Indeed, 140-residue hormones have been synthesized and partially purified. The ability to synthesize peptides is useful in several respects. First, peptide fragments from proteins can be used to generate specific antibodies, which can occasionally bind to the parent protein from which the fragment was derived. Thus peptide synthesis can generate useful analytical reagents, or in the clinical realm, diagnostic or therapeutic reagents. Second, peptides can be useful in determining which amino acid residues are key for particular functions (for example, hormones can be synthesized with amino acid modifications at interesting sites). New techniques for protein synthesis, purification, and the joining of peptide fragments will enhance our ability to carry out structure-function studies and scale up the synthesis of valuable peptides.

DNA Sequencing. Two methods are available for DNA sequencing—a chemical and an enzymatic method. Both methods generate nested sets of radioactively labeled DNA fragments starting at a single fixed point and terminating at every A, every T, every C, and every G base in four distinct reaction mixtures. These mixtures are run separately on gels that can resolve fragments differing by a single nucleotide, and from the resulting patterns visualized by radioautography on film, the DNA sequence can be detected. Recently, a fluorescent chemistry for separately labeling each of the four bases has been developed and the reading of the separated fluorescent-colored fragments automated. This instrument can simultaneously sequence 16 DNA fragments for 300 nucleotides each over an 8-hour cycle time. Accordingly, under ideal conditions 15,000 nucleotides can be sequenced per 24-hour period. In the near future, machines should be developed that are capable of sequencing 150,000 or more nucleotides per day with acceptable error rates. This type of DNA sequencing instrumentation raises interesting questions about the feasibility of sequencing the human genome, a topic that will be discussed elsewhere.

DNA Synthesis. A solid-phase method for DNA synthesis has been automated, and a machine is now available that can synthesize hundreds of nucleotide linkages per 24-hour period. Typically, oligomers 30 to 50 nucleotides long are synthesized. Oligonucleotides can readily be joined to synthesize entire genes chemically and those can serve as probes for genes of interest. Oligonucleotide primers can be synthesized rapidly to facilitate DNA sequencing and complementary DNA synthesis. Genes can be mutated readily and specifically by oligonucleotide-directed mutagenesis. Finally, in conjunction with the protein sequencer, a powerful strategy is available to clone rare message genes that produce very little RNA and correspondingly small amounts of protein. Once a small amount of amino acid sequence is determined, it can be translated by the genetic code dictionary into a DNA sequence. This sequence can in turn be synthesized as an oligonucleotide fragment that can be used in conjunction with routine recombinant DNA techniques to clone the corresponding gene.

FLOW CYTOMETRY

Flow Cytometry Is Used to Sort Cells

A flow cytometer, also known as a fluorescence-activated cell sorter, is an instrument that quantifies fluorescent molecules bound to individual cells or subcellular particles. Many fluorochromes can be used to obtain information about cellular structure and function. Monoclonal antibodies specific for cell surface, cytoplasmic, or nuclear antigenic sites can be coupled to fluorescent molecules such as fluorescein, phycoerythrin, or Texas red. Fluorescent dyes such as propidium iodide and acridine orange can be used to bind to DNA. Because this binding is proportional to the amount of DNA per cell, it reflects the percentage of cells in a population undergoing mitotic division. A number of fluorescent molecules can be used to quantify cell functional activities such as changes in membrane potential, calcium uptake, and intracellular pH. The repertoire of reagents is growing daily. The flow cytometer is capable of looking at multiple fluorochromes simultaneously. Therefore a combination of monoclonal antibodies can be used to label cells, allowing subpopulations of cells that were previously indistinguishable to be identified by means of their unique combination of cell surface markers. The flow cytometer is also capable of sorting populations of cells on the basis of any of the above variables under aseptic conditions that retain cell viability.

Cytometric analysis is done by injecting a suspension of cells into a fine stream that passes through a finely focused halogen light source. As a cell passes through the light beam, light-scatter signals are collected by photodetectors. The light-scatter signals give morphological information that can identify some cell subpopulations and separate viable from dead cells. In addition, fluorescence emissions are detected by very sensitive photomultiplier tubes. All signals derived from one cell are digitally processed simultaneously, and the values from that cell are stored by the computer as frequency histograms. The fluorescence intensity can be empirically related to the number of fluorescent molecules bound per cell, which in turn quantitates the number of antigenic sites or dye-binding sites per cell (for example, receptor density or RNA levels).

Cell sorting is carried out by oscillating the stream of cells so that droplets are formed, most of which contain only one cell. When the sort begins, cells pass through the light beam. If the cell falls within the defined parameters, an electrostatic charge is placed on the droplet that contains the cell and the charged droplet is deflected into a test tube. Cells can be sorted at rates up to 5,000 cells per second on commercial instruments. Sort rates as high as 100,000 cells per second have been obtained with specially designed cytometers.

MICROSCOPY

A Revolution in the Application of Light Microscopy Has Occurred

Advances in light microscopy have resulted from an integration of the fields of microscope optics, video technology, digital image processing, biology, and chemistry. The remarkable advances in the past few years have been driven by the need for biologists to define the dynamics of the chemical constituents in living cells. Cells perform a variety of cellular functions such as growth, division, movement, intracellular transport, and communication by the coordination of hundreds of complex chemical reactions occurring at distinct times in different regions of cells. Therefore, a complete understanding of the molecular basis for normal, as well as abnormal, cell functions requires methods that can yield both temporal and spatial information about the chemical constituents and chemical reactions in living cells.

Modern light microscopy has emerged as a tool that can yield such information about the chemistry of living cells when used in conjunction with state-of-the-art photodetectors and computers and with specially designed chemical and biological probes. The most important recent advances in optical microscopy have been in two major areas: video-enhanced contrast microscopy and low-light-dose microscopy.

Video-Enhanced Contrast Microscopy

Video-Enhanced Contrast Microscopy Combines the Technologies of Modern Light Microscopy, Video Imaging, and Digital Image Processing

Video-enhanced contrast microscopy results when differential interference microscopy uses a high-resolution video camera instead of photographic film to record images. Stray light, which limits contrast, can be suppressed by electronically changing the black and white levels (video offset) and the gain (sensitivity) of the video camera. The resulting improvement in contrast has permitted the camera to detect fine structures that were invisible by conventional light microscopy to the human eye or film. Biological structures as small as 24 nanometers in diameter or 1/10 the resolving power of a light microscope can be detected in living cells with a time resolution of one video frame (1/30 second).

Digital image processing—including averaging a few video frames, subtracting a background image, and increasing the contrast by computer enhancement methods—has improved the image quality even further. This imaging system is capable of recording the movements of biological structures in living cells that had been seen previously only by electron microscopy of killed cells.

Video-enhanced contrast microscopy has had an immediate impact on neurobiology. Nerves consist of a series of elongated cells that propagate electrical impulses and transmit chemical signals between nerve cells and finally to targets such as muscle cells. Many membrane-bound organelles such as synaptic vesicles, which carry chemical signals to the nerve cell termini, are transported inside the long extensions or axons of these cells. The transport of these organelles is critical for the normal functioning of nerves, but the mechanism for this transport was unknown until recently. The video-enhanced microscope recorded the directions and rates of movement of very small vesicles in squid axons.

Low-Light-Dose Microscopy

Coupling Biological Chemistry with Advanced Image Processing Has Permitted Low-Light-Dose Microscopy to Evolve as a Powerful Method of Investigation

Low-light-dose microscopy allows the analysis of very weak light signals from luminescent (light-emitting) molecules in or on cells and tissues.

Quantitative fluorescence microscopy, when combined with the multitude of biological probes and fluorescent probes now available, offers an approach for gathering complex chemical and molecular information from living cells and tissues. The power of quantitative fluorescence microscopy results from its high sensitivity and specificity, combined with spatial as well as temporal resolution.

Fluorescence spectroscopic measurements extend the power of fluorescence microscopy to the molecular level. For example, resonance energy transfer allows the distance between two suitably labeled molecules to be determined over the range of 1.0 to 7.0 nm. By the interaction of fluorescent analogues, the assembly of subunits of actin or microtubules can be analyzed in the smallest regions of single cells that can be resolved spatially. Measurement of fluorescence anisotropy of fluorescent analogues can yield rotary diffusion coefficients, which can be used to determine whether the analogues are free to diffuse or are bound to other structures. Because some fluorescent probes change their spectroscopic characteristics in response to the chemical environment, changes in the pH, the free calcium ion concentration, or other measures can alter the excitation or the emission properties of environmentally sensitive probes. Measuring the fluorescence at two wavelengths can be used to rapidly quantify the pH, the calcium ion concentration, or other specific variables, allowing temporal and spatial changes in these variables to be determined in living cells.

Scanning Acoustic Microscope

The Scanning Acoustic Microscope Measures the Elastic Properties of the Cell

With the advent of the scanning acoustic microscope as a commercial instrument, it became possible to study the elastic properties of cells and biological material on a scale that is similar to that of the optical microscope. The acoustic microscope uses sound waves propagating through liquids rather than optical waves, so that images reflect elastic properties rather than changes in the index of refraction. The relative changes in elastic properties of cells and organelles are often larger than the change in the index of refraction, which means that the contrast may be enhanced in the acoustic images. With living cells attached to substrates, it is easy to monitor the attachment sites, the contour of the cells, and the intercellular networks of fibrils and microtubules. Furthermore, it is possible to monitor the changes in these elastic features in living cells.

For the future, acoustic microscopes are being developed that can operate in a cooled liquid such as helium. In such a liquid, the wavelength of sound can be much shorter than that of optical wavelengths. Research instruments have been constructed with a resolution of 200 angstroms, which approaches the resolution of the scanning electron microscope.

Scanning Tunneling and Atomic Force Microscope

The Scanning Tunneling Microscope Allows One to Image Surfaces with the Resolution of a Few Angstroms

The functioning components of the scanning tunneling microscope consists of a sharp tip, usually tungsten, which is mechanically scanned over the surface of the specimen. The scanning tip is prepared in such a way that it consists of a single atom. The size of this atom sets the resolving power. The tip is placed within a few angstroms of the surface of the specimen. In this position, the electrons can move, or tunnel, across the barrier between the tip and the specimen, which allows the examination of the density of electrons on the surface being scanned. In the usual case, the electrons are concentrated near the atomic nuclei, and, therefore, the measurement of the electron density gives us a measure of the position of the individual atoms. Since electrons are involved, the specimen used must be conductive.

A large amount of work has been done with this new instrument in studying the atomic arrangement on surfaces of single crystals of such materials as silicon and graphite, but the imaging power of the scanning tunneling microscope extends beyond such observations into the realm of biology. Primitive images showing some of the components of a bacteriophage, as well as DNA strands, have been recorded. Such results are more a demonstration of future potential, however, than an informative study.

The more useful and definitive work accomplished so far with the scanning tunneling microscope has been with organic molecules. Monomolecular layers of hydrocarbon chains, for example, have been imaged with definition sufficient to resolve the shape and spacing of the periodic array. Molecules of sorbic acid deposited on graphite substrates have been imaged, and their vibrational spectra have been identified.

The Atomic Force Microscope Holds Great Promise for Analyzing Biological Specimens

Even more informative images of biological molecules may come from the atomic force microscope. In the atomic force microscope, the tip is placed on a cantilever beam, which can deflect as the force on the tip is changed. In practice, the tip is scanned over the specimen, and the variation in force between the atoms on the tip and the atoms in the specimen gives information that is displayed in the image. These forces are small and the deflection of the cantilever minute. Nevertheless, several techniques can be used to measure these small deflections with great accuracy. Because tunneling electrons are not used with the atomic force microscope, it is possible to examine nonconducting samples on the atomic scale. This property indicates the great potential of the instrument for biological studies.

MAGNETIC RESONANCE

Magnetic Resonance Spectroscopy Is Becoming an Invaluable Tool for Determining the Structures of Complex Molecules

Magnetic resonance continues to grow as an increasingly powerful technique for gaining insights into a wide variety of structural and dynamic aspects of biologically important processes; much of this knowledge is unavailable from any alternative technique. Some of the specific present and future applications include (1) determination of structures of nucleic acids, proteins, and complex oligosaccharides; (2) measurement of dynamic aspects of these molecules and of interactions between them (for example, interactions between enzymes and substrates or between antibodies and antigens); (3) observations of metabolic events with cells, tissues, and isolated organs; and (4) medical applications, including noninvasive high-resolution imaging of humans and observation of metabolic activity in various organs.

Knowledge of the three-dimensional structures of proteins and nucleic acids has enormously advanced our understanding of biology; x-ray diffraction of single crystals has contributed greatly. In the past several years advanced techniques of magnetic resonance (such as two-dimensional spectroscopy) have allowed major structural questions to be resolved for molecules in solution (under conditions in which they exist in their native environments). These approaches reveal subtle, but biologically significant, deformations of DNA duplexes and important aspects of protein folding. Insights have also been gained into structures of complex oligosaccharides that, because they are generally not a single molecular species and do not crystallize, are not acessible to x-ray diffraction techniques.

Proteins and nucleic acids interact dynamically to accomplish their specific functions. Static structural techniques, such as x-ray crystallography, reveal little of these dynamic events; in contrast, magnetic resonance provides a powerful tool for studying dynamics: how enzymes bind substrates, how antibodies bind antigens, how receptors bind ligands.

Metabolic events within cells of all types can be directly observed by magnetic resonance; it is a completely nondisruptive technique for studying the complex interrelations among metabolic pathways, which is not possible with any other approach. These techniques have also been extended to studies of tissues and even intact organs such as the heart.

Magnetic resonance can yield three-dimensional images of human patients of the quality of drawings in anatomy texts; it has no requirement for injection of radio-opaque dyes. It is also becoming possible to observe the metabolic state of various internal organs by extensions of these imaging techniques that allow, for example, noninvasive monitoring of kidney function.

Both in areas of basic understandings of the structures and functions of biological molecules and in areas of diagnosis of human disease, magnetic resonance technologies offer versatile and powerful methods of gaining crucial knowledge inaccessible by alternative approaches.

COMPUTERS AND DATA ANALYSIS

Computers Are Coming to Play a Central Role in Modern Biology

To date, more than 15 million nucleotides and 1 million peptide linkages have been determined for genes and proteins. Computers play a central role in data-base management and pattern recognition. In modern biology it is now possible to take a newly isolated gene (or protein) and search through the entire DNA or protein data banks to determine whether it resembles any known gene (or protein). These so-called homology relationships can provide critical insights into the possible function of the corresponding gene product. Moreover, it is possible to take protein sequences and search for patterns of amino acid subunits that correlate with various forms of secondary structure. With the advent of automated DNA sequencing, the DNA data bank will enlarge rapidly. Hence, there will be a compelling demand for better methods to search for sequence patterns in large data bases. Clearly, these demands will require large and more rapid computers (for example, parallel or concurrent processing) as well as better software for pattern searches. It is important to stress the role large and fast computers will play in deriving the rules of protein folding: that is, how the primary sequence of amino acid subunits directs the three-dimensional folding of the polypeptide chain. Computers will also play a critical role in correlating three-dimensional structures with function. In addition to computer applications in structural biology, the use of expanding ecological data bases will require increasing sophistication as interests in modeling increase.

As biology moves toward an ever more detailed analysis of the chemistry of life, computers will play an ever-increasing role in data management, data analysis, pattern recognition, and imaging. The training of computer-literate biologists will be essential. Conversely, the training of computer scientists with greater understanding of chemistry and biology presents an immediate and compelling need.

BIOLOGY AND THE FUTURE

Synergistic Interactions of the New Biology Have Shortened the Time Between Fundamental Observations and Applications

The interactions among the various biotechnologies (such as recombinant DNA techniques, monoclonal antibody techniques, and microchemical instrumentation) are striking. Given a small amount of protein, the corresponding gene can be cloned through the use of protein sequencing, DNA synthesis, and recombinant DNA techniques. Alternatively, for a gene of known sequence, peptides matching part of the gene product can be produced that can in turn be used to generate antibodies to assay the corresponding gene product. If the gene is unusually short, its entire protein product can be synthesized. Moreover, given the protein sequence of a particular gene product, the corresponding gene can be altered for optimum protein production in bacteria, yeast, or mammalian cells.

Such techniques and instrumentation will help stimulate biological research into the next century. Much progress will come from the interaction of types of scientists who are working together for the first time. The new synergistic interactions will present new challenges, however, in areas including methods of information analysis and dissemination, funding strategies, training and education, and methods of balancing single-investigator research with that of large research centers. Our success in addressing these challenges will largely dictate the effectiveness of biology research programs in the United States, and hence the contributions of American biologists to the development of the global data base for the field.

3

Molecular Structure and Function

Biological Macromolecules Are Machines

All biological functions depend on events that occur at the molecular level. These events are directed, modulated, or detected by complex biological machines, which are themselves large molecules or clusters of molecules. Included are proteins, nucleic acids, carbohydrates, lipids, and complexes of them. Many areas of biological science focus on the signals detected by these machines or the output from these machines. The field of structural biology is concerned with the properties and behavior of the machines themselves. The ultimate goals of this field are to be able to predict the structure, function, and behavior of the machines from their chemical formulas, through the use of basic principles of chemistry and physics and knowledge derived from studies of other machines. Although we are still a long way from these goals, enormous progress has been made during the past two decades. Because of recent advances, primarily in recombinant DNA technology, computer science, and biological instrumentation, we should begin to realize the goals of structural biology during the next two decades.

Much of biological research still begins as descriptive science. A curious phenomenon in some living organism sparks our interest, perhaps because it is reminiscent of some previously known phenomenon, perhaps because it is inexplicable in any terms currently available to us. The richness and diversity of biological phenomena have led to the danger of a biology overwhelmed with descriptions of phenomena and devoid of any unifying principles. Unlike the rest of biology, structural biology is in the unique position of having its unifying principles largely known. They derive from basic molecular physics and chemistry. Rigorous physical theory and powerful experimental techniques already provide a deep understanding of the properties of small molecules. The same principles, largely intact, must suffice to explain and predict the properties of the larger molecules. For example, proteins are composed of linear chains of amino acids, only 20 different types of which regularly occur in proteins. The properties of proteins must be determined by the amino acids they contain and the order in which they are linked. While these properties may become complex and far removed from any property inherent in single amino acids, the existence of a limited set of fundamental building blocks restricts the ultimate functional properties of proteins.

Nucleic acids are potentially simpler than proteins since they are composed of only four fundamental types of building blocks, called bases, linked to each other through a chain of sugars and phosphates. The sequence of these bases in the DNA of an organism constitutes its genetic information. This sequence determines all of the proteins an organism can produce, all of the chemical reactions it can carry out, and, ultimately, all of the behavior the organism can reveal in response to its environment.

Carbohydrates and lipids are intermediate in complexity between nucleic acids and proteins. We currently know less about them, but this deficit is rapidly being eliminated.

The central focus in structural biology at present is the three-dimensional arrangement of the atoms that constitute a large biological molecule. Two decades ago this information was available for only several proteins and one nucleic acid, and each three-dimensional structure determined was a landmark in biology. Today such structures are determined routinely, and we have begun to see structures of not just individual large molecules, but whole arrays of such molecules. The first three-dimensional structures were each consistent with our expectations based on fundamental physics and chemistry. Most of the structures determined subsequently, however, were completely unrelated, and a large body of descriptive structural data began to emerge as more and more structures were revealed by x-ray crystallography. From newer data, patterns of three-dimensional structures have begun to emerge; it is now clear that most if not all structures will eventually fit into rational categories.

The Main Theme of Structural Biology Is the Relation of Molecular Structure to Function

Since biologists are ultimately interested in function, structural biology is often a means toward an end. The role played by structural biology differs somewhat depending on our prior knowledge of the function of particular molecules under investigation. Where considerable knowledge about function already exists, the determination of three-dimensional structure has almost inevitably led to major additional insights into function. For example, the three-dimensional structure of hemoglobin, the protein that carries oxygen in our blood stream, has helped us understand how we adapt to changes in altitude, how fish control their depth, and how a large number of human mutant hemoglobins relate to particular disease symptoms.

Often knowledge about structure can provide dramatic advances in our understanding about function even when prior knowledge is sketchy. For example, early biological experiments had shown that DNA contained genetic information, but these experiments offered no real clues to how a molecule could store information or how that information could be passed from cell to cell or from generation to generation. The structure of DNA, with bases paired between two different chains, led immediately to the correct conclusions about the mechanism of information storage and transfer. The information resided in the sequence of the bases; the apparent redundancy of two strands with equivalent (complementary) information meant that each could serve to pass the information onto a daughter strand. Furthermore, the redundancy offered a natural defense against loss of information. Even if one strand is damaged (as by chemicals or radiation), in the vast majority of cases the information on the other strand can be used to recover the missing information. Indeed, cells have evolved truly elegant mechanisms to determine which strand contains the original undamaged information; such models could provide useful paradigms for the current human preoccupation with electronic information handling.

The ultimate challenge for structural biology occurs when we have a structure but no clues at all about its function. Because of dramatic advances in our ability to determine structures, this challenge is likely to occur with increasing frequency. There have been a few remarkable cases in which limited structural information, such as a knowledge of the sequence of amino acid residues in a protein, without any three-dimensional structural information, has led to significant insights into function. In general, however, our current ability to predict function from structure in the absence of prior biological clues is limited, and one of our major needs is to improve our predictive abilities.

Biological Structure Is Organized Hierarchically

The structures of large biological molecules such as proteins and nucleic acids are complex. It is not practical or useful to describe these structures in words. In fact, highly specialized computer-driven graphics systems have been especially created to display molecular structures visually. An example of the output from one of these display systems is shown in Plates 1 and 2. Such devices are an invaluable aid to today's structural biologist, and future advances should make such devices cheaper, easier to use, and thus more readily available to all biologists.

Because of the complexity of biological structures, it is frequently convenient to deal only with certain aspects of these structures. It is common practice to describe structure at a series of hierarchical levels, called primary, secondary, tertiary, and quaternary structure. This hierarchy reflects some of the types of information provided by particular experimental techniques used to determine the structures of biological molecules.

The *primary structure* is the covalent chemical structure, that is, a specification of the identity of all the atoms and the bonds that connect them. The major molecules with which we work—proteins, nucleic acids, and carbohydrates— usually consist of linear arrays of units, each of which has a similar overall structure; they differ only in certain details. The types of units are limited in numbers: 4 common ones in typical nucleic acids, roughly a dozen in typical carbohydrates, and 20 in proteins. Thus, the primary structure can be specified almost completely by naming the linear order, or sequence, of each type of unit of the chain. The primary structure is given by the sequence plus a description of any additional covalent modifications or crosslinks.

The sequence of proteins, nucleic acids, and carbohydrates is determined principally by chemical methods. This is understandable since it is, in fact, the chemical structure. These methods have advanced tremendously in the past decade, and the implications of these advances constitute the second section of this chapter.

The *secondary structure* refers to regular patterns of folding of adjacent residues. Most secondary structures are helices. Some of the most frequent and best-known helices are the alpha helices found in many proteins and double helices found in virtually all nucleic acids. Carbohydrates also form helices. Helices are convenient structural motifs: They are easy to recognize by inspection of a known three-dimensional structure, they are relatively easy to detect experimentally by physical techniques, and their appearance within many structures is relatively easy to predict just from a knowledge of the primary structure.

The *tertiary structure* is the complete three-dimensional structure of a single biological unit. Until recently the only available method for determining this structure was x-ray diffraction studies of a single crystal sample. Now electron and neutron diffraction have become available as tools for solid samples, and nuclear magnetic resonance spectroscopy has been developed to the point where it can be used to determine the tertiary structure of small proteins and nucleic acids in liquid solution, that is, close to the state in which they are usually found inside living cells. The tertiary structure usually provides the starting point for studies that attempt to correlate structure and function.

Quaternary structure describes the assembly of individual molecular units into more complex arrays. The simplest example of quaternary structure is a protein that consists of multiple subunits. The units may be identical or different. The arrangement of the subunits frequently has important functional implications. Some quaternary structures have been determined by experimental methods that reveal not only the arrangement of the subunits but also their individual tertiary structures. However, many quaternary structures are too complex to be addressed by existing techniques. Here a variety of methods ranging from electron microscopy to neutron scattering to chemical crosslinking can still provide information about the overall shape of the assembly and detailed arrangement of the components.

In the sections that follow, we will first explore the levels of biological structures; our concerns will be improved methods for revealing these structures and the application of the resulting information to solving biological problems. We will then consider the current and future prospects for predicting the higher order structure of biological macromolecules from more readily available information on lower order structure. Finally we will consider the power of our newfound abilities to alter macromolecular structure more or less at will.

PRIMARY STRUCTURE

Nucleic Acid and Protein Sequence Data Are Accumulating Rapidly

The amount of available information on the primary structure of biological polymers is increasing at an astounding rate. Two decades ago we knew the nucleotide sequence of only a single small nucleic acid, the yeast alanine transfer RNA. We knew the amino acid sequence of fewer than 100 different types of proteins.

Today more than 18 million base pairs of DNA have been sequenced, and the data are accumulating at more than several million bases a year. The first completed sequences were research landmarks. Now sequences are appearing so rapidly that many research journals refuse to publish such information unless it has some particular novel or utilitarian aspects. Indeed, sequence data are currently accumulating faster than we can analyze them, and even faster than we can enter them into the data bases by existing methods.

The longest block of continuous DNA sequence known is the entire primary structure of Epstein-Barr virus. This 172,282-base-pair genome is responsible for a number of human diseases including infectious mononucleosis, Burkitt's lymphoma, and nasopharyngeal carcinoma. Knowledge of the DNA sequence potentially unlocks for us all of the secrets of the virus. The challenge now is to use this sequence information to learn how to prevent or control the diseases caused by the virus. Other landmarks of recent DNA sequencing include the complete DNA sequence of the maize (corn) chloroplast DNA (about 130,000 base pairs) and the complete sequence of the gene for human factor VIII, one of the proteins involved in blood clotting, which is defective in certain hemophilias. We know the complete sequence of many other important proteins, RNAs, and viruses. Perhaps what is most important is that we have the technical ability to determine the sequence of virtually any piece of DNA, RNA, or protein.

Sequence Comparisons Lead to Structural, Functional, and Evolutionary Insights

Much valuable comparative sequence information awaits us as the data accumulate and as analytic methods become more reliable and informative. Already, one can do much using the data bases to help interpret any DNA sequence plucked more or less at random from a genome. The patterns of sequence in the regions that code for the amino acid chains of proteins differ enough from the noncoding regions that the former can usually be identified. For example, we know about types of sequences that are required for efficient synthesis of proteins in many different types of organisms. We know about some general types of control elements for certain genes important in developmental pattern formation or in an organism's response to environmental stress.

When the protein sequence predicted from a gene is compared with all known protein sequences, there is about one chance in three that it will be similar enough to one or more of them to be recognized in a match. This provides an immediate clue to the function of the previously unknown protein. Perhaps the most spectacular example of such a match was the discovery that the product of the *sis* oncogene, a protein of unknown function that is associated with some cancers, was extremely similar to a blood protein that promotes normal growth, the platelet-derived growth factor. As the data base grows in size, and as our general knowledge about the function of its constituents does likewise, the probabilities of informative matches should rise steadily. One can anticipate the growth of a new speciality, molecular archeology, that resembles the field of archeology itself. Protein and gene sequences are old. They have been rearranged and altered much as the residual artifacts from a town or fortification have partially deteriorated and become dispersed. The components that remain, however, when properly viewed, provide clues to the function of the whole.

An archeologist might conclude that a room full of amphoras was likely to have been a storage room and not sleeping quarters. In the same way, we can already look at some protein sequences and gain clues about functions, even about functions that we have never observed in detail in the laboratory. Proteins that have transmembrane domains will sit in membranes, proteins with nucleic acid binding sequences will bind DNA or RNA; a protein with both would probably bring a nucleic acid into the vicinity of a membrane and keep it there. The analysis can be carried further because the details of the protein sequence can provide even greater clues, just as the details of the decoration on a piece of pottery or the shape of an arrowhead can identify the geographic origin of the people who produced it.

Many proteins with related functions have probably evolved from common ancestors. Thus receptors—proteins designed to sit at the cell surface and detect the environment—may represent one or more fundamental families of structures and sequences. For example, the sequence of the beta-adrenergic receptor, which binds the hormone adrenalin, and the sequence for rhodopsin, which detects light, are sufficiently similar that we can tell both were once related through a common progenitor. In the same way, proteases often resemble other proteases and structural proteins resemble other structural proteins.

Many proteins are modified chemically after they are synthesized. Proteases may remove one or both ends of the initial chain as well as make cleavages in the middle: Carbohydrates may be added to form glycoproteins. Some of these modifications occur as the protein travels from its initial site of synthesis to its final location in the cell. Others, such as phosphates, are added and removed repeatedly as part of the functioning or regulation of the protein. The enzymes that perform these modifications frequently do so by recognizing particular signal sequences. Because we now know some of these signals, a search of protein sequences can frequently reveal potential modification sites and in turn provide additional clues to the function of the protein.

Three-dimensional structure is better conserved in evolution than sequence is. Apparently there are severe constraints on folding a protein to make a compact three-dimensional array that is stable in the aqueous medium of a cell and resistant to proteases. Once we have mastered techniques for estimating possible folded structures from amino acid sequences, we will enhance our ability to explore molecular archeology. Mastery of these techniques itself will probably require an examination of many more three-dimensional structures by x-ray diffraction. What we still cannot do with much success is predict the function of an arbitrary protein without some molecular archeological clues.

When inspected by eye, a three-dimensional protein structure is complex and confusing; about the best most trained observers can do is identify potential binding sites as clefts or pockets and find potential sites of flexibility, such as connectors, between domains. Clues to functional regions can emerge from amino acids that are found in places other than their usual locations. For example, in typical soluble proteins, hydrophilic residues (which have an affinity for water), such as charged residues, reside on the surface, whereas hydrophobic (which avoid contact with water) residues, such as those with hydrocarbon side chains, are found buried in the interior. A buried charged group, particularly if it is not paired with an opposite charge, can be a clue to a functional site. Similarly an exposed hydrophobic group may reveal a binding site for a hydrophobic small molecule. A whole set of such groups may indicate a surface of the protein that interacts with another protein or a membrane.

One can go only so far with visual inspection. Methods for systematic analysis of three-dimensional protein structures are needed that can extract, from the structures, as many clues as possible about protein function. Such procedures are still in their infancy; the next decade should see rapid growth in such techniques now that a sufficient library of known structures and functions exists on which to develop, test, and refine these methods.

The DNA Sequences of Entire Genomes of Some Simple Organisms Will Soon Be Known

The explosion in sequence data has just begun. DNA sequencing is far easier than protein sequencing, and the tools already available for cloning and efficient sequencing of 500-base-pair blocks of DNA will ensure that the current stream of new sequence data will become a torrent.

The ultimate target would be to determine the sequence of all the DNA in an organism, that is, to sequence an entire genome. Genomes range in size from 750,000 base pairs (a mycoplasma) to more than 3 billion base pairs.

Such large-scale sequencing programs are feasible by today's technology, but they are expensive in both manpower and actual dollar cost. Automated DNA sequencing techniques have begun to be developed, which should markedly diminish manpower requirements and decrease costs. It now seems likely that in the next few decades we will determine the complete DNA sequence of the bacterium *Escherichia coli*, the yeast *Saccharomyces cerevisiae*, the human genome, the fruitfly *Drosophila*, the mouse genome, the nematode *Caenorhabditis elegans*, and possibly even a number of plant and other bacterial and yeast genomes. The resulting information will stimulate future generations of biologists as they explore the functions of the tens of thousands of genes that will be revealed for the first time by such sequencing programs.

The major issue facing us today is how to stage the process of large-scale sequence determination. One set of concerns related to this issue deals with the optimal scientific strategy and the selection of targets for sequencing. Another set of concerns deals with attempts to organize and accelerate this work by mechanisms other than the types of investigator-initiated individual research projects typical in current biological science.

Most investigators favor making a physical map of a genome before commencing really large-scale sequencing. This physical map will consist of an ordered set of large DNA fragments that covers the entire genome. From each large DNA fragment, smaller pieces can be isolated (or cloned) and used as source material to perform the actual DNA sequencing. Some workers favor constructing the ordered set of fragments by isolating individual ones at random and then determining which fragments are neighbors in the genome. Others favor dividing the genomes into successively smaller pieces—first chromosomes, then chromosome fragments, then very large DNA pieces—until an ordered set of DNA fragments is created. At present there are good arguments for attempting both approaches simultaneously.

The first physical maps of genomes or segments of genomes are almost completed. In principle, one could use these and simply start large-scale sequencing now, with existing approaches. However, the likelihood of major improvements in automated technology over the next 5 to 10 years leads many people to favor concentrating current efforts on speeding the development of that technology and delaying most massive sequencing until the technology is available. As automated DNA sequencing machines become common, the rate-limiting step in obtaining data will shift to the production of the DNA needed for sequencing. Thus, we need to enhance our ability to prepare large numbers of discrete DNA fragments (preferably kept in linear order from some larger starting fragment). Robotics seems an attractive and useful new technology. At present, most sequencing methods are limited to a maximum of about 500 base pairs per DNA fragment. Every significant increase in the size of the fragment that can be sequenced will improve the overall efficiency of the process. Multiplex methods, in which numerous different fragments are handled in parallel or in series, offer another way to accelerate and expedite the entire process.

A third set of scientific concerns deals with the choice of species, individual organisms, and genes to sequence. At one extreme are those who believe that current efforts should be restricted to sequences tied to existing biological problems. For example, in the pursuit of human disease genes, it might be far more important to determine the DNA sequence of the same gene in many individuals than to extend a given sequence into neighboring regions to see what is there. Similarly, comparisons of different species frequently provide biological insights that would not have been possible if studies had been restricted to a single organism.

The advantage of this traditional problem-oriented approach is obvious: The sequences obtained are more or less guaranteed to be interesting and useful. However, the disadvantage is also obvious: As interesting regions are sequenced, it will become more difficult to motivate people to risk explorations of regions of genomes for which little or no information is available. While explorations of these regions have the potential to make major advances through finding completely unexpected genes and functions, the work is also risky since some regions may yield no rewards at all. The realities of tight funding and frequently competitive review for renewed funding militate against such work; if it is to be encouraged, new support mechanisms may need to be created, with longer term commitments and rewards for more risk-taking.

The final set of concerns deals with whether genomic sequencing should be organized in ways similar to those in which "big science" has been dealt with in other disciplines. The actual process of sequence determination is boring. It seems to require more dedication and large-scale organization than most typical biology projects. The intellectual rewards of obtaining sequences of entire genomes are likely to be missed by most of those involved in the massive effort to accumulate the data. Much of the data may not result in publications in the primary scientific literature, and some publications that do result may have very large numbers of authors. Thus special efforts may be needed to maintain investigators' morale.

Structural and Computational Methods Need to Advance to Keep Pace with the Explosion in Sequence Data

As the acquisition of sequence data continues to accelerate over the next few years, the problem of managing these data will become increasingly severe. Considerable thought and resources will be needed to optimize the collection of data in consistent formats, the entry of data into computerized data bases accessible to all investigators, and the refinement of computer algorithms for all sorts of sequence and structure analysis. The anticipated size of the data base—100 million base pairs in the next few years, 10 to 100 billion base pairs eventually—is not staggering even by today's standards. However, the way the data are being accumulated, by efforts in hundreds of different laboratories, each with its own computer systems and idiosyncracies, poses a serious problem. What would help is a relatively uniform system of data annotation and transmittal. If this can be done by translation programs that accept a wide variety of inputs and return them to the data base and to the investigator in the standard format, it would probably win broad acceptance by the community because each laboratory could then maintain its own style.

A second complexity of the existing data bases is that there are three independent repositories for nucleic acid sequence data: GenBank, operated by the Los Alamos National Laboratory; the EMBL data base, operated by the European Molecular Biology Laboratory in Heidelberg; and Protein Identification Resource, operated by the National Biomedical Research Foundation in Washington, D.C. The multiplicity of data bases poses severe problems for current and potential users. Ideally, the three should be combined into one.

Nomenclature is a major problem for all three data bases. The names of molecules, species, and genes are constantly changing, and the data bases also change the cryptic names that they use to identify entries. The various data bases are also not cross-indexed. A major research problem is to determine what data are common to two or all three data bases. Moreover, any such cross-comparison is outdated as soon as one of the data bases is updated. International responsibility for entering data into major data bases and greatly expanding the use of electronic communication for this purpose is badly needed. Someone will have to construct and maintain a cross-index of related biological data bases. Possibly a direct-access system will be set up. In any case, the availability of a periodically updated cross-index will allow other installations to provide an integrated retrieval system to their users more easily.

The third major complexity of the sequence data base is the sophistication of many of the interrogations that will be made. Today each new sequence is almost automatically run through a comparison program to see what matches can be found with preexisting sequences. At the most trivial level, this procedure may reveal that the sequence has already been reported by someone else, and perhaps the same gene will have been reported under another name. At a more profound level, the comparison may reveal functional and structural insights. These comparisons can consume large amounts of computer time if they are carried out with algorithms that try to detect even very slight degrees of sequence similarity.

In the near future, we should begin to see many more attempts to use the data bases to refine methods for structure prediction, studies that will consume enormous amounts of computer time. It seems prudent to plan ahead and support research and development of better computer algorithms and better computer hardware to optimize the biologist's use of the DNA data base. Among the needs are database management systems designed to keep track of inquiries and results, so that insights gained by different inquiries can be synergistic and so that unwarranted duplicate inquiries can be short-circuited. Other needs are for improved analytical tools for predicting structure and for comparing structure and sequence. These tools may take the form of new chips, parallel processors, or more powerful algorithms.

Carbohydrate Research Is Gaining Momentum

In the past decade, structural studies on carbohydrates have begun to approach the capabilities of more developed areas of protein and nucleic acid structure. Techniques have been developed to deduce the complete structure of complex oligosaccharides, including oligosaccharides found in scarce glycoproteins, such as cell-surface molecules.

Glycoproteins are proteins containing covalently attached sugars, usually short carbohydrate polymers attached to the side chains of the amino acids asparagine, serine, or threonine. Glycoproteins are found throughout nature, from simple single-celled organisms to humans, and they play critical roles in these organisms. Glycoproteins are usually, but not exclusively, found on the surfaces of cells and in cellular secretions. For example, almost all of the human blood proteins and all of the well-characterized eukaryotic cell-surface macromolecules are glycoproteins. In addition, glycoproteins are key components in the outer coatings of a number of pathological agents, including viruses and parasites. Many of the molecules used by the immune system to combat these pathogens are also glycoproteins. Recently, important roles have been identified for some glycoproteins that remain in the cell's interior, such as the proteins that form the pores in the nuclear membrane.

The new techniques in carbohydrate research include nuclear magnetic resonance (NMR), fast-atom-bombardment mass spectrometry, and metabolic labeling with radioactive sugars combined with stepwise degradation with a battery of purified glycosidases. The consequence has been the elucidation of hundreds of oligosaccharide structures, which has enhanced our understanding of the actual repertoire of structures synthesized by various organisms and cell types. These data form the basis for studies of the biological role of oligosaccharides.

The techniques have also been invaluable for studies of the biosynthesis of complex oligosaccharides. One important discovery is that a lipid-linked oligosaccharide serves as the precursor of asparagine-linked oligosaccharides. It is noteworthy that this lipid-linked precursor oligosaccharide structure has been highly conserved; it is the same in yeast as it is in mammals, and presumably occurred in a common ancestor of fungi and animals.

Several Major Advances in Studying Carbohydrates Can Be Anticipated

New NMR methods will enable the determination of the three-dimensional structures of oligosaccharides. This information will greatly enhance studies of the interaction of oligosaccharides with carbohydrate-binding molecules, such as receptors and lectins, and with glycosyltransferases. In 1986 the first partial complementary DNA (cDNA) clone for a glycosyltransferase was obtained. That a number of laboratories are focusing on this area suggests that the genes for many of these enzymes will be cloned soon. This achievement will provide valuable information about the structures of the molecules that synthesize complex carbohydrates and whether or not they exist as gene families. The clones will also be used for studying the catalytic function of the enzymes and for manipulating their cellular levels.

The organic synthesis of complex oligosaccharides has lagged behind their biochemistry. Improvements in synthesis have been made, and it seems likely that progress will continue. The ability to synthesize large quantities of complex oligosaccharides of known structures will be of considerable value to the investigators. Many laboratories are trying to implicate oligosaccharides in a variety of biological interactions, ranging from adhesion between the cells of sponges to sperm-egg interactions to the inhibition of growth that occurs when certain cells contact each other. The availability of synthetic oligosaccharides, of inhibitors that alter oligosaccharide biosynthesis, and of endoglycosidases that remove oligosaccharides from glycoproteins will be useful in these studies.

Trying to understand the features of glycoproteins that are distinct from those of unglycosylated proteins is of current interest. Contributions of sugars to protein folding and macromolecular assembly might be fundamentally different from those of amino acids. The physical properties of sugars differ from those of amino acid side chains. They can confer special mechanical and other properties, such as those of fish-antifreeze glycoproteins, mucins, or connective tissue proteoglycans. The vast diversity in the types of sugars and the ways they are connected suggest that oligosaccharides on glycoproteins (and glycolipids) can potentially encode large amounts of information, which may play an important part in the development of multicellular organisms, recognition in immune systems, and information storage and processing in nervous systems.

The main outstanding question for the future is, What function, if any, does the carbohydrate play in determining glycoprotein function? With the exception of the specialized physical properties imparted to mucins and proteoglycans, we do not know why most glycoproteins are glycosylated. In some studies of the functions of oligosaccharides on glycoproteins, the proteins appear to function perfectly well without the carbohydrates, whereas in others glycosylation is essential. We do not understand the basis for these different results; no underlying rules have been uncovered.

The main obstacle to understanding function is the lack of proper assays. For instance, the finding that oligosaccharide structures are dramatically altered during embryonic development has led to the proposal that surface oligosaccharides function as recognition molecules during development. However, there is no direct evidence for this. What is lacking is the ability to manipulate the oligosaccharides in these systems and to devise assays to test for effects. The same type of problem faces investigators studying the potential roles of oligosaccharides in other systems. One way to improve this situation would be to develop more inhibitors that are specific for particular glycosyltransferases or processing enzymes. This would make it possible to alter the oligosaccharides that cells synthesize. Another approach is to transfect active genes for particular glycosyltransferases into cells that normally do not express these genes. The objective would be to see how cell behavior is affected when the cells synthesize oligosaccharides with altered structures.

THREE-DIMENSIONAL STRUCTURE

The Three-Dimensional Structure of Biological Macromolecules Determines How They Function

It is the three-dimensional shape of proteins and nucleic acids that endows them with their biological activities. Structural molecular biology uses x-ray diffraction, nuclear magnetic resonance, and other techniques to determine the three-dimensional arrangement of the atoms in biological molecules. Studying these detailed atomic arrangements provides insight into how the molecules fold and how their folded surfaces can act as biological catalysts, as recognition and adhesion devices, as architectural elements in cells, and as the storage libraries of genetic information.

X-Ray Diffraction Provides Highly Revealing Images of Molecules

By recording the scattering pattern produced by shining x-rays on a crystal, it is possible to compute an atomically detailed image of the molecule forming the crystal. The x-ray diffraction method, pioneered in the early part of this century, can now be applied to crystals as simple as those of table salt or as complex as those formed from polio virus.

The particular power of x-ray diffraction is that, because x-rays are just a very short wavelength form of light, the method produces a three-dimensional image of molecules in the same sense that our eyes produce images from light. Thus we can literally see the arrangement of atoms and distribution of electrons in biological molecules, rather than have to infer such information from indirect experiments. The experimental difference between x-ray diffraction and seeing (for example, in a microscope) is that there exist no x-ray lenses. This lack has been overcome by exacting experimental methods involving binding heavy metal atoms to proteins as atomic markers and by using a computer in place of an x-ray lens to transform the raw x-ray scattering pattern into an image. The latter operation is possible because we know the mathematical equations governing the actions of lenses.

The three-dimensional atomic structures of more than 300 proteins are now known. X-ray diffraction continues to unveil between one and two dozen new protein structures each year. The distribution of atoms in any given protein is bewilderingly complicated, difficult to describe, and satisfactorily revealed only in three-dimensional stereoscopic drawings. However, now that so many structures have been determined, structural themes are seen to recur, allowing recognition and categorization of structural motifs.

Structural Motifs Are Repeatedly Used to Carry Out Similar Functions

A structural motif composed of three extended strands and two helical coils of protein, named the nucleotide-binding domain, is found as part of many enzymes that bind nucleotides, such as adenosine triphosphate (ATP). Recently, from the nucleotide sequence of the gene encoding a protein associated with bladder cancer in humans, the nucleotide-binding domain was correctly predicted to be part of this protein's three-dimensional structure.

Another structural motif has been seen in some of the proteins that recognize specific sequences of DNA and consequently regulate genes by turning them off or on. In this instance, two helical coils of protein connected by a short bend form a module that can plug into the major groove of a DNA double helix. The atomic surface of this recognition-helix motif is different in different proteins, imparting to them the ability to recognize and bind tightly to different specific sequences of DNA. As a result, one protein turns on one specific gene, whereas another might turn off another gene. Understanding how this structure functions has allowed scientists to synthesize novel regulators of genetic information.

Three-Dimensional Structures Give Insight into How Proteins Fold and How Groups of Proteins Assemble

A number of metabolic enzymes have similar structures that look strikingly like a barrel. Each stave of the barrel resembles a hairpin composed of one extended strand folded back against one helix. Genetic studies have shown that each stave is encoded by a unit of genetic information—an exon—and that the barrels are formed by stringing together eight of these genetic units. Correlating the genetic map of the staves with about a dozen three-dimensional structures of various barrels has led to the suggestion that the staves may derive from independent folding units. The discovery of such substructures may help unlock the mystery of how proteins fold by letting us see the primitive folding units that first appeared as life evolved. Maybe we can learn how to fold proteins by seeing how nature learned to do it.

The Major Goal of Protein Crystallography Is to Show How Proteins Function

The majority of the proteins of known structure are enzymes. When the unique three-dimensional spatial information from the structure is combined with a vast array of observations on the properties of the protein, many of the secrets of how enzymes catalyze chemical reactions are revealed. For example, we know how enzymes can bind certain substrate molecules specifically, how certain amino acid side chains are positioned to act as catalysts, and how the enzyme can change its shape in response to binding the substrate or regulatory molecules. We know that sometimes these shape changes can be transmitted through the structure. We also know how the same structural motifs can be varied in different proteins to produce a series of enzymes with similar mechanism but different substrate specificities. For example, a whole set of proteases have extremely similar active sites, all of which contain an activated serine residue.

Photoreaction Center. In 1987, the structure of the first membrane assembly involved in photosynthesis, the photoreaction center, was determined. This complex of four proteins converts light energy into an electrical gradient across a membrane (Figure 3-1). The structure immediately showed the path through the protein that electrons traverse to cross photosynthetic membranes, a key initial step in the conversion of light into chemical energy.

This structure also gave major new insights into the exact nature of packing of alpha helices in a detergent micelle that is thought to mimic the lipid bilayer environment in a real membrane. It revealed a logical relationship between the length of particular hydrophobic sequences and the angle at which they cross the membrane. It also showed that other helices were actually located at the surface

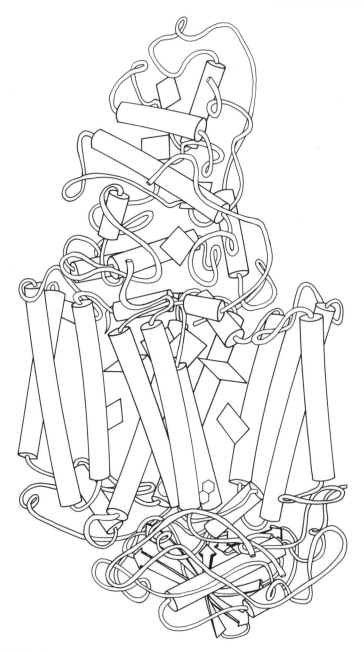

FIGURE 3-1 Cartoon of the three-dimensional structure of photoreaction center. [L. Stryer, Biochemistry (Freeman, ed. 3, New York, 1988)]

of the micelle and that these had arrangements of hydrophobic and hydrophilic residues that enabled them to interact favorably with both the lipid in the micelle and water outside it. Thus the structure of the photoreaction center provides the first understanding at high resolution of the interface between a protein and the lipid bilayer.

It is possible to diffuse small molecules (enzymatic substrates, inhibitors, reaction products, or drugs) into protein crystals because the spaces between proteins in protein crystals are large, water-filled channels. A herbicide was soaked into the crystals of the photoreaction center to see how it bound to the protein. The three-dimensional atomic image of a photosynthetic protein being inhibited by a herbicide provides a new level of understanding for developing agricultural products.

Antibodies. Antibodies are protein molecules that play a key role in an organism's defense against foreign molecules called antigens. The first step in this process is specific complex formation between an antibody molecule and a molecule of antigen. The structural basis of this antigen-antibody recognition is beginning to be elucidated. Structures of certain antigen-binding fragments of antibodies were determined some time ago. These structures demonstrated directly that certain regions of antibodies were extremely variable in conformation from one antibody to another. Six stretches of amino acids, containing these hypervariable sequences from the heavy and light chains (See Chapter 7) of the antibody, are adjacent in space. Together these describe a surface believed to be complementary to the recognized region (epitope) on the antigen used to raise the particular antibody. Binding studies with small-molecule fragments of antigens (haptens) revealed the importance of these complementarity-determining regions in forming a complex of small molecules, and by implication, macromolecular antigens. This work was responsible for providing an intellectual framework for understanding and investigating antigen-antibody interactions. In 1986, the first structures of complexes between protein antigens and antibodies of the immune system were determined. These direct pictures of how antibodies recognize an enzyme and a protein from the surface of influenza virus are providing fundamental information about how our immune systems recognize and destroy foreign molecules.

Future challenges will be to characterize the complex antigen-receptor interactions by which different types of cells participate in and regulate immune responses. The physiological-cellular description of the process is already giving way to a molecular characterization. Genetic engineering, cell culture, and hybridization techniques will allow production of molecules directly involved in the cell communication process and in the elaboration of the response to antigenic simulation.

Knowledge of How Structures Function Is Being Used to Design Drugs

The three-dimensional atomic structure was recently determined for a complex between a protein on the outside of the 1968 Hong Kong influenza virus and the receptor molecule on human cell surfaces to which the virus binds to initiate an infection. By observing the atomic details of how the virus binds to a cell, scientists hope to design and synthesize drug molecules that can mimic the cell receptor. Such inhibitors would prevent infection by binding to a virus, interfering with its ability to attach itself to and to infect a human cell.

Dihydrofolate reductase is an essential enzyme for cell growth. It is the target both for antibacterial drug design and for chemotherapeutic agents that arrest human cancers. So far, studies of the three-dimensional structure of the enzyme complexed with various inhibitors have resulted in the development of the new antibacterial drug trimethoprim.

The antihypertensive drug captopril lowers blood pressure by inhibiting angiotensin-converting enzyme that normally produces a substance that constricts blood vessels and raises blood pressure. Captopril was designed by studying the three-dimensional structure of a digestive enzyme related in its chemical activity to the angiotensin-converting enzyme and by synthesizing a compound that would fit tightly to and block the active site of a converting enzyme.

Drugs that bind to the blood protein hemoglobin, preventing it from aggregating in individuals with the hereditary sickle-cell trait, are also being studied. Knowledge of the structure of β-lactamase, an enzyme that bacteria use to destroy penicillin and similar antibiotics, has opened the way for foiling the method by which bacteria resist drugs.

Designing drugs by understanding the atomic details of how inhibitors fit onto the surfaces of proteins and block normal activities is just beginning. Many believe we are entering a new era of drug discovery based on designing molecules for stereochemical fit to their targets by actually seeing the target molecules with substrates and the drugs that are bound to them.

DNA Is a Dynamic Molecule That Can Switch Between Different Structural States

When the double helix structure for DNA was first announced, it was an instant public success. It represented a neat solution to a number of chemical and biological problems, and it was easy to describe and to remember. The importance of pairing between bases on the two DNA strands and stacking of adjacent bases along each individual DNA strand is overwhelming in nucleic acid structures. In terms of relative importance to the overall structure, there are no counterparts in proteins. However, with time, the structure of DNA has been found to be much more complex than was originally thought, since there are a variety of different double helical structures. The diversity of such structures has dramatically altered our thinking about the DNA molecule.

To date, the folding of DNA has been largely thought of as the assembly of the double helix through formation of successive base pairs. The insertion and deletion of extra helical twists in circular DNA molecules has forced attention on the topology of these complex systems and has presented a massive mechanistic problem at the enzyme level. The bending of the double helix and its control by sequence variation is also under intensive investigation. The double helix was initially thought to be rigid. In view of the compact packing required in the nucleus of the cell, bending was obviously essential. However, the structural details of the contortions that the double helix can actually undergo have only recently been recognized. The structure of the nucleosome, now known at high resolution with its coiled double helix and protein core, is a beautiful example of the biological importance of bending. Other proteins that interact with DNA can also induce bending.

The dynamic aspects of the equilibrium structures of DNA have become clear with direct experimental measurement of the swinging in and out of individual bases to and from the axis of the helix. Larger scale motions on a much longer time scale are revealed by pulsed field gel electrophoresis, which separates molecules of enormous molecular weight.

RNA Structure Is a More Challenging Area of Research

The structure of RNA has been even more difficult to deal with than that of proteins and DNA. Only the smallest class, transfer RNA, has yielded any solved crystal structures. All the transfer RNAs turn out to be similar L-shaped molecules. This similarity is reflected in the cloverleaf model for secondary structure, originally derived by searching for similar base pairing possibilities within the single chains. Although no other RNA structures are yet available through diffraction procedures, the extensive use of sequence data and sequence homology has led to a large array of secondary structure predictions that will almost certainly be retained in the three-dimensional structures eventually determined. Nuclear magnetic resonance (NMR) is starting to provide a substantial amount of structural information on RNAs, but diffraction-quality crystals would be enormously useful.

One of the most unexpected discoveries of the past few years is the catalytic activity of certain RNAs in RNA splicing reactions (See Chapter 4). The enzyme RNA polymerase produces an RNA copy, called the transcript, from the gene in the chromosome. In eukaryotic cells, such a transcript frequently has several long stretches (introns) that interrupt the functional portion, that will form a messenger RNA (mRNA), a transfer RNA (tRNA), or a ribosomal RNA (rRNA). These introns are precisely cut out of the transcript and the functional structures (exons) are rejoined to yield the mature RNA in what is called an RNA splicing reaction. In some cases, this reaction can be carried out by the intron RNA itself without the help of any proteins. These RNA molecules are the first known examples of true, nonprotein biological catalysts. The details of the highly organized three-dimensional structures of these catalytic RNA molecules have not yet been unraveled.

Much Remains to Be Learned About the Structures of Carbohydrates

Significant by its absence in the above discussion is any mention of the three-dimensional structure of polysaccharides (a carbohydrate made up of a large number of sugar molecules). As mentioned earlier, these substances have been particularly intransigent in yielding high-resolution structural data. Only the smallest compounds have provided truly crystalline material. Most studies have been chemical or spectroscopic. In view of their unquestioned biological importance, much greater effort on the three-dimensional structure of this class of polymers is indicated. We do not even know whether such molecules have unique three-dimensional structures.

A Technical Breakthrough Promises Information About Dynamic Processes in the Function of Proteins

The massive electron-storage rings that physicists use to probe the fundamental components of matter also emit x-ray beams high in power. These synchrotron x-ray sources have recently been used to study large biological molecules. The beams of x-rays are thousands of times as strong as those from conventional laboratory x-ray sources, reducing x-ray data-collection time from months to hours. An experimental breakthrough in the application of multiple-wavelength x-ray diffraction now provides exposure times of milliseconds. The biochemical events on the surface of a protein can therefore be studied by a series of snapshots of the structure every few milliseconds. This should allow the sequence of events that constitute a chemical reaction or protein conformational change to be understood in atomic detail. Examining the dynamics of fundamental biological reactions will deepen our understanding of how proteins work, provide insight into normal functions, and raise the possibility of understanding abnormal functioning in disease.

Crystallography Will Continue to Increase in Importance

The future for structural biology is particularly bright at present because two factors have coincided. First, the recent explosive growth in the power of molecular biology, as a result of gene cloning and recombinant DNA technology, suddenly provides a large amount of any given macromolecule and the ability to modify these at will, to test or alter their functions. This brings the fundamental molecules at the basis of almost every process in living systems into the range of structural study.

Second, as the discovery of new molecules has accelerated, the technology by which x-ray structures are determined has undergone a rapid evolution. New methods and algorithms have made determining x-ray structures easier, but most important, because x-ray crystallography is highly technical, it has benefited enormously from the recent leap in computational power and computer-controlled instrumentation.

Nuclear Magnetic Resonance Is the Technique of Choice for Studying Molecular Structures in Solution

Recently, NMR, a structural and analytical tool used by chemists for many years (Chapter 2), has made rapid progress in providing information in structural biology. NMR is a spectroscopic technique in which the absorption of radio-frequency energy is measured for the nuclei of molecules placed in highly magnetic fields. Because the absorption frequency is related to the chemical environment of a nucleus, NMR measurements provide structural details such as inter-atomic distances and conformational angles from samples in aqueous solution.

The spectra are very complex. The recently developed procedures for spreading out the absorption peaks in two dimensions have dramatically improved both resolution and the ability to assign each peak to specific atoms in the macromolecule. Intricate patterns of radio-frequency pulses can be used to collect information on which atoms are close to which others. As with all spectroscopic techniques, each absorption process occurs on a different, characteristic, time scale, and it is sensitive to events that occur on that same time scale. Thus, the dynamic behavior of many parts of the molecule can be directly measured over a broad frequency range.

Nuclear Magnetic Resonance and X-Ray Diffraction Form a Strong Partnership

NMR and x-ray diffraction provide both overlapping and distinct information about molecules. Recently, the chain-folding of a small protein was determined from an analysis of interatomic distances provided by NMR. X-ray diffraction simultaneously verified the structure, confirming as a side benefit that the structure of a protein in solution as seen by NMR is the same as that in a crystal as seen by x-ray diffraction. We can now confidently predict that NMR will make it possible to determine a series of structures of small proteins in solution.

The two techniques are complementary in the nature of the information that they can best supply. X-ray crystallography can provide precise atomic positions for almost all the atoms in a macromolecule, and it can be applied to very large molecules with no size limit yet found. However, it requires crystals of excellent quality, poorly ordered regions cannot be defined, and no estimates of the rates of certain types of molecular motion can be inferred. NMR can provide structural information, but not yet with the precision obtainable from x-ray structures. So far, NMR studies have produced atomic level data for only very small proteins. However, the procedure uses solutions rather than crystals, it can provide information on flexible regions, and it can reveal the times required for many dynamic processes. Both techniques are rapidly improving their capabilities and are likely to continue to dominate structural biology in the near future.

MOLECULAR ASSEMBLIES

The past decade has seen major advances in our ability to study the structure of molecular assemblies. These are aggregates of individual macromolecules, most frequently complexes between proteins or proteins and nucleic acids. Among the major triumphs have been solution by x-ray methods of large structures such as the nucleosome, the photosynthetic reaction center, an antibody-antigen complex, and spherical viruses. Another highlight was the solution of the structure of a protein within its native membrane by electron microscopy; this structure is still at relatively low resolution, but it should be possible to extend it to 3 angstroms.

Understanding of the molecular architecture and function of some key filamentous complexes and organelles—such as actin filaments decorated with the myosin head, the ribosome, and clathrin-coated vesicles—has significantly advanced through the use of three-dimensional electron image reconstruction combined with in vitro reassembly. Also important has been the use of deuterium labeling and neutron scattering to derive three-dimensional maps specifying the relative locations of macromolecular components in large complexes such as RNA polymerase and the *E. coli* ribosome. Crucial insight, paralleling physiological and biochemical data, has been gained into dynamic processes such as muscle contraction and microtubule assembly through the use of time-resolved synchrotron x-radiation.

New Technology Has Improved Our Ability to Determine Complex Structures

These successes have stemmed in the main from technological developments and breakthroughs. In x-ray crystallography, particularly of viruses, the progress can be traced to four advances: (1) the introduction of methods to acquire and process high-resolution data from very large repeating units (such as crystals with very large unit cells); (2) the development of methods that take advantage of the symmetry of many molecular assemblies in solving the structure; (3) the tremendous advances in the speed, size, and affordability of computers; and (4) the development of computer graphics that efficiently communicate the information obtained.

An additional major breakthrough was learning to crystallize membrane protein assemblies in the form of protein-detergent micelles. Crystallization has been achieved now for several such membrane proteins, including the photosynthetic reaction center, *E. coli* matrix porin, and bacteriorhodopsin. Also of particular significance for membrane structure has been the progress made in electron microscopy. Preparative methods that preserve specimens have been developed as have mathematical analyses of the image that can provide three-dimensional details. The advent of cryo-electron microscopy, in which wet specimens are prepared by being frozen so rapidly that the water remains amorphous rather than crystalline, has made it possible to see functional molecules in action for the first time.

Nucleosomes. The nucleosome is the fundamental repeating structural unit that makes up the chromosomes of all eukaryotic cells. The elucidation of the molecular details of the nucleosome necessitated the use of a wide range of newly developed physical and biochemical approaches. The existence of nucleosomes was first recognized by electron microscopy and from the finding of a unit of histone organization that could explain the nuclease cutting pattern of chromosomal DNA. The nucleosome was perceived and eventually proven to consist of a histone octamer about which is wrapped approximately 160 base pairs of DNA, with a single molecule of an additional histone bound on the outside. Electron and x-ray crystallographic analyses showed that the DNA is coiled in two left-handed turns around the central histone octamer. With the methods used to analyze nucleosome digestion, and with the use of cloned DNAs, the chromatin organization of many genes has been studied. It is now understood that transcriptionally inactive genes are packaged in nucleosomes, but the active genes are organized differently, with regulatory sequences in special exposed regions and with gene and flanking sequences in altered nucleosomes or in novel particulate structures. Challenges for the future are to determine how chains of nucleosomes are further coiled or folded in condensed states of chromosomes, to elucidate the mechanism of unfolding that accompanies gene activation, and to solve the structure of transcriptionally active genes.

Membrane Proteins. Major insights into the three-dimensional structure of membranes came from the electron microscopic analysis of bacteriorhodopsin—a light-driven proton pump—in the purple membranes isolated from the cell membrane of the bacterium *Halobacterium halobium*. The structural maps provided the first case in which a membrane protein was shown to be composed of a bundle of alpha-helical segments. There were seven such segments, closely packed in a left-handed configuration extending roughly perpendicular to the plane of the

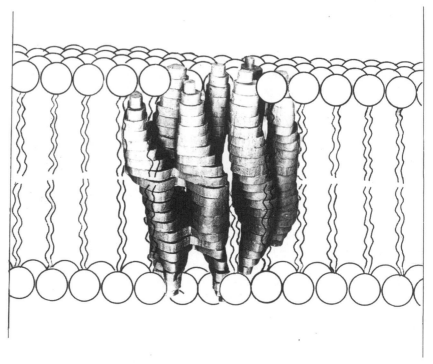

FIGURE 3-2   Bacteriorhodopsin as it sits in a bilayer [R. Henderson and P. N. T. Unwin, Biophys. Struct. Mech. 3:121 (1977)]

membrane bilayer for most of its width (Figure 3-2). The amino acid sequence for a number of cell-surface receptors such as rhodopsin, the beta-adrenergic receptor, and the muscarinic acetylcholine receptor have recently been determined. Analysis of the patterns of secondary structure and distribution of hydrophobic residues predicted from these sequences indicate that the 7-alpha-helical bundle is a recurring motif among cell-surface receptors. The quaternary structures of protein oligomers that form membrane channels, such as the nicotinic acetylcholine receptor and the gap-junction connection, have become better understood through the same approach.

Viruses. Viruses consist of nucleic acid (RNA or DNA) packaged in a multi-subunit protein coat. The three-dimensional structures of a number of plant and animal viruses have been determined by x-ray diffraction; they show assemblies of as many as 180 proteins forming icosahedral shells that package the virus's genetic information (Figure 3-3). Each determination of a viral structure has been a triumph of both persistence and innovation, and we have learned much from every new step. Determining the structures of the simple plant viruses was a major advance. These structures were far larger than any previously determined by crystallography; they provided our first real insights into viral architecture at the molecular level. The viral images also led to insights into how viruses assemble themselves. One day we may learn to use this insight to design strategies to prevent viruses from assembling as part of an effort to control the infections they cause. Recently, the structures of two animal viruses (rhinovirus and poliovirus) and the structure of the adenovirus hexon have revealed that the molecular topology of their coat proteins is essentially the same as that of the simple plant viruses. Thus all such viruses may have a common evolutionary ancestor.

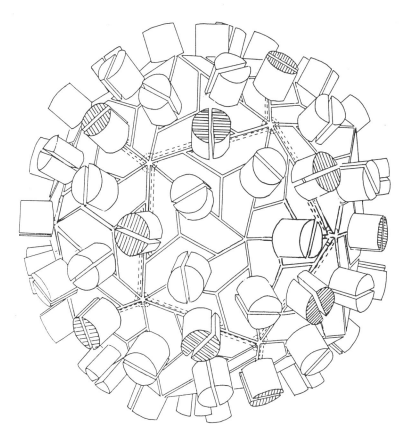

FIGURE 3-3   A schematic diagram of the location of the 180-protein subunits on the surface of the tomato bushy stunt virus, as determined by x-ray diffraction. [S. C. Harrison et al., Nature 276:368-373 (1978)]

The structures of the human rhinovirus (common cold virus) and poliovirus have especially important implications for medical research. From these structures it has been possible to ascertain the sites of attachment of various types of neutralizing antibodies and the sites of binding for a series of experimental antiviral drugs that are suspected of inhibiting virus replication by preventing the low pH-mediated uncoating of the viral RNA. The structural results show that these antiviral compounds insert themselves into the hydrophobic interior of one subunit of the protein coat, suggesting how the drugs may inhibit the disassembly of the virus.

In the near future, this newly developed technology should allow a fairly rapid survey of the structures of numerous viral pathogens or their components and thereby provide a good deal of information about how they work. We should come to understand the mechanism of neutralization of viruses by antibodies by looking at the structure of neutralizing monoclonal antibodies and their complexes with the virus. Moreover, it should become possible to design more powerful antiviral drugs that interfere with attachment, penetration, uncoating, or assembly.

How Our Immune System Recognizes Viruses. In the past five years, the determinations of the crystal structures of poliovirus, rhinovirus, and both of the surface proteins of influenza virus have allowed us to visualize those parts of the viruses recognized by the human immune system. Monoclonal antibodies and the amino acid sequences of many strains of viruses have made it possible to map the regions of each virus that are attacked by human antibodies. Plate 3, for example, shows five sites (A-E) on an influenza virus that bind antibodies. How these antigenic sites vary in structure every few years, resulting in new epidemics of "the flu," is now under study.

Even Larger Structures Will Be Solved During the Next Decade

Research on molecular assemblies over the next decades is likely to involve extending x-ray analytical methods to still larger aggregates, such as ribosomal subunits, and to more difficult materials, such as membrane complexes and networks of peptides and sugars called proteoglycans, which are materials that play important roles in the interactions between cells. There will be greater use of clusters of heavy atoms (rather than simply individual atoms) to determine the structures. Area detectors and synchrotron radiation will be used to collect x-ray data. The increase in power and accessibility of computing resources will greatly benefit the data processing. With symmetrical assemblies, new methods may eventually allow the determination of new high-resolution structures without the need to resort to heavy-atom methods. Since progress on each of these points is already being made, one can anticipate that early in the next decade determining the structure of small viruses will be routine and that a number of larger viruses will also be solved.

Electron microscopy of rapidly frozen specimens, in the form of two-dimensional crystals or isolated molecules and in conjunction with heavy-atom cluster labels, should provide a wealth of new information on molecular assemblies trapped in different conformational states in nearly physiological conditions. Methods of three-dimensional image reconstruction from crystalline specimens will be further refined by computational procedures. Averaging methods to extract details from isolated molecules will become more powerful. With further improvements in methods for electron microscopy at very low temperatures, we can expect images of two-dimensional crystals in some instances to yield atomic resolution and, more generally, resolution at the secondary structure level. These images, when combined to derive three-dimensional maps, will provide essential frameworks upon which other diverse information can be added to build up detailed pictures of molecular structure and action. Oligomeric membrane proteins may become fairly well understood by such an approach, since the lipid bilayer imposes stringent constraints on possible transmembrane structures. The identification of residues exposed at one or the other surface of the bilayer can readily be accomplished by labeling, for example, with antibodies to specific strings of amino acids.

The ability to capture a particular conformation by rapid freezing should soon make it possible to visualize the configuration of the myosin head attached to actin at different stages in the contraction-relaxation cycle and to visualize membrane channels in their closed, open, and inactivated states. Closer ties with physiology can also be expected to emerge as a result of further development of microscopic and time-resolved techniques. Computer-aided light microscopy, for example, has facilitated the discovery of the mechanisms of microtubule-based motility and is beginning to reveal the fate of very small populations of molecules in cells. Similarly, time-resolved x-ray diffraction has been developed on the basis of very few model systems (muscle contraction, microtubule assembly), but the applicability of these methods goes far beyond the original aims.

Crystallization of the Sample Is Still a Major Hurdle

The major obstacle to the structural analysis of molecular assemblies has been and will continue to be in preparing suitable crystals. The growth of two-dimensional crystals has become a key aspect of electron microscopic structure determination, and new general approaches are urgently needed. Thus, the lipid-layer crystallization technique (in which macromolecules are bound to a lipid ligand in mono- or multilayers in order to facilitate high lateral concentration and oriented binding), if successfully developed, would play a critical role. The difficulties encountered in three-dimensional crystallization, as needed for high-resolution x-ray analysis, will depend on the type of assembly in question. With membrane complexes, the crystals must be grown from precise mixtures of detergents, amphiphiles (polar molecules that have an affinity to both aqueous and nonaqueous areas), protein, and lipid; the process of crystallization has an additional dimension compared with that of soluble proteins. A major difficulty at present, therefore, is in obtaining sufficient commitments of financing and time to support such crystallization efforts. The first such crystallizations were carried out only after many years of trials in Europe, where the support of science can maintain a constant effort in a high-risk, long-term endeavor. Because risks have been demonstrably reduced, considerable weight must be given to early successes in growing crystals of sufficient quality for high-resolution analyses.

The crystallization of viruses is often made difficult by the small supplies available for systematic experimentation. Thus, large cell-culture laboratories with suitable biohazard containment are required. New methods of crystallization may also be needed, particularly for lipid-enveloped viruses such as rubella or measles. The forming of homogeneous complexes of viruses with antibodies, drugs, and receptors will call for considerable effort. An alternative approach, which has proven successful in the past, consists of crystallization of components of the assembled structure, determining the three-dimensional structures of each of the components, and then using electron microscopy studies to provide the architectural details of how the components are arranged in the assembly. Complementary analyses of this nature will also, in many instances, provide the most appropriate pathway toward understanding the details and action of the large intracellular organelles, such as the nuclear pore and the various types of cytoskeletal filaments. An exciting result to be expected along these lines over the next 10 years is the merging of the available low-resolution picture of myosin-actin filament interaction with that of the high-resolution structures currently being determined for the actin monomer and the myosin head.

Complex Biological Structures Can Assemble Themselves

Researchers have begun to unravel how molecular assemblies are formed. In living cells, the production of the components destined to be assembled is often coordinated tightly both spatially and temporally. This coordination is revealed by studies with mutants, in which individual components are defective or are not synthesized in the proper amounts. Sometimes assembly occurs just by spontaneous association of individual proteins and nucleic acids, but steps in assembly are frequently accompanied by the covalent modification of key proteins and nucleic acids. Such modification can make the assembly irreversible—in essence, to lock the pieces into place. Other assembly mechanisms have been found to make use of scaffolding molecules. These molecules are present at intermediate stages in the assembly to help align critical components, but they disappear before the final structure is formed, just as a scaffold is taken down as a building is finished.

Studies that attempt to assemble biological structures in vitro have been particularly fruitful. These allow the timing of particular steps to be controlled at will, and particular components can be added sequentially or simultaneously, which in turn allows detailed study of assembly pathways and direct tests of the function of specific components of the assembly by single-component-omission experiments. Such studies are not always possible in vivo. For example, if a protein has functions critical to the cell, it will be difficult to see its effect on the structure or function of an assembly by simply preventing its being synthesized.

In vitro assembly is also useful for many structural studies. For example, neutron-scattering measurements usually require the creation of an assembly in which some components contain the normal isotope of hydrogen whereas in others the hydrogen is substituted with deuterium. Such manipulations can be carried out only by starting with isolated components in vitro. Complex biological structures successfully assembled in vitro include ribosomes, microtubules, nucleosomes, and even many viruses.

DIRECTED MODIFICATION OF PROTEINS

We Can Now Design and Construct New Molecular Machines

Until recently, the experimental strategies available to structural biology were largely limited to examining naturally occurring biological structures. Testing specific hypotheses by altering structures was limited to observing naturally occurring biological variants when they could be identified, as in the numerous mutant hemoglobins. This approach is limited in having no systematic way to search for a particular desired variant. Furthermore, one was restricted to those variants that had no lethal consequences for the organism and variants that had a significant chance of arising by natural biological mutation or evolution.

The development of recombinant DNA technology has dramatically altered our study of the structure and function of proteins. The major breakthrough lies in our new ability to modify or synthesize de novo genes (DNA) that, when introduced into cells, direct the synthesis of modified or new protein molecules. What was only a fantasy a few years ago is today a routine procedure: We can produce protein molecules of any desired sequence. We can produce altered proteins in bacteria, yeast, or plant or animal tissue-culture cells, which makes it possible to isolate large enough quantities for structural and functional studies. In addition we can produce the altered proteins in vivo in transgenic animals to gauge the effect of the altered protein on complex biological processes.

The Future Will See a Heightened Interdisciplinary Cooperation Between
Structural and Molecular Biology

The techniques of traditional structural analysis and of recombinant DNA when combined increase the value of both. Such integrated approaches will allow more rapid and informative studies of the structures of proteins and how these structures determine function. The future will see ever-closer working relations among scientists expert in these different disciplines.

The potential to alter proteins at will is remarkable since it transforms structural biology from a science limited to strictly descriptive observations to an experimental science in which specific hypotheses can be tested with appropriate controls in specifically modified molecules. Our ability to do this is still in its infancy; much experience will be needed before the strategies in routine use approach optimal design. However, it is already clear that the ability to alter the sequence of proteins and nucleic acids systematically has revolutionary applications for structural biology. The importance of these new technologies is twofold, the first of which is widely appreciated, the second of which is perhaps less often noted.

First, by altering protein structure, or by creating new proteins, we are able to produce improved or even new proteins of value to human welfare, such as new pharmaceuticals. We are already using recombinant DNA technologies to produce human growth hormone, the blood-clotting factor VIII, the anticoagulant tissue plasminogen activator, interferons, and several lymphokines (which regulate development of various cells in the body). Variants of these proteins are being made and tested for improved properties such as increased heat stabilities, or increased lifetime in the blood. We are limited in these efforts by the fact that in most cases we do not yet sufficiently understand the structures of these proteins or the relations of structure to function to know what changes to make. As we learn more about the structures of these molecules from x-ray crystallography and other techniques, the successful production of useful variants will increase.

Second, determining protein structure will be enhanced by our ability to modify proteins. These modifications will result in proteins that crystallize better, that are designed for easy insertion of heavy metals needed in x-ray crystallography, and that have specific perturbations introduced to test hypotheses. In parallel, as the body of defined structures grows, we will be in a better position to design rationally modified proteins. We will know more about possible protein-folding motifs, domain attachments, and the effects of certain kinds of single-residue modifications. Today, in the absence of a known three-dimensional protein structure, the behavior of site-directed mutants can be unpredictable. The cellular location of the new product, its stability, and its properties can frequently differ from our naïve predictions. This situation should change markedly as we gain more experience with site-directed protein modification and its structural consequences.

Methods for Designing New Proteins. The first approach to protein design is site-directed mutagenesis. Here one usually alters a single amino acid by changing one or two nucleotides in the gene at the point coding for that amino acid. The result is a site mutant, which may resemble natural mutants, except that the experimenter can choose the site and the replacement.

The second approach is to make larger alterations. Usually this involves interchanging segments of two or more different proteins. Domains are segments of sequence often associated with particular functions of a protein. Therefore, by appropriate switches in domains, one can rationally create proteins likely to have desired hybrid functions.

The creation of such chimeric proteins in the laboratory mirrors the events that seem to occur in protein evolution. The genes for proteins in most organisms occur in blocks of coding regions (exons) and blocks of noncoding regions (introns). The introns are removed from the message by RNA splicing before a final transcript is used to direct the synthesis of protein (See Chapter 4). The exons frequently appear to correspond to functional or structural motifs in the protein, and often they correspond to actual three-dimensional structural domains. Many new protein functions may have evolved by exon shuffling—rearrangements among pre-existing exons of proven functional capability. Such exon shuffling provides a rapid way to create proteins with new, hybrid functions. It also provides a rationale for the presence of interrupted genes. An organism that has such a pattern of genetic information is likely to be more able to cut and paste its genes in a meaningful way and thus should have a selective advantage.

Domains are also regions that appear to fold independently into three-dimensional structures. Switching pre-existing domains maximizes the likelihood that the new chimeric protein will still be able to achieve a stable, well-ordered, three-dimensional structure.

Genetically Engineered Proteins Reveal Much About How Proteins Function

The use of site-directed protein modification offers great promise for answering some of the fundamental questions in contemporary biology. For example cell-surface receptors must migrate throughout the cell from one organelle to another, moving from the endoplasmic reticulum (site of synthesis) to the Golgi complex (site of carbohydrate modification) to the plasma membrane (site of clustering in specialized regions of the cell surface called coated pits). Once inside a coated pit, these proteins are taken inside the cell in a coated vesicle and then recycled back to the cell surface in a recycling vesicle. All of these movements seem to be dictated by signals contained within the structure of the protein itself. What are these targeting signals? Are they simply short, continuous stretches of amino acids or are they determined by the three-dimensional structure of the protein? Are protein modifications, such as phosphorylation or fatty acylation of the protein, required for any of these targeting signals?

The use of chimeric proteins has made it possible to define the functions of linear sequences responsible for protein translocation into the endoplasmic reticulum, mitochondria, and nucleus. However, signals that are defined by noncontinuous amino acid sequences are more difficult (if not impossible) to define functionally with chimeric proteins. Incorrect protein folding becomes a major obstacle when the function of an internal sequence or domain is examined by this approach.

Once a targeting signal for a given movement is identified, the scientist is in a superb position to study biochemically how the proteins of the cell interact with the cell-surface receptor to affect the desired targeting event. All the potential questions can now be answered in model systems, in which cloned genes for cell-surface receptors are transfected into cultured cells and then studied functionally and biochemically.

The targeting problem is related in a major way to another crucial problem in biology: protein folding. Many of the rules for protein folding can be derived from site-directed mutagenesis studies of proteins such as cell-surface receptors.

For example, the folding of a cell-surface receptor in the lumen of the endoplasmic reticulum depends in large part on the arrangement of cysteine residues and other amino acids in the primary structure of the polypeptide. By use of site-directed mutagenesis, one can begin to vary the position and number of cysteine residues to determine their effects on the interaction of the protein with the cellular machinery of the folding process.

Cell-surface receptors are key molecules that mediate a variety of physiologically important processes, ranging from the regulation of blood glucose and cholesterol levels to the control of body iron and vitamin B_{12} stores. Fundamental research on these molecules should shed light not only on basic science but on medicine as well.

For example, it has recently been possible to create functional, chimeric cell-surface receptors. In a chimera, the extracellular domain of epidermal growth factor receptor can stimulate the tyrosine kinase activity of an attached, insulin receptor—intracellular domain. This result shows that the insulin and epidermal growth factor receptors use a common mechanism for signal transduction across the plasma membrane (See Chapter 7). Future applications of this type of approach include studying the function of new receptors for unknown ligands by activating the new receptor's cytoplasmic domain with a heterologous ligand-binding domain derived from an already characterized receptor.

Some oncogenes represent naturally occurring receptor mutants. These will help us to understand normal mechanisms of receptor function and should lead to an understanding of how normal signaling processes are subverted to result in tumorigenesis.

Functional Protein Molecules Can Also Be Synthesized Chemically

For peptide chains of fewer than about 100 amino acids, chemical (as well as biological) synthesis is now possible and will frequently be the method of choice for shorter chains. By synthesizing chains in blocks, much longer chains will become practical synthetic goals. Chemical synthesis permits the insertion of isotopes, either stable or radioactive, at specific single sites in the chain. The peptide bond itself, can be replaced in selected locations to render the product totally resistant to proteolytic degradation at that position. In general, such products are extremely difficult or impossible to prepare biologically. Such a chemical approach is particularly valuable in testing hypotheses related to small structural and functional domains and the possible refolding of these isolated units. Single- and multiple-site mutagenesis experiments are equally easy chemically since any amino acid can be selected for any position with the automatic equipment for synthesis that is now available. The simultaneous use of recombinant DNA techniques and sophisticated polypeptide chemical synthesis will create new approaches to both the understanding of protein structure and the development of specific reagents and functions.

FOLDING

Initial reaction to the appearance of the first high-resolution crystal structure of a protein was one of shock at the complexity and the total absence of obvious symmetry. This reaction was conditioned, in part, by the earlier appearance of the model for DNA, in which the base-paired double helix, once revealed, seemed elegant and simple. During the past two decades, the successful determination of many protein structures has led to the realization that there are, indeed, underlying substructural motifs in these molecules. These motifs are themselves complex and asymmetrical, but they are repeated in many structures.

The properties of the chemically bonded atoms in the peptide chain severely constrain the possible conformations the chain can assume, and only a small number of secondary structures are possible regardless of the sequence of amino acids. The structural motifs are made up of various combinations of these secondary units. These supersecondary units, in turn, are packed together into structural domains. A domain may be the whole molecule, but, in larger proteins, the native molecules are frequently composed of several domains.

We Do Not Yet Understand How Proteins Assume Their Intricate Three-Dimensional Forms

The biosynthetic machinery that synthesizes peptide chains is the same for all proteins. As far as is known, the peptide emerges as an essentially straight chain having no intrinsic biological activity. During, or shortly after, the completion of this synthesis, this chain folds up spontaneously to give the final unique, compact, biologically active structure, which is characteristic of the native protein. In the early 1960s this process was shown to occur in vitro. By now a large number of proteins have been shown to undergo this reaction without any apparent help other than the correct solvent environment. The mystery of how a biologically active protein is formed from an inert disordered chain is known as the folding problem. Research in this area, a major interface between the fields of biology and chemistry, has rapidly expanded during the past five years.

The basic question has always been, From chemical theory, can the three-dimensional structure of a protein be derived solely from its known amino acid sequence? Since folding appears to be a purely spontaneous process, prediction of the structure is a severe test of the level of our understanding of the chemistry of polypeptide chains. The attack on this problem is both experimental and theoretical. Its solution is not only essential as a basic underpinning for all of molecular biology, but would also be of great practical importance in the industrial application of genetic engineering.

Experimental Studies Search for Folding Intermediates

On the experimental side, much work has centered on the search for folding intermediates. Do the secondary structural elements in native proteins exist, as such, in small purified peptides apart from the rest of the structure? In the past it was thought that such structures would be so unstable that they would not be found, but recent evidence, based largely on optical spectroscopy, suggests a positive answer to the question, at least for certain sequences. Is there a definable folding pathway along which such structure intermediates can be found (Figure 3-4)? This question is controversial. Thermodynamic measurements can frequently be fitted to a two-state model reflecting only the native and the unfolded forms. Kinetic data, on the other hand, are often difficult or impossible to interpret without the assumption of one or more intermediate states. The methods are invariably indirect and the interpretation non-unique. Since crystals cannot be obtained for the unfolded state, x-ray diffraction data are not even potentially available. Detailed NMR studies on long peptides are still on the horizon, but may in the future provide direct structural information on these intermediate states.

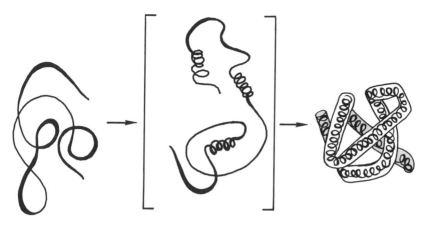

FIGURE 3-4   Schematic diagram of the folding process for an all helical protein. The reaction starts with an extended chain containing no permanent intrachain interactions. This proceeds to a hypothetical intermediate with fluctuating helical segments that occasionally associate. The end point is the folded, compact, biologically active structure.

A Diverse Range of Theoretical Studies Is in Progress

Theoretical approaches to folding have been proceeding along three lines. The first two are fundamental procedures for which, in principle, we do not need to know the final structure. The third is a collection of ad hoc procedures with which we try to produce useful generalizations by examining the known structures.

Energy Minimization. Minimization of the conformational energy of peptides is perhaps the oldest of the three procedures. The goal is to predict the native folded structure of the protein by assuming that it is actually the most stable structure. This requires computing a potential energy function with many terms representing possible conformational changes: bond stretching, angle bending, torsional rotations, van der Waals interactions, and various electrostatic terms. The minimization of this potential energy with respect to the locations of each of the atoms in the protein should lead to the observed native structure. The difficulties are formidable. The largest hurdle one must overcome is the problem of multiple local energy minima. The potential energy function is like the surface of the earth. Energy minimization corresponds to finding the lowest point on the surface. Wherever one starts on the surface, one can find the lowest point nearby. But how does one know if this is the lowest point on the whole planet? How can one tell if a locally stable protein structure is actually the most stable possible structure? Although intense efforts are under way, it is not yet possible to consistently derive the correct native structure by starting with an unfolded chain and attempting to minimize the potential energy.

Molecular Dynamics. In the second theoretical approach, one actually simulates the motions of the atoms of a protein. The potential energy function (in principle, the same one used for energy minimization) can be used to obtain the forces on each atom. The movement of the atoms in response to these forces is then calculated. The applications of this powerful procedure are under intensive development. As with energy minimization, the fidelity of this approach depends on the quality of the potential energy function and on the proper modeling of the solvent. Molecular dynamic simulations have not yet successfully folded a protein in the absence of additional structural information. The latter can, in principle, be provided experimentally by NMR or optical spectroscopic procedures. When such data can be supplied, some recent tests have shown notable success in carrying out the folding simulations.

The Protein Data Bank Is a Rich Resource for Predicting Structure

In the ad hoc approaches, the protein data bank is searched for patterns and statistical correlations. For example, probabilities based on the occurrence of each amino acid in various types of secondary structure differ and can, in turn, be used predictively to estimate probable regions of alpha helix, beta strand, and beta turn structures in any sequence. In parallel efforts, combinatorial algorithms aimed at packing assigned secondary structures into supersecondary and larger tertiary units have been developed. Most recently, combinations of such secondary and tertiary prediction schemes that show great promise in providing probable domain structures have been worked out. Whether the resulting models are close enough to converge to the native structure through molecular dynamic or energy minimization procedures is not yet known. Although all ad hoc approaches are implicitly based on the underlying chemistry through the use of known structures, only a few explicitly refer to these properties in the algorithm itself.

New Experimental Tools Will Aid Studies of Protein Folding

Genetic Approaches. The power of modern genetics is being brought to bear on the problems of folding. Some mutants seem to be clearly deficient in the folding process, and yet the final folded protein does not seem abnormal in any way. The discovery of other systems of this sort and their detailed analysis may provide a great deal of information on folding pathways that would not be found by other procedures, or even suspected.

Polypeptide Synthesis. The chemical and recombinant DNA approaches have been discussed in an earlier section. These complementary approaches to providing peptides of known sequence will play major roles in the future study of protein folding. At this time, the behavior of peptides at membrane interfaces has been studied in detail by chemical synthesis; general specifications of such interactions are starting to appear, and marked improvement in our understanding of electrostatic interactions in alpha helices seems imminent.

Through recombinant DNA approaches, many different molecules appropriate for structure-function investigations are already being created, largely through single-site mutagenesis. Estimates of hydrogen bonding energies are being derived from comparisons between carefully planned and constructed mutants of proteins of known structure, and factors affecting protein stability are being outlined.

The Folding Problem Now Seems Ripe for Major Advances

The immediate future for the folding problem looks remarkably (and unexpectedly) bright. The development of both fundamental and ad hoc theoretical approaches is advancing rapidly. The correlation and interactions between theory and experiment will be much closer than has generally been true in the past. Combined approaches, with various levels of theory or theory and experiment, seem likely to be the most fruitful. The ability to easily synthesize specific polymers, themselves specifically designed to test theoretical predictions or to provide missing values for parameters, seems particularly promising.

Instrumentation. The solution of the structures of new proteins, and of mutant versions of older proteins, will continue to be of major importance. Thus the development and implementation of new and improved x-ray and neutron diffraction procedures is as important to the folding problem it is as to other areas in structural biology. Improvements in both solid-state and high-resolution NMR will be central to the specification of the unfolded state and the search for definable folding intermediates. Proteins that are isotopically labeled at specific sites will be essential in this process, and they will also permit the study by NMR of substantially larger proteins than can currently be tackled.

NEW TECHNIQUES AND INSTRUMENTATION

Improvements in Analytical Techniques and Instrumentation Are Necessary

Better methods for automated X-ray diffraction are critical to our increased understanding of molecular structure and function. In addition, more general and effective methods are needed for direct analysis of x-ray data without the need for preparing many heavy metal derivatives. Many of the most interesting biological molecules have not been crystallized. Systematic studies are needed, aimed at producing crystals and other ordered arrays suitable for high-resolution structural determinations. Funding mechanisms must be adjusted to allow the long-term support of this speculative but extremely critical area.

Methods are also needed to extend two-dimensional NMR to larger structures and to automate its analysis. The development of instruments operating at higher magnetic fields will certainly play an important role in this work.

Advances in Computation Will Revolutionize the Study of Molecular Structure and Function

Improved methods are needed for collecting and transmitting DNA sequences including a single, international data base. Improved methods are also needed for extracting more biological information directly from sequence data. We must ensure that continuing advances in computer science are made available, rapidly and broadly, to the field of structural biology.

More accurate protein folding calculations must be developed, including better methods for refining x-ray structures and improved semiempirical methods based on the ever-increasing data base of structures. In addition, uniform, inexpensive devices to display three-dimensional structures are needed so that, ultimately, every biologist can view any known structure directly and accurately.

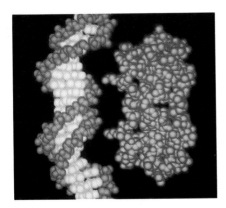
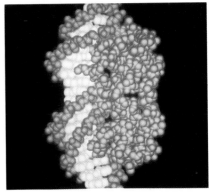

PLATES 1, 2   A repressor protein (from bacteriophage 434) is shown (left) approaching DNA and (right) bound to DNA in a crystal of the repressor-DNA complex. This binding turns off expression of a bacteriophage gene. [J. E. Anderson et al., Nature 326:846-852 (1987)]

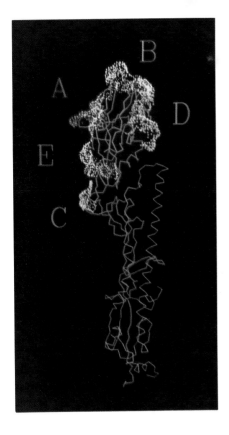

PLATE 3   Sites A-E on the surface protein of the flu virus are recognized by our immune system. Variation in the structure of these sites results in the recurrence of epidemics in the human population. [Based on work of D. Wiley, Harvard University, and J. Skehel, National Institute of Medical Research, London]

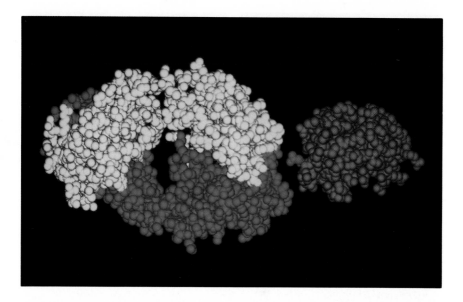

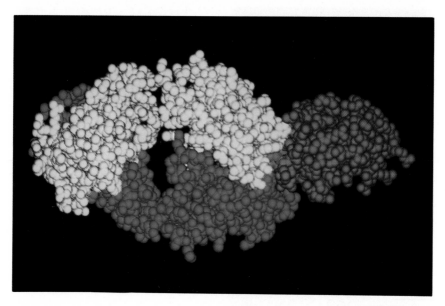

PLATES 4, 5 Space-filling representation of Fab (antibody fragment containing antigen-binding sites) of an anti-lysozyme antibody and lysozyme. The antibody heavy chain is shown in blue, the light chain in yellow, lysozyme in green, and glutamine 121 of lysozyme in red. [A. G. Amit *et al.*, Science 233:749 (1986), figure 2]

4

Genes and Cells

The Fundamental Biological Unit of All Living Organisms Is the Cell

To a surprising degree, all cells are similar in design and function, whether in human beings, in plants, or as simple single-celled organisms such as bacteria. One major difference, however, is the presence or absence of a distinct compartment, the nucleus, for the genome. Cells with a nucleus, called eukaryotes, are found in advanced single-celled organisms and multicellular organisms. Those without nuclei, called prokaryotes are the simplest single-celled organisms, the bacteria.

The importance of the cell as a biological unit is made clear when we consider the life cycle of advanced multicellular organisms such as human beings. The cycle begins with the fusion of egg and sperm, themselves single cells of specialized types, to form the one-celled embryo. At the earliest stage of our life cycle, therefore, we exist as a single cell. This cell divides into two cells, each of those into two more, and so on, to give rise to the adult organism, which may contain as many as 1 million billion cells (see Chapter 5). Every one of these cells is autonomous in some functions and dependent or interdependent in others.

The entire developmental process is regulated by the myriad interactions within and among individual cells. These interactions regulate the capacity of cells to multiply, to differentiate into the hundreds of different cell types that make up our bodies, and to organize themselves into tissues, organs, and finally the human body itself, according to a specific, well-defined architectural plan.

This cycle of activity characterizes normal healthy individuals, but the process can go awry, leading to abnormal states characterized by disease in humans, animals, and plants. Since both normal and abnormal progression of the life cycle is governed in a fundamental way by cells, it is logical that to understand and exert control over these phenomena and, in a real sense, over our own lives and biological environment, it is necessary to learn all we can about the cell and how it conducts its activities.

RESEARCH STRATEGIES

Since all organisms are related through evolution, we can use information from simple organisms to determine protein and gene functions in higher and more complicated organisms. Furthermore, new technologies allow us to test these inferences by expressing the genes of more complicated species, such as humans, in the cells of simple organisms such as yeast or bacteria.

The identification of the genes and the proteins involved in various cellular functions is only part of the story. A deep understanding of the roles played by individual proteins requires detailed characterization of their mechanisms of action. Ultimately this requires knowledge of the three-dimensional structure of each protein and of the rates of its reaction with other molecules. This characterization requires the development of test-tube (in vitro) assays of function. Genetics can contribute to this effort by providing components from mutant cells that can be used to identify essential components and to test the molecular roles inferred from the physiology of live cells. Likewise, components identified as part of in vitro mechanisms can be used to isolate the corresponding genes and produce defective mutants to test in the live cell the functions inferred in vitro.

Many Cellular Functions Are Carried Out by Macromolecular Assemblies That Can Be Isolated and Reconstituted from Their Constituent Molecules

Some biological functions such as the fermentation of sugar to alcohol are carried out by individual proteins, alone or in a series of catalytic reactions, but many others including photosynthesis and the beating of sperm tails require complex assemblies consisting of many different types of proteins and other molecules such as lipids and carbohydrates. These subcellular molecular machines include the organelles that can be seen by light microscopy as well as many smaller structures.

Perhaps the most successful research strategy in cell biology for the past 30 years has been to purify these molecular machines so that their functions can be studied outside the living cell. Fortunately, the organelles and many smaller macromolecular complexes are so well constructed that they emerge from the trauma of separation still able to perform complex, highly integrated functions such as contraction, transport of molecules across membranes, photosynthesis, the production of adenosine triphosphate (ATP), protein synthesis, and gene transcription. This strategy has yielded a biochemical inventory and, in addition, a rather detailed functional characterization of each organelle showing that functional specialization and division of biochemical and biophysical labor prevail at the subcellular level.

At the same time, studies of reconstituted cell fractions have demonstrated that most macromolecular machines are formed by the self-assembly of their component molecules. More recently, this approach has been used to demonstrate that cell organelles can interact in vitro to reconstitute a specific intracellular process. Whole organelles such as the nucleus can now be reversibly disassembled and reassembled in the test tube. A striking feature of the current state of research in molecular and cell biology is the remarkable degree to which the effectiveness of the reductionist strategy has been confirmed by experiments.

In Each Field of Cell Biology, Particularly Favorable Cells Have Been Identified for Experimental Work

To a great extent, progress in molecular and cellular biology depends on finding cell types that are suitable for experiments. Fortunately, because of the parsimony of nature, mechanisms for determining the most important cellular processes are shared by cells all along the phylogenetic tree. For example, the force-producing protein myosin from muscle can bind functionally to actin filaments from all eukaryotic organisms. Similarly, protein synthesis can be reconstituted in vitro from different components from plant and animal cells.

Fruit flies, sea urchins, and chickens have all been excellent sources of model systems for analyzing cell behavior during embryonic development. Amphibian eggs, because of their large size, are ideal for the production of foreign proteins by microinjection of messenger RNAs (mRNAs). Highly motile amoebas that can be grown in large quantities have yielded much of the basic information about cytoplasmic contractile proteins. *Chlamydomonas*, an alga with two flagella and well-characterized genetics, has provided much of what we know about the molecular biology of flagella and cilia.

The occurrence in yeast of three specific cell types, each of which plays a distinctive role in the cell's life cycle, makes the organism suitable for investigations in several important areas of cell biology, including genetic programming for cell differentiation. Yeasts are also especially valuable for combined biochemical and genetic studies of control of the cell cycle, exocytosis (secretion), endocytosis (uptake), and biogenesis of such cell organelles as mitochondria.

THE NUCLEUS

The Nucleus Is Both the Warehouse and the Factory for Most of the Cell's Genetic Material and Activity

The nucleus has three main functions: the storage, replication, and expression of the genes. The nucleus in most human cells is a sphere roughly 10 micrometers in diameter that contains the almost 2 meters of DNA that makes up the human genome.

The DNA contains the code for all cellular proteins and RNAs, as illustrated in the pathway represented by the "central dogma":

$$DNA \rightarrow RNA \rightarrow protein.$$

This information is processed with exquisite precision. In most cells only a tiny fraction of the total number of genes are actually expressed at any given time (or ever). Furthermore, the patterns of gene expression change in precisely programmed ways during development. Thus, cells in kidney and brain express independent repertoires of genes even though they originate from a single fertilized ovum and have identical copies of the genetic material. The cells of a developing embryo act like tiny computers that accumulate and remember information concerning their past and present locations relative to other cells and express genes appropriate to this information. The study of the control of gene expression is an important focus of modern cell biology research. How do the cells of complex multicellular organisms turn on and off their genes? And how is the position of a cell "read out" by the cell to control specific genes? These are difficult but central questions to cell biology that are being actively pursued in many laboratories.

DNA is complexed with proteins to form chromosomes. All human body cells (somatic cells) contain 46 chromosomes. The germ cells (sperm in the male and eggs in the female) that contribute to the embryos of the next generation contain only half that number—23. In addition, the nucleus contains an abundance of proteins and RNAs, representing various structural elements and the enzymes and products of replication, transcription, and RNA processing. How all of these components are organized within the nucleus is unknown, and research on the three-dimensional structure of chromosomes and other nuclear components is needed to better understand how genes function and are controlled.

Nuclear Envelope

All Molecular Traffic Between the Nucleus and Cytoplasm Is by Way of the Nuclear Envelope

Replication of chromosomes and synthesis of most mRNA molecules is restricted to the nucleus, whereas translation of messenger RNA into protein occurs only in the cytoplasm. Separation of chromosomes from the cytoplasmic space is thought to ensure proper regulation of nucleic acid and protein synthesis in the cell. Except at the time of mitosis, when the nuclear envelope breaks down, all molecular traffic between the nucleus and cytoplasm is by way of this envelope, which consists of two parallel membranes joined at regions called nuclear pores.

Nuclear pores consist of a precise geometrical arrangement of structural elements. The transport of large molecules across the pores and the direction in which individual molecules are transported (cytoplasm to nucleus or vice versa) seems to be selective. Selective proteins are actively moved from their site of synthesis in the cytoplasm into the nucleus by means of an active transport system in the nuclear pores. Selective RNAs are pumped out of the nucleus through the same pores. For the transported proteins, the specificity is known to reside in a short sequence of their amino acids that seems to be recognized by a component of the nuclear pore. A number of pore complex proteins have been identified. However, the organization of these proteins in the pore complex and their role in regulating traffic in and out of the nucleus are largely unknown. Future analyses of the structure, molecular composition of the pore complex, and the genes encoding its proteins will undoubtedly reveal how this gatekeeper of the nucleus works.

Attached to the inner surface of the nuclear envelope with its pores and pore complexes is a fibrous network known as the nuclear lamina. The major proteins of the lamina, called lamins, have been identified and shown to be related to the intermediate filaments found in the cytoplasm. (See the section on cytoskeleton in this chapter.) The nuclear envelope seems to be a major site for anchoring chromosomes and may also facilitate the packing of DNA in the nucleus. The reversible assembly of the lamina may help control the breakdown and reformation of the nuclear envelope during mitosis.

Chromosomes

Chromosomes Are the Structural Units That Contain the DNA

Each chromosome consists of one extraordinarily long DNA molecule complexed with a multitude of proteins. An orderly condensation of long DNA molecules into much smaller chromosomes is mediated by nuclear proteins. The first level of folding involves a set of five proteins called histones. DNA interacts with histones to form a regular beadlike structure, the nucleosome. Approximately 160 nucleotide pairs of DNA are wrapped around a core formed from two copies each of four histone molecules (the histone octamer). This basic structure is repeated over and over to give a beads-on-a-string appearance when chromosome fibers are viewed in the electron microscope.

DNA complexed with histones is generally referred to as chromatin. Chromatin fibers consisting of nucleosome beads are folded into still more complex structures known as superbeads or supercoils, depending upon one's view of their organization. For technical reasons this higher-order folding of chromatin has been hard to study, and we are only beginning to understand how chromatin folding may be related to important processes such as the turning on and off of genes.

Beyond superbeads or supercoils, one major level of chromosome folding can be recognized, the so-called loops. Loops have been studied in the lampbrush chromosomes of oocytes for more than 100 years; the giant lampbrush chromosomes consist of several hundred individually recognizable loops of chromatin extending laterally from the main axis. These lampbrush loops are regions of unusually intense RNA synthesis that provide a unique opportunity to visualize active gene transcription. When RNA synthesis is completed, the loops contract back into the main axis of the chromosomes. The loops are thought to retain their individuality even during mitosis, the time of maximal chromosome condensation. Loops of chromatin similar to those of the lampbrush chromosomes can be seen when normal mitotic chromosomes are chemically treated to loosen their structure.

We know very little about the composition (or intranuclear location) of the macromolecular complexes that anchor the DNA into loop-domains inside nuclei. Fortunately, however, the genome does not remain in the dispersed state characteristic of physiological activity during the entire cell cycle. At cell division (mitosis), the nuclear envelope disassembles and the genome condenses into discrete chromosomes. These mitotic chromosomes provide an opportunity to study the organization of the DNA loop-domains in the absence of the many soluble components that participate in transcription and replication. Methods are currently being developed to identify proteins that regulate mitotic chromosome architecture. Recently one putative component of the loop-domain anchor complex was identified as the enzyme DNA topoisomerase II, which had previously been identified as able to knot and unknot DNA molecules in vitro. Genetic analysis shows that this activity is required at mitosis if the two sets of intertwined DNA molecules are to be successfully partitioned to daughter cells. Future studies should identify other components of the anchor complex and eventually enable us to determine what role they may play in the regulation of a gene's activity.

The domain idea has matured sufficiently during the past decade that the genome can now be conceived of as being constrained into loops whose average size is roughly 100,000 base pairs. In theory, the loop-domain model permits the coordinate control of complex arrays of genes, since regions that may be very distant in the DNA sequence may actually be physically adjacent (at the base of a loop) in the nucleus. The nucleolus is an example of the clustering of dispersed DNA regions into a single-functional domain where the genes (often found on different chromosomes) encoding ribosomal RNAs associate physically and are expressed together.

It is a striking coincidence that DNA replicates in multiple independent blocks, which again are about 100,000 base pairs long. Thus, loop-domains (or clusters of loop-domains) may constitute the control units for both transcription and replication. This model predicts the existence of a new set of nuclear components that control DNA function—the structural components that anchor the loop-domains; at present, however, these ideas have yet to be demonstrated.

Centromeres and Kinetochores Play a Key Role in the Migration of Chromosomes During Mitosis

During the separation of chromosomes at mitosis, microtubules of the mitotic spindle attach to a specific site on each chromosome known as the centromere or kinetochore. (As now used, centromere refers to the DNA sequences at this site and kinetochore to a rather complex structure of unknown composition visible by electron microscopy.) The centromere is fundamentally important because movement of the chromosome at mitosis and meiosis into daughter cells depends on this region; chromosomes without a centromere fail to move normally and are eventually lost from one of the daughter cells. Centromeric DNA has been cloned and sequenced from the yeast *Saccharomyces*. Remarkably, the DNA region necessary for accurate segregation of a chromosome is no more than a few hundred base pairs long in this organism. The centromeres of higher eukaryotes are larger and their characterization will be difficult, but should prove of great interest for comparison with the presumably simpler condition in yeast. The use of antibodies produced by patients with an autoimmune disease has made it possible to identify and clone the DNA sequence for a centromeric protein. Characterization of this human protein and others that mediate association of the centromere with the spindle microtubules should give insight into critical questions of chromosome movement during mitosis and meiosis.

Telomeres Maintain the Structural Integrity of Chromosomes

The ends of eukaryotic chromosomes, the telomeres, are special in several ways. Telomeres stabilize chromosomes and prevent their fusion with other broken or natural ends. In addition, their structure allows replication without loss of DNA. The ends represent a vanishingly small amount of the total DNA in a typical chromosome. For this reason telomeric sequences were first recognized in certain ciliates, which have thousands of extremely small chromosomes and hence thousands of telomeres. Telomeres have subsequently been identified in yeast and several other organisms, including humans. In all cases they consist of hundreds of nucleotides of simple repeated DNA (such as CCCCAA and CCCCAAAA repeats) associated with unique proteins. Telomere sequences are added to the ends of chromosomes at the time of chromosome replication by a special enzyme or enzymes without the need for a DNA template, and in this way they form a protective cap at each end of a chromosome.

Artificial Chromosomes Are Valuable Research Tools

The identification of centromeres and telomeres as well as sequences that initiate DNA replication now permits the synthesis of artificial "minichromosomes" by genetic engineering techniques. Such minichromosomes have already been introduced into yeast cells, where they function normally during both mitosis and meiosis. Work on artificial chromosomes in yeast will undoubtedly lead to increased knowledge about chromosome structure and mechanics; eventually similar studies will be possible in higher eukaryotes, simplifying the introduction of specific genes or gene combinations into experimental organisms. One important use for the artificial yeast chromosomes is in the cloning of very large fragments of DNA (as many as a million nucleotide pairs).

The Nucleolus

The Nucleolus Is the Site in the Nucleus for the Transcription of Ribosomal RNA

The nucleolus is the major structural differentiation seen in nondividing nuclei. It is formed from a specific chromosomal locus, the nucleolar organizer, which contains the genes coding for ribosomal RNA (rRNA). When rRNA is synthesized, it first accumulates in the nucleolus in association with a large number of ribosomal proteins. Eventually the rRNA and proteins are transported to the cytoplasm, where they constitute the mature ribosomes. Although the major features of rRNA synthesis and its relation to the nucleolus were worked out more than 15 years ago, the nucleolus continues to be of interest as a model for RNA transcription and processing. In particular, we are just beginning to learn about the ribosomal proteins and the ways they interact with rRNA to form the ribosomes. Ribosomes themselves are crucially important to cell function, since they are the machines that catalyze all protein synthesis.

GENES AND GENE ACTION

The Primary Questions That Have Been Asked for the Past 100 Years by Geneticists Are Still Inspiring Research Innovation

Geneticists have always wanted to know how traits are passed from one generation to the next. We now seek to answer these questions at higher levels of resolution. For example, we can ask, What is the structure of genes? How do they replicate? How are they organized on chromosomes? How do they mutate, recombine, and repair themselves? What controls the timing of gene expression and repression? What mechanisms control tissue-specific and cell type–specific gene expression? The answers that we anticipate are at the level of nucleotide sequences, the three-dimensional structure of chromatin, the mechanism of action of enzyme complexes and specific DNA binding proteins. As answers to these questions emerge, we can use the resulting picture of how genes act to ask even more sophisticated questions about their products and functions.

Genetic Analysis

The Combination of Classical Genetics and Biochemistry Has Resulted in an Explosion of Genetic Understanding

The field of genetics has made dramatic advances within the past 25 years largely because biologists learned that genes are made out of DNA. In classical genetics, the arrangement of genes on the chromosomes was inferred by analysis of crosses between organisms. In modern genetics the order of genes can in principle be determined directly by sequencing the DNA molecules. The rapidity and certainty with which genes can be sequenced by biochemical procedures, even in organisms having no convenient sexual cycle or, as in humans, having a long cycle, have come to give the illusion that classical genetics has been subsumed by biochemistry. The most elegant insights, however, have emerged in the collaboration between classical genetics and biochemistry, which has created the field of molecular genetics. In this collaboration, the goals of genetic research have not changed. Geneticists still seek to learn the set of instructions that specify the architectural plans encoded in the DNA for building a functional organism.

Mutations in Essential Metabolic Pathways Have Contributed to Our Understanding of Both Genetics and Biochemistry

Classically, genetic analysis starts with mutations. These are alterations in a single gene that result in changes in an organism's appearance or biochemistry (phenotype). For example, normal yeast cells can grow on a simple medium with no amino acid supplements. Mutations can be found that lead to the requirement that an amino acid, say histidine, which the cells would normally produce itself, be supplied in the growth medium. Mutations of this type are extremely useful because the cells can be maintained and genetic crosses can be made on complete medium, yet the mutation can be detected at any time by plating a sample of cells onto medium that lacks the required nutrient. Such conditional and selectable mutants have been the basis for learning the fundamentals of molecular genetics, the physical nature of the gene, and the structure of chromosomes. These mutations have also been important to our current understanding of biochemistry, The analysis of many mutants, each blocked in the same biochemical pathway, but at different steps, can lead to an understanding of how complex molecules could be formed by sequential chemical reactions.

When coupled with genetic crosses, mutant analysis becomes a valuable tool for dissecting some of the most complex biological functions by correlating genotype (the nucleotide sequence) with phenotype (how the organism looks). A decade of intensive research correlated gene structure and function with biochemical activities of simple organisms. In some cases a single base-pair change in the DNA can be correlated with the loss of the catalytic function of an enzyme in a biochemical pathway. The metabolite required in the diet of the mutant organism reveals the identity of the defective biochemical pathway. The loss of enzyme function easily explains the phenotype of the mutant organism—the failure to grow without a metabolite in the medium—since the mutant cannot carry out one of the steps in the biosynthesis of the metabolite. Thus, classical genetic analysis allowed the elucidation of biochemical pathways because the essential steps in the pathways were easy to eliminate: The genes were discovered through mutations with a metabolite-requiring phenotype, the functions were inferred by inspection of the phenotype, and the proteins were identified by application of chemical and biochemical analysis of the mutants in comparison with the normal.

The Analysis of Mutations in Other Types of Genes Requires a Different Approach

Although these approaches permitted an understanding of metabolic pathways in simple organisms, they are difficult to apply to the direct analysis of cellular structure and to questions involving the complex interactions that determine protein structure and function in higher organisms. The major difficulty lies in our inability to connect genetics and biochemistry at higher levels of complexity. Very often the function that we wish to study is part of a subcellular structure and cannot simply be added back from the outside to correct the effect of the mutation. Furthermore, a component of the cell's architecture may be essential for viability, and then a mutation in this function would be lethal. Undaunted, geneticists still made considerable progress through the mutation analysis approach by using conditional lethal mutations (for example, mutations that permit growth at one temperature, but not at another). Despite many imaginative efforts, it became clear that major cellular processes could not be analyzed simply by examining phenotypes in the old ways that had elegantly sufficed for determining basic biochemical pathways.

To associate a gene with its function in one of these complex cellular processes required development and exploitation of new experimental strategies that can permit the enumeration of genes controlling the synthesis and function of the cell's architecture. The new methods enable scientists to carry out a new chain of discovery that allows them to associate the gene with a cellular function, with a gene product (usually a protein), and ultimately with a mechanism by which the gene product executes the function.

Even with the latest in technologies, one still faces problems in associating genes with products and the products with essential functions related to cellular structure, progress of the cell division cycle, or other functions carried out by macromolecular assemblies. The geneticists can identify candidate genes by observing mutant properties that suggest failure in cell architecture or cell cycle. Biochemists and cell biologists, on the other hand, can find proteins (such as actin and tubulin) that are abundant in structures implicated in basic cell function. The challenge to the molecular geneticist is to find ways to bring these lines of endeavor together so that the geneticists' genes can be associated with the biochemists' proteins and the cell biologists' structure and function.

A Gene Isolation Experiment Usually Begins with the Construction of a Gene Library

The gene library is a collection of DNA fragments carrying every gene from the organism, each recombined with a carrier DNA molecule called a vector. The vector contains genes that permit replication and transmission of the vector and any DNA joined to it when the DNA is transformed into the appropriate cloning host (usually a bacterial or yeast cell). The library is made by chopping the chromosome into gene-sized pieces of DNA and then joining each piece to its vector. The size of the library depends on the size of the organism's genome.

Isolating a Specific Gene from All the Rest of the Recombinant Molecules Can Occur in Two General Ways

The more classical route begins with mutations, which define a gene and whose properties (phenotype) indicate failure in a particular cellular function. One can readily isolate the gene as a DNA fragment by using DNA transformation with gene libraries to complement a mutation. Restriction fragment length polymorphisms (RFLPs), to be discussed later, can also be used to localize genes or chromosomes. The isolated gene can be analyzed as a physical entity, after which one can find the gene product and determine, in favorable cases, something about the way in which this protein contributes to cellular function. By this route one eventually obtains all the elements: the gene, the function, and the protein product. This route works well with bacteria, yeast, and cultured cells of higher organisms if the gene desired has a strongly selectable phenotype.

The second approach begins with a gene product, perhaps a protein involved in some cellular process. Isolation of the gene for that protein depends on what is already known about that protein and often on the ingenuity of the investigator. If the protein sequence is known, oligonucleotide probes can be chemically synthesized on the basis of information contained in the protein's amino acid sequence. These probes can be used to screen the library by hybridization.

Other techniques involve isolating mRNA from the cell type of interest, transcribing it into DNA, and cloning it into a vector that is designed to direct transcription of that mRNA when it has been transformed into a bacterial host. Colonies of bacteria producing the desired protein can be identified by antibody binding (if an antibody is available) and then grown to yield quantities of the desired cloned genes. These clones can be used to screen a library of genomic DNA sequences to obtain the gene encoding the mRNA.

Once the gene is cloned, much can be learned from its sequence. Much more can be learned if the gene can be put back into the species of origin. In many organisms it is now possible to introduce DNA, usually by injection, that will become stably integrated into chromosomes and frequently be normally expressed. It is often possible to use such transformation to show that a cloned gene will repair a mutation.

Specific features of transformation vary significantly with the organism. In yeast and in the slime mold *Dictyostelium*, the introduced gene can displace the resident gene by homologous recombination (recombination that requires sequence similarity). This feature makes it possible to analyze the new gene without interference by the preexisting one and also ensures that only one copy of the transformed gene will be added. Homologous gene replacements are not yet routinely possible in other organisms, although techniques are being actively sought. In spite of some limitations, gene transformation has proved to be a powerful research tool that has opened analysis of complex function by direct genetic approaches.

The Genome

The Genome Refers to the Assemblage of All of the Genes of an Organism

The size of the genome (that is, the total amount of DNA) varies widely among organisms (Figure 4-1). Even though the number of genes in different multicellular animals is probably reasonably constant (totaling perhaps tens of thousands), the total amount of DNA in their genomes varies by a factor of 100 or more. Paradoxically, the size of the genome is not directly correlated with the complexity of the organism. Thus, much of the DNA in organisms with high DNA values (such as mammals, including humans) is noncoding DNA. What is the function of this DNA? Some of it may have no function at all, although we are beginning to learn about functions of DNA segments that do not contain conventional genes.

One intriguing class of "extra" DNA includes transposons (transposable elements). These are short pieces of DNA that have the ability to move from one place in the genome to another or from one genome to another. The transposon usually contains sequences that specify the machinery required for its own mobility and perhaps one or two other genes that can affect the host. Although the transposons may appear to be innocuous passengers in the genome, they play a significant role in generating evolutionary diversity by introducing chromosomal rearrangements and changes in gene expression. Some of these transposons are the inserted DNA copies of RNA retroviruses.

Another fraction of the extra DNA almost certainly plays a role in maintaining the higher-order structure of chromosomes and ensuring their appropriate segregation and disposition in the nucleus. The role of extra DNA is an aspect of genetics that is only beginning to be explored.

The Physical and Chemical Characterization of Bacteriophages and of Bacterial Genomes Has Revealed a Remarkable Array of Molecular Structures

Bacteriophage genomes (the genomes of viruses that attack bacteria) range from single-stranded linear RNA and DNA molecules to small double-stranded linear and circular DNA molecules. By comparison, the double-stranded linear or circular molecules that compose the entire genomes of various bacteria are enormous. The quest for correlations between the physical map of the genome and the genetic map obtained by recombinational studies led to the complete sequencing of bacteriophage genomes and is currently the driving force behind the attempt to completely sequence bacterial genomes. The first genome to be sequenced, that of bacteriophage ϕx-174, resulted in the surprising finding that some of the stretches of nucleic acids encode overlapping genes.

Although much has been learned about the DNA of bacteria, less is known about the proteins that are bound to that DNA. Specific proteins have been identified that bind adjacent to and perhaps within the coding region of active genes and are necessary for specific gene transcription and function, but many such proteins are as yet uncharacterized. Existing techniques of genetic, chemical, and physical analyses need to be pushed further, and other techniques will need to be invented to yield a dynamic picture of the changes in the structure of the genome during the cell cycle. Both the overall structure and the disposition of specific proteins and nucleic acids within these structures need to be determined.

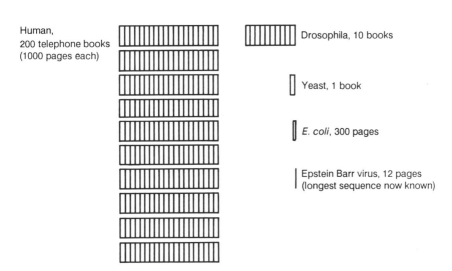

Human, 200 telephone books (1000 pages each)

Drosophila, 10 books

Yeast, 1 book

E. coli, 300 pages

Epstein Barr virus, 12 pages (longest sequence now known)

FIGURE 4-1 Genome sizes. Shown are the number of telephone books needed to display the DNA sequence of various genomes, ranging from that of a simple virus to the entire human genome.

DNA Replication

The study of DNA replication promises at long last to yield information important for understanding both normal proliferative responses, such as those basic to the immune response, differentiation, tissue regeneration, and abnormal processes such as tumorigenesis and carcinogenesis.

The Major Components of DNA Replication Have Been Determined

Although the problem of how a DNA sequence is transmitted in dividing cells was solved conceptually by Watson and Crick in 1953, the elucidation of the actual molecular basis of replication has required considerable time and effort. At many points in the history of the development of the field it appeared that replication was "solved" and that all that was left was to work out the details. For instance, in 1959, the discovery of DNA polymerase I of *Escherichia coli* led to 10 years of study based on the assumption that replication was catalyzed by this single enzyme. Then, in 1969, it was shown that polymerase I mutants were viable and that DNA polymerase I was, thus, not required for replication. This finding led to the search for and discovery of DNA polymerase III. Polymerase III mutants that were replication-defective identified this polymerase as the authentic replicative enzyme. Additional replication mutants showed that more than DNA polymerase was required. What was needed was a strategy to identify the products of the genes defined by the replication mutants. The complexity of replication forced biochemists to first work out the number of components required in prokaryotes.

By using a variety of DNA viruses as model systems, biochemists have been able to identify six essential classes of proteins in addition to DNA polymerase that are required in virtually all replication systems.

The Details of DNA Replication Can Now Be Investigated

A current model of how the purified proteins interact to carry out movement of the replication fork along both strands of DNA simultaneously is shown in Figure 4-2. Rather than carry out a set of sequential individual reactions with one class of protein carrying out its function, dissociating from the DNA, and the next protein further processing an intermediate, the proteins form a stable complex that remains associated with the DNA during growth. Fork movement of the chain of newly synthesized DNA may be facilitated by ordered conformational changes in the proteins in the complex by the hydrolysis of bound nucleotide triphosphate molecules. Thus the replication complex resembles a machine, with proteins as the moving parts.

Good evidence exists that these replication complexes are stable; despite the conceptual attractiveness of the machine model, however, we actually know very

LAGGING STRAND

LEADING STRAND

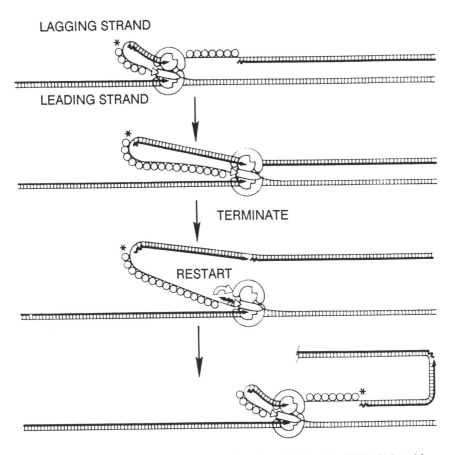

TERMINATE

RESTART

FIGURE 4-2 The trombone model for the propagation of a replication fork of DNA. [Adapted from B. M. Alberts, Cold Spring Harbor Symp. Quant. Biol. 59:1-12 (1984)]

little about how the proteins move along the DNA. In the future, many experimental disciplines will be brought to bear on this problem. What is the detailed architecture of the complexes? What is the driving force for the set of coordinated protein movements, and what is the inherent timing mechanism of the polymerization cycles? How are error rates minimized? How do accessory proteins increase the time of association of DNA polymerase with the template? How is the size of intermediate fragments determined? Are all of the components of the machines defined or have we missed some by our search methods? The key enzyme remains DNA polymerase, for which an x-ray crystal structure is now available, allowing more precise questions about polymerization mechanisms. Sequencing of cloned polymerase genes and in vitro mutagenesis, based on the crystal structure, offer fundamental new insights into how these enzymes work, especially with respect to mechanisms of fidelity (ability to copy accurately) and mutagenesis. Comparisons between cellular and viral polymerases are defining differences that can be exploited for antiviral therapy by rational drug design.

The original idea that replication was carried out by a single enzyme may seem naïve to us now, but it would also be naïve to think that DNA replication is the only process mediated by such complex protein machines. Impressive evidence has already accumulated that transcription, translation, RNA processing, and muscle contraction are also organized in this way. Yet, only for bacterial DNA replication is the field so advanced as to have all the components not only defined, but purified in abundance, and in a form allowing reconstitution of the entire process in the test tube. Although the current description of the replication apparatus in itself represents an enormous achievement, the most exciting era is just beginning. The basic principles that emerge from future enzymatic replication studies will have a general biological significance far beyond the specific light they shed on DNA synthesis.

The Regulation of Chromosomal Replication Is Poorly Understood

The regulation of chromosome replication will be emphasized in research in the coming years because of its important implications for cellular and developmental biology. In both prokaryotic and eukaryotic cells, regulation of replication occurs when replication is initiated rather than when the DNA chain is elongated. Therefore, the first step toward understanding regulation is the identification of proteins, genes, and DNA sequences involved in initiation. Apparently only one more protein is necessary for initiation than for elongation in bacteria [although this may not be the case for the replication of simian virus 40 (SV40)]. The extra protein is one that selects the sequence of nucleotides in the origin of replication, binds there, and mediates unwinding of the DNA, creating a nucleoprotein structure that directs binding and further unwinding by the enzyme DNA helicase. Interestingly, only months after this simple mechanism was established for bacteria, an identical process was reconstituted from partially purified human proteins in the eukaryotic DNA virus SV40. This finding suggests that initiation mechanisms may have been conserved over long periods of evolutionary time; although the SV40 system is undoubtedly simpler in detail than eukaryotic chromosome replication, the basic enzymology of replication may confidently be assumed to be the same in prokaryotes and eukaryotes.

Since the systems for the study of initiation of replication are new, much less is known about initiation than elongation. In the future, this recent work on initiation will be extended to address the role of transcription in activating DNA replication. Are there additional proteins? What are the primary, secondary, and tertiary structures of the DNA in specific areas where replication begins? Most importantly, how is the process controlled? The prokaryotic systems will continue to be important for enzymological studies because of the ease of genetic and biochemical analysis and because of their recently proven relevance for higher cells. In addition, new technological advances should permit the testing of principles learned from prokaryotes in the more complicated eukaryotic systems.

A second aspect of the regulation of eukaryotic DNA synthesis needs to be understood. The eukaryotic cell apparently retains the ability to initiate new replication units throughout the S phase, but a region once replicated cannot be replicated again in the same cell cycle. This block to reinitiation is essential to ensure that each region of a chromosome is replicated only once per cell cycle. The next few years will see major efforts to try to understand the molecular basis for this aspect of replication regulation.

Initial Efforts Will Emphasize the Further Development of Eukaryotic Experimental Systems for the Study of Replication

Two of the systems that seem to offer the most promise for study of eukaryotic replication are yeast and amphibian eggs. Yeast has been uniquely successful in contributing to our knowledge of eukaryotic DNA replication in at least two ways: (1) in the isolation of the DNA sequences essential for chromosomal replication and segregation: origins of replication, telomeres and centromeres; and (2) in the isolation of genes and mutants that have helped characterize several of the proteins involved in replication. Recently it has been found that extracts of amphibian eggs (*Xenopus*) initiate replication, with two unexpected features: Replication initiation requires the formation of an intact nuclear envelope around the added DNA; and synthesis is periodic, even though there is no periodic decondensation and condensation of the chromosomes. Intensive study of the proteins that catalyze replication in these eukaryotic systems and others should lead to a complete molecular understanding of the process.

Recombination

Genetic Recombination Is the Rearrangement of DNA Sequences to Create New Genetic Information

Genetic recombination is one of the few ways to introduce variability into the genome. Recombination is most important in the generation of diversity among individuals in a population, although it is also a mechanism for generating antibody diversity in vertebrates (see Chapter 7). Recombination is also important in pathological processes such as cancer. Many viruses, including the cancer-causing retroviruses, rely on recombination for an essential part of their life cycle: the joining of the viral genome with the host genome. In addition, many parasites rely on recombination to alter the genes that code for components of their surface coats so as to escape the host's immune defenses. A final area of importance for genetic recombination centers on the use of recombination by researchers to artificially introduce gene segments into the genome of cells or whole organisms.

General Homologous Recombination Is the Simplest and Most Studied Category of Recombination

General homologous recombination is the exchange of DNA strands that have a certain degree of complementarity. In homologous recombination the two DNA segments need not be completely identical, but only homologous, to recombine. Studies of homologous recombination in bacteria have identified a key protein, RecA, that catalyzes the exchange of genetic information. Many of the essential features of homologous recombination have been reconstituted with purified RecA protein. These biochemical studies have brought researchers to the point of asking two fundamental questions. First, how does RecA locate the homologous segments of DNA and bring them together to initiate the exchange? Second, how does RecA catalyze the extension of the initial point of DNA exchange so that hundreds of thousands of nucleotides of DNA are recombined? We know that this process requires hydrolysis of the high-energy compound adenosine triphosphate.

In eukaryotic systems, the stage is set for major research advances. Molecular genetic techniques should expedite the identification, isolation, and characterization of the genes that encode the recombination apparatus. Because of the ease of genetic manipulation, research on the lower eukaryotes may lead the way in this effort. One hopes, however, that conservation of gene structure will permit identification of related genes in higher eukaryotes. The biochemical study of meiotic components—the complex that aligns homologous chromosomes, DNA strand transferases that initiate exchange, and DNA resolvases that terminate exchange—is in its infancy, but recent progress in yeast should attract more workers to the field.

In addition to understanding the basic mechanism, we need to learn more about important control features of meiotic recombination. For example, failure of homologues to separate at meiosis is the basis of trisomy diseases like Down's syndrome. A second item for further research is the means by which cells prevent recombination between repeated sequences. Such repetitive DNA is a hallmark of higher eukaryotic cells and presents a potentially devastating opportunity for a homologous recombination system. If left unchecked, recombination between repeated sequences could scramble the genome. We need to know how this is limited and whether breakdowns in this control are responsible for some human disease.

A separate issue in the study of general recombination of eukaryotes concerns somatic cells. When recombination has been studied in living organisms (usually *Drosophila* and yeast), somatic recombination is less frequent than recombination in germline cells and seems to be primarily involved in DNA repair. Nevertheless, many scientists have demonstrated the efficient recombination of homologous segments of DNA when they are artificially introduced into somatic cells in culture. Although ongoing studies attempt to define the mechanism of this recombination, the most exciting possibility is its potential use to manipulate the genetic content of cultured cells. Research will focus on controlling the frequency and fidelity of recombination so that investigators can introduce genes into cultured cells and replace existing genes with modified ones. This sequence of manipulations is readily accomplished in bacteria, yeast, and slime molds, where the technique has been of tremendous value in the genetic analysis of complex loci. Formidable problems arise in extending this technology to higher organisms because of the complexity of their genomes. However, several initial successes in targeting have opened the way to a thorough attack on the problem.

Site-Specific Recombination Has Been Studied for the Past 20 Years and Many Systems Have Been Studied in Depth

In the 1960s it became clear that efficient recombination could take place between segments of DNA that lacked sufficient homology to undergo general homologous recombination. Two features are required for such nonhomologous recombination—the presence of special DNA sites in at least one partner in the recombination process and the activity of specific enzymes (recombinases) that act at these sites. The rearrangements of the genome promoted by site-specific recombination are mechanically fascinating. One wants to know how small segments of the genome are identified, brought close together, and rearranged with high efficiency and fidelity. Moreover, site-specific recombination is a critical step in many important biological processes. As a result, the study of site-specific recombination has emerged as a major area of concentration of biological research that will continue to grow rapidly in the next 5 to 10 years.

The molecular mechanism of these reactions has been worked out by using the molecules and mutant strains identified by genetic analysis to reconstitute the recombination event in a cell-free (in vitro) system. Site-specific recombination can be divided into broad mechanistic categories according to the degree to which both partners must have specific sites and on the amount of DNA synthesis that accompanies the reaction. In some systems, recombination uses a completely conservative mechanism in which the participating DNAs are simply broken, rearranged, and rejoined. In other systems, complete replicas of the DNA must be synthesized during the recombination event. The integration of the bacterial virus called lambda into the *E. coli* genome is a reaction with high target specificity and strict conservation of the parental DNA. This complete reaction has been reconstituted in vitro. By combining topological investigations with biochemical and molecular genetic studies, the field is reaching the goal of elaborating plausible models for complex processes such as synapsis, the juxtaposition of DNA segments that is a prerequisite for efficient, precise recombination. In the future, structural studies using x-ray diffraction and nuclear magnetic resonance techniques should be able to add atomic details to the understanding of these processes that is being deduced from molecular biological studies.

In contrast to the conservative reactions (no DNA replication), study of the mechanism of the replicative site-specific recombination (which requires DNA replication) is in its infancy. The transposition of the genome of the bacterial virus mu, is the only example that has been reconstituted in vitro. Several features of the DNA sequence of donors and targets suggest close parallels between mu transposition on the one hand and other rearrangements such as the integration of retroviruses (RNA viruses that replicate by way of a DNA intermediate that is inserted into the host's chromosome) and the recombination of antibody genes. The pioneering work on mu transposition has thus opened the way for rapid progress in the biochemistry of these eukaryotic recombination events.

Another important research area, control of the recombination process, has already been extensively studied in prokaryotic and lower eukaryotic systems. Here, the key event seems to be control of the synthesis of a specific recombinase. In some systems, such as the integration of lambda and sexual differentiation of yeast, such control is elicited through a complex chain of events that couples the synthesis of the recombinase to related events in the life cycle of organisms. In higher eukaryotes, we know almost nothing about the control of recombination. We need to learn whether these events are regulated and, if so, by what mechanism. For the immune system, some experiments suggest that accessibility of the recombining sites rather than the availability of the recombinase may be critical. The finding that several human tumors result from an inappropriate rearrangement of immune elements highlights the importance of understanding the control of site-specific recombination in higher organisms.

DNA Repair and Mutagenesis

Organisms Correct Mistakes by DNA Repair Mechanisms

It is critical for organisms to maintain and protect the fidelity of the information encoded in their DNA so they can pass this information on to their descendants and so they can access that information during their own lifetimes. However, the integrity of the DNA itself and the information encoded therein are undergoing constant challenges from agents in the environment. Furthermore, mistakes can be introduced into the information by normal cellular metabolic events such as DNA replication and recombination. The study of mechanisms of DNA repair and mutagenesis is of direct relevance to human health, since mutations in DNA are responsible for both genetically inherited diseases and the somatic cell diseases known as cancer.

The Molecular Mechanisms of DNA Repair and Mutagenesis Need to Be Studied in Prokaryotes, Yeasts, and Mammals

Around 1975 it was generally thought that DNA repair and mutagenesis were understood in bacteria and that this knowledge made it possible to approach these issues in eukaryotic organisms. However, new discoveries of a variety of complex independent enzymatic systems specializing in repair of different kinds of damage over the past 10 years have revealed that these processes are much more complex, even in bacteria, than was assumed in 1975. These new revelations about the complexity of bacterial systems explain the relatively modest progress in dissecting DNA repair in mammalian cells. It is important that major studies of DNA repair and mutagenesis be continued at several levels at once: studies involving at least prokaryotes, yeasts, and mammals. The main reason for lack of progress in mammalian systems has been the lack of genetic tools similar to the ones used in work on prokaryotes.

Opportunities Abound for Studies of DNA Repair and Mutagenesis in Prokaryotes such as E. coli

Recent studies show that the complexity of the repair systems demand the power of *E. coli* molecular genetics if we are to obtain a framework within which to understand these processes in all organisms. To begin with, the molecular mechanism by which damaged DNA is processed to give rise to mutated DNA ("error-prone repair") is not understood in any organism, and *E. coli* is likely to be the first organism in which this fundamental problem will be solved. The existence of regulatory networks of genes induced by DNA damage was discovered in *E. coli*. Studies of two such gene systems that are induced by DNA damage, the SOS repair system and the adaptive response, have led to the discovery that some proteins (such as the proteins RecA and Ada) have both a regulatory and a mechanistic function in DNA repair. The complexities of the regulation of these systems need further study, as do the functions of all of the induced gene products. Detailed analysis of complex proteins such as these are needed to provide models for systems that lack sophisticated genetics.

Major Opportunities Exist to Study DNA Repair in Eukaryotes

Yeast can be used to explore uniquely eukaryotic aspects of DNA repair, such as the relationship of DNA repair processes to the cell cycle and the effect of chromatin on DNA repair. In yeast, many genes affecting survival or mutagenesis after DNA damage have been identified, and many of these have been cloned. However, very little is known about the biochemistry of DNA repair in yeast. Moreover, although a few yeast DNA repair genes have been shown to be induced by DNA damage, virtually nothing is known about the mechanism of their regulation.

The study of DNA repair in mammals, especially humans, is a challenging problem of great significance, but one that needs to be approached with greater sophistication. Many studies to date have been primarily descriptive and have allowed only low-resolution inferences concerning the mechanism. Another problem has been that humans with inherited diseases caused by mutations in DNA repair, such as xeroderma pigmentosum and ataxia telangiectasia, have complex phenotypes that make analysis of their deficiencies difficult. Furthermore, to date, no one has succeeded in cloning genes that complement the DNA repair deficiencies of these naturally occurring mutants. Problems unique to higher organisms that will need to be addressed include the relative DNA repair capabilities of different types of differentiated cells and the DNA repair capabilities of germline cells versus somatic cells.

Gene Expression

Genes Provide the Information Required for the Formation, Function, and Reproduction of Cells and Organisms

The information in the genes specifies the sequences of RNAs and proteins that have the biochemical capacity to synthesize other cellular components (sugars and lipids) as well as the DNA, RNA, and proteins themselves. The process that converts information stored in selected genes into RNA and proteins is called gene expression.

Specific gene expression leads to the selected transcription copies of appropriate genes at particular times in the life cycle of the cell or organism. Although we now understand some of the strategies involved, our knowledge of the control of transcription still contains enormous gaps, for both prokaryotic and eukaryotic organisms. An understanding of the regulation of gene expression lies at the heart of understanding how the genome functions to guide the development, growth, and differentiation processes that generate a complex organism from a single-celled fertilized ovum.

Gene expression must be properly controlled during the life cycle of any organism. For example, hemoglobin is produced in red blood cells but not in other cells. Yet the hemoglobin gene is present in all cells of the body. Similarly in plants, proteins of the photosynthetic apparatus are expressed in specialized leaf cells but not in cells of the root. By what mechanisms are the genes in individual cells differentially expressed?

At least two levels of regulation exist: large-scale activation or inactivation of groups of genes and the precise control of individual genes. Since eukaryotic development is characterized by changes in the expression of complex cohorts of genes, these cohorts may be regulated in parallel. This procedure would be straightforward if the structure of the nucleus could in some way regulate the activity of the DNA. For example, groups of genes that show coordinate expression might be packaged in discrete domains or compartments that could be regulated at one or a few key control sites. However, whereas coordinately regulated genes in prokaryotes are frequently linked and controlled by a single promoter (nucleotide sequence to which RNA polymerase initiates transcription), such linkage is not usually found in eukaryotes. In most instances, members of coordinately controlled gene sets in eukaryotes are broadly scattered over the chromosomes, implying that transcription is controlled by *trans*-acting regulators, which can affect gene regulation at widely separated chromosome sites.

However, large-scale gene control is found in several notable exceptions to the general picture of eukaryotic gene regulation. These include inactivation of the X chromosome in mammals and dosage compensation of the X chromosome in insects, in which the transcription of genes on an entire chromosome or a large chromosome segment is regulated as a unit. In this latter case, regulation is associated with a general change in chromatin structure; for example, the inactive X in mammals can be distinguished by its pattern of staining. Genes that are moved onto the inactivated X chromosome are also inactivated: This type of gene inactivation does not generally occur when a gene is simply moved near a member of a coordinately controlled gene set.

Formation of Primary RNA Transcripts Depends on Specific DNA Signals

For a gene to be transcribed, it must have certain nucleotide sequences arranged in serial order along the DNA chain. The minimal elements are— starting at one end and reading from left to right—a promoter sequence, which binds an enzyme (RNA polymerase) that facilitates transcription; a coding region that specifies the mRNA transcript; and terminal sequences that terminate and stabilize the transcript. Experiments in bacteria showed that the promoter region also serves as a binding site for specific gene-regulatory proteins that regulate the transcription activity of the gene. New techniques, especially gene cloning and DNA sequencing, have enabled investigators to demonstrate that a similar system of transcriptional regulation exists in higher organisms, including humans. The promoter region of each gene in a higher organism has sites for several different DNA-binding proteins. Many of the specific proteins are now being characterized, and their genes cloned, so that the mechanisms by which they work can be studied in detail at the molecular level.

One class of binding site that need not be located precisely at the promoter region of genes is called an enhancer. Enhancers can function from distances of nucleotide pairs away from the RNA start site they affect. Enhancers were originally given this designation because gene activity was enhanced when the element was present and diminished when it was absent; we now know that an enhancer can also repress transcription, depending on the exact proteins bound.

An important property of many enhancers is their tissue-specific action. Sequences in the promoter region of the human insulin gene, for example, bind proteins found in insulin-producing cells but not in cells that do not produce insulin. Such results indicate that tissue-specific gene expression is at least in part caused by DNA-binding proteins that bind at specific sites next to genes.

The demonstration of enhancers helps explain how cells can coordinately express large sets of genes physically separated on different chromosomes. That is, specific DNA-binding proteins may recognize and bind with a similar nucleotide sequence present in the promoter region or enhancer of a number of different genes. In this way, a particular set of genes typical of a certain cell type would be activated or inhibited in a coordinated, programmatic fashion by the activation of a specific type of DNA-binding protein. Many such DNA-binding proteins exist, and a catalogue of many of them is needed before we can understand how the network of regulatory proteins affects cell differentiation in higher organisms. These networks are fundamental to the processes of normal cell differentiation and tissue morphogenesis. Moreover, the inappropriate expression of *trans*-regulatory proteins could lead to the abnormal gene expression associated with some forms of cancer.

Primary RNA Transcripts Are Modified by Additions at Their Ends and Deletions of Internal Segments

Modifications of RNA transcripts have an important influence on gene expression. When a transcript is first formed, its free end (the 5' end) is chemically modified, or capped. Such capping is believed to stabilize the transcript. In addition, immediately after completion, most transcripts are modified by having a string of adenylnucleotides added to their other end (the 3' end); this polyadenylated, or poly(A), modification is believed to influence their life expectancy.

Most transcripts are also extensively modified by the precise deletion of internal segments, a process called mRNA splicing. The splicing mechanism must proceed with high fidelity, since an error of even a single nucleotide could change the reading frame (nucleotide sequence that codes for a specific protein), leading to translation of an altered protein. Variations in splicing may lead to greater flexibility in gene formation and in the expression of genes.

RNA splicing is required because of the existence of introns in eukaryotic genes. Introns are segments of DNA that are not used to specify protein sequences but are interspersed among the coding sequences. Their presence explains why long nuclear RNAs are produced as precursors for shorter cytoplasmic mRNAs. The intron sequences are excised from the precursor by RNA splicing, and the mature mRNA is transported to the cytoplasm. The length of an intron can vary from 60 or 70 to more than 100,000 nucleotides, and a typical vertebrate or plant gene might contain several introns.

Since the coding sequences used for translation of a protein are formed during splicing, variations in the pattern of splicing of a precursor RNA permit one gene to specify the formation of multiple proteins. Not only does such variation in splicing exist, but more importantly, the splicing pattern of a precursor can vary from one cell type to another. This means that splicing is regulated by factors specific to cell type.

Several questions are central to the study of RNA splicing. What specifies the set of intron sequences to be excised by splicing? Small nuclear ribonucleoprotein (snRNP) particles are crucial to sequence recognition during splicing. The splicing of the shortest intron requires the activities of at least four different particles, whereas longer introns may require even more. Changes in such factors could regulate different splicing patterns in different cell types.

Another important question concerns the chemistry of the splicing reactions. What are the mechanisms of the endonucleolytic cleavage and ligation steps, and are these reactions catalyzed by RNA or protein components?

The study of splicing is a new frontier in molecular biology. The most startling finding from this research has been that of the self-splicing of particular types of RNA sequences. As the name suggests, this splicing reaction occurs in the absence of protein and is catalyzed by conserved sequences within the intron. When analyzed in detail, the RNA sequences within this intron have a catalytic activity similar to that commonly associated with several different types of protein enzymes. Catalytic RNAs are now thought to have been common in primitive biological systems and still to be integral to many contemporary processes (such as protein synthesis). In the past few years, they have been discovered in viruses, bacteria, fungi, protozoa, animals, and plants. It is even possible to imagine that the first cells on earth may have contained only RNA enzymes, and that cells existed before there were proteins.

Precursors to mRNAs are transcribed at many sites in the nucleus, processed by splicing and polyadenylation, and then transported through nuclear pores to the cytoplasm. Nothing is known of the mechanism of transport. These precursors are bound to a highly conserved set of basic proteins. In the cell, these basic proteins are retained in the nucleus and may play an important role in both splicing and polyadenylation. The complex steps of RNA processing offer several points at which gene expression could be modulated, such as by differential splicing to yield different mRNAs. The field of RNA processing is young and has only begun to be explored.

Specific DNA Modification May Influence Gene Expression and Confer Heritable Traits

Specific DNA base pairs are modified most frequently by methylation (the addition of methyl groups to DNA). In prokaryotic organisms it was thought that modification was required only to prevent DNA degradation by restriction enzymes. Recently, however, direct effects of methylation on the regulation of gene expression have been found. In mammalian genomes it has long been thought that methylation might correlate with the heritable control of specific gene expression. A variety of indirect experiments suggests that demethylation can derepress gene activity (activate a gene from a repressed state). In mammalian embryos, the expression of certain genes depends on whether the genes have been transmitted by the egg or by the sperm (see Chapter 5). The basis of this "imprinting" is not known, but data suggest that specific methylation changes that occur during the formation of egg and sperm may be involved.

Although DNA methylation could explain the heritable patterns of gene expression displayed by various cell lineages in multicellular organisms, we do not know how large a role methylation plays, nor do we know what controls the patterns of DNA modification. In some complex multicellular organisms (for example, *Drosophila*), the absence of evidence of DNA methylation suggests that it evolved rather late in evolution as a back-up system. Clearly we have much to learn about the role of methylation and other heritable modifications in the control of gene expression.

The Study of the Regulation of Gene Expression Offers Great Potential for New Research Opportunities and Practical Applications

At each level—DNA structure, transcription, RNA processing, mRNA translocation, translation, and the determination of the three-dimensional structure of the protein product and its introduction to the appropriate cellular compartment— new genetic techniques can be used to probe and understand the controlling processes. It is these processes that constitute "gene action" and determine cellular function. We know very little about how RNA polymerase reads the various promoter and enhancer sequences that determine the genes to be transcribed. It is not at all clear that the presence of transcription factors is sufficient to generate a regulatory system that can persist even after cells replicate, yet we know that cell determination results in patterns of gene expression that are inherited by daughter cells. We need to know the details of these and associated processes to understand the mechanisms that control cellular differentiation.

The ability to intervene and affect any of these processes specifically would make it feasible to design a host of new therapeutic drugs. For example, if we understood how genes such as those that control the production of interferon are induced, we might be able to mimic such a process selectively to enhance resistance to viral infections. The same kind of arguments apply to the regulation of growth in cancer cells and of immune responses.

Gene Transfers Provide Many Opportunities to Study in Detail How Genes Function

Using gene transfer technology, one can remove the control regions (enhancer or promoter) of a gene a bit at a time to identify segments that are critical for normal function. One can also switch control regions from one gene to another, so as to cause genes to be expressed under circumstances in which they would normally be silent. Although such studies have been possible for only a short time, they have provided much information (and some surprises) about gene regulation. In many organisms it is now possible to insert the modified gene directly into germline cells so that entire organisms carrying the new DNA are obtained. Studies of transgenic animals are now providing important information on both the mechanisms that regulate tissue-specific gene expression and those that cause cancer (for a discussion of transgenic animals, see Chapter 2).

Genome Organization

The Organization of Complex Genomes Can Be Studied by Gene Mapping

Every cell of the human body contains an identical set of an estimated 60,000 to 100,000 genes. Most genes are present only once (single-copy genes). Each gene can be assigned to a particular chromosome site. This activity, called gene mapping, is improving our understanding of cells.

Gene mapping in humans and selected mammalian species has advanced impressively during the past decade. In humans, approximately 1,400 genes of known function have been mapped to some degree of precision, and another 1,000 DNA markers of unknown function have been localized. A preliminary low-resolution genetic map of the human genome has already been constructed.

Restriction Fragment Length Polymorphism DNA Markers Have Increased the Power of Mendelian Techniques for Mapping Genes

RFLPs are naturally occurring variants in the nucleotide sequence of DNA that can be used as genetic markers. They are transmitted from one generation to the next in the same way as the genes that govern eye or flower color. The advantage of RFLPs is that they can be detected through the use of only a few cells of the organism by the techniques of molecular biology. In addition, RFLPs relate the genetic linkage map to the physical (DNA) map. In essence, RFLPs add a large set of new markers for mapping any genetic locus of interest and for obtaining a high-resolution genetic map of any chromosomal region of particular interest.

The Value of a Human Gene Map Has Become Apparent in the Past Several Years

The map serves as a body of data that can be used to address a broad range of biomedical questions. It becomes more useful in generating and resolving hypotheses as the data base grows. For example, the map has been useful in establishing the connection between chromosome rearrangement and proto-oncogene activation. Proto-oncogenes are normal genes generally involved in cellular growth control, but they have the potential to cause cells to become cancerous when inappropriately expressed. In some cases inappropriate activation has been associated with rearrangement of the chromosome bearing the proto-oncogene. The gene map has also enabled investigators to detect linkage relationships between RFLP markers and specific genes that carry a risk for genetic disease. This can now be accomplished for many genetic diseases, such as various anemias, phenylketonuria, hemophilia, and Huntington's disease.

The identification of linkage relationships between genes causing genetic disease and RFLP markers is an important first step in physically isolating genes whose mutants cause genetic disease. A start in this direction has been made for a number of human heriditary diseases, such as Duchenne's muscular dystrophy, Huntington's disease, and cystic fibrosis. Isolation of these disease-causing genes will be the first step in analyzing the molecular basis of these diseases. The use of this approach to isolate genes and determine their products can reveal the underlying molecular basis of more than 3,000 human genetic conditions and provide crucial insights into human developmental processes.

If the number of human genes is approximately 100,000, only 1 percent of them have been mapped to date. Currently, new genes are being mapped at the rate of one every day. Two important new mapping activities are just now beginning: the formulation of maps based on overlapping cloned DNA segments and the DNA sequencing of the nucleotides that constitute these segments (see Chapter 3).

CYTOPLASM: ORGANELLES AND FUNCTIONS

In a eukaryotic cell, the nucleus is surrounded by the cytoplasm, which in animal cells usually accounts for nine-tenths of the cell's volume; the cytoplasm is surrounded in turn by a cell membrane, or plasmalemma. The cytoplasm contains within a protein polymer matrix several types of minute, functionally specialized cell organs, or organelles, most of them present in multiple copies. In prokaryotic cells (bacteria), by contrast, the genome is not separated from the cytoplasm, there are no membrane-bounded organelles, and the plasma membrane subserves many of the specialized functions of eukaryotic organelles.

Prokaryotic and Eukaryotic Cells Use Related Strategies to Compartmentalize Their Biosynthetic Reactions

Prokaryotic cells have only one membrane, located at the cell surface (although some bacteria have a double membrane). In contrast, eukaryotic cells, such as those of animals and plants, have, in addition to their plasmalemma, at least a dozen different types of chemically specific membranes that create separate intracellular compartments with different microenvironments required by various processes, such as respiration, photosynthesis, protein synthesis, and intracellular digestion.

In prokaryotic cells, the plasmalemma is the site of a number of important biosynthetic activities, including the synthesis of membrane lipids, adenosine triphosphate, and cell-wall components. This membrane also contains a multitude of transporters, receptors, and sensors for chemotactic movements, which create the interface of the bacterium with its extracellular environment.

In eukaryotic cells, comparable activities are differently distributed. Some of them (such as those of the transporters and receptors) remain in the plasmalemma, but others are relocated to different intracellular membrane systems. Lipid synthesis, for instance, in animal cells occurs only in the endoplasmic reticulum (ER), a network of membrane-bounded channels that pervades the cytoplasm. In plant cells, lipid synthesis occurs in plastids. From the ER or plastids, newly synthesized lipids are distributed to all other cellular membranes. The ER also contains a complex set of enzymes that modify lipid-soluble aromatic compounds imported by the cell from the environment. The modifications increase the water solubility of these compounds, thereby facilitating their elimination. Since sterols, drugs, herbicides, toxins, and chemical carcinogens are among these compounds, the relevant enzymes constitute an intracellular detoxifying system.

The ER membrane is also the site of important steps in protein traffic regulation. Issues awaiting resolution include the means by which proteins and lipids are transported from their site of synthesis in the ER to a multiplicity of destinations and the way differences in lipid chemistry are established and maintained in different membranes. We must also advance our understanding of the ER detoxifying system to shed light on problems related to chemically induced cancers and toxic effects of chemical pollutants in the environment.

Protein Synthesis and Regulation

Protein Synthesis and Regulation of Protein Traffic Take Place in the Cytoplasm

Protein Synthesis. For all cells, protein synthesis is a major, continuous activity needed for the production of intracellular enzymes, contractile and cytoskeletal assemblies, membranes, ribosomes, chromosomes, and many other functionally important macromolecular assemblies. In more complex multicellular organisms, it is also needed for the production of proteins destined for export out of the cell, such as enzymes, hormones, growth factors, blood proteins, antibodies, or components of the extracellular matrix.

In all cells, proteins are synthesized on ribosomes, which translate into amino acid sequences the instructions received from active genes in the form of mRNAs. The ribosomes themselves are macromolecular assemblies of ribosomal RNA and protein molecules. The two sets of components are produced separately in the cytoplasm (proteins) and in the nucleus (ribosomal RNAs) and are modified and assembled in a special compartment in the nucleus (the nucleolus) before being transferred to the cytoplasm as part of a functioning ribosome.

In both prokaryotic and eukaryotic cells, ribosomes are basically similar and protein synthesis proceeds by similar steps. In eukaryotic cells, however, the ribosomes are larger and require more factors for their activity. These differences are probably related to the emergence of more versatile regulatory processes in eukaryotes.

Protein Traffic Control. In eukaryotic cells, ribosomes and protein synthesis occur primarily in the cytosol, to which mRNAs have direct access from the nucleus and in which the pool of amino acids and all ancillary factors required for protein synthesis are located. Only 2 percent of the total protein production is accounted for by small mitochondrial ribosomes whose products are used exclusively in mitochondria. In plant cells a somewhat larger fraction of the cell's protein is produced by chloroplasts (or other differentiated forms of plastids).

From the cytosol, proteins are accurately directed to more than 20 different sites of final functional residence. These sites are membranes or compartment contents, each endowed with chemical specificity.

ER Targeting System. Many proteins are directed to the ER membrane as they are being synthesized. The selectivity is based on mutual recognition between signals (called signal sequences) within the amino acid sequence of the protein to be transported and a signal recognition complex. This complex involves a ribonucleoprotein particle (called a signal recognition particle) and at least one transmembrane protein (its receptor) in the target ER membrane.

The ER targeting system recognizes and processes proteins destined for secretion and lysosomes (storage compartments for degradative enzymes), as well as membrane proteins for many intracellular compartments. Secretory and lysosomal proteins are fully translocated across the ER membrane into the ER cisternal space. Membrane proteins are partially translocated and remain anchored in the membrane. Recent experiments indicate that a membrane protein can be converted into a secreted polypeptide, and conversely, a secretory protein can be converted into a membrane protein by deleting or adding the information for the membrane anchoring sequences from their mRNAs.

The mode of operation of the ER targeting system has been elucidated in reconstituted in vitro systems in which ribosomes are allowed to translate into proteins the genetic information encoded in specific mRNAs in the presence or absence of ER-membrane vesicles in vitro. The results show that many components of the systems are functionally equivalent in different species, phyla, and even kingdoms, which implies that this part of the traffic regulation system originated early in evolution and has been conserved.

Post-ER Steps in Traffic Control. Once past the ER, proteins are moved within the cell through a specialized membrane-bounded compartment called the Golgi complex, where they are further modified by glycosylation, sulfation, and proteolysis; they are sorted while in transit to lysosomes, secretion vacuoles, or different membranes. The Golgi complex itself contains at least three subcompartments. Transport from the ER to Golgi subcompartments, from one subcompartment to another in the Golgi complex, and finally, from the last Golgi elements to the plasmalemma requires energy and is effected by vesicular carriers; thus, past the ER, protein traffic can be regulated at least in part by controlling the movements of vesicular carriers. These carriers apparently recycle continuously among compartments. The best known among these vesicular carriers are the secretion granules or secretion vacuoles of various glandular cells. They transport products to the cell surface and discharge them into the extracellular medium by a process known as exocytosis. Sorting of the proteins to their ultimate destinations probably involves mutual recognition between a signal and a receptor, but so far only the signal for lysosomal proteins has been chemically defined. Its receptor has been isolated and partially characterized and its gene sequenced. Reactants involved in the sorting of other proteins remain unknown, as are the signals and receptors that regulate the traffic of vesicular carriers. Studies to identify them are being actively pursued.

Other Traffic-Control Systems. The ER targeting system (which includes the Golgi complex) is undoubtedly the most complex component of the overall protein traffic-control system of the cell. The other components are simpler, and most of them probably transport the protein in a single step: from the cytosol directly to the final destination. The amino acid sequence of the signal that directs certain proteins to the nucleus is known in a few cases, but the corresponding receptor remains to be identified. A substantial body of information already exists about traffic regulation of proteins targeted to mitochondrial and chloroplast membranes. Among the protein products of plant nuclear genes are some that function in the mitochondria and some in chloroplasts; it is not known how the systems differ.

In a simpler form, protein traffic regulation occurs in prokaryotic cells and has been studied extensively in gram-negative bacteria, which are provided with two concentric membranes at the cell surface. The number of final destinations is considerably fewer—the two membranes and the intervening space and perhaps the outside of the cell. A signal, generally comparable to that found in eukaryotic proteins targeted to the ER membrane, has been identified and analyzed in detail by sequencing and by extensive genetic modifications. This line of work has led to the recognition of functionally critical residues in the signal sequence, but the other components of the system are still unknown.

We can anticipate considerable activity in this fertile and exciting area, especially in relation to the identification and characterization of signals and their receptor partners and to the intracellular location of receptors and sorters (molecules that control the selection and movement of proteins from one compartment to another). Although the picture is already rich in detail, many uncertainties remain to be resolved by further research. The reasons for removing the signal sequence are not clear, nor are the reasons for the complexity of the enzyme that effects the removal. The enzyme may have additional functions since it consists of six different proteins. The translocation process itself is still a mystery.

Structural biology is likely to yield three-dimensional information on such interactions if large enough quantities of relevant proteins can be obtained. The main goal is to understand how cells process their many proteins and how they achieve and maintain the chemical specificity of their membranes.

Another basic process that remains to be understood in molecular terms is membrane fusion. The process is critical for cell division, cell fusion in egg fertilization, and vesicle fusion along different pathways of intracellular transport. Membrane fluidity is a prerequisite for membrane fusion. It is also a prerequisite for membrane growth, which appears to proceed by expanding preexisting membranes. At the same time, intact diffusion barriers need to be maintained in highly dynamic membrane systems. Much remains to be discussed about how these processes are controlled.

MITOCHONDRIA: FUNCTION AND BIOGENESIS

Mitochondria Produce Most of the Cell's Main Energy Source, ATP

The mitochondrion is characterized as the power plant of the cell, because it performs the enzyme-catalyzed, stepwise oxidation of nutrients (such as sugars, fats, and amino acids) in a process called respiration. The most interesting product of this process is ATP, the direct source of the energy required for most of the chemical work the cell must perform to power its growth, movement, synthesis of new components, and other functions. A relatively small amount of ATP is produced in the cytosol during sugar catabolism (glycolysis), but by far the largest amount is generated in mitochondria. The mechanism of mitochondrial ATP synthesis has stubbornly resisted full elucidation, but progress continues to be made as a result of our insistent probing into this fundamental energy-transducing process.

Mitochondria are also of interest because they contain their own complement of DNA, which cooperates with the DNA of the nucleus in the control of mitochondrial formation. The origin and evolution of the mitochondrion are linked to the origin and evolution of eukaryotic cells. Understanding of mitochondrial function in turn sheds light on a wide array of fundamental and practical issues, ranging from certain metabolic and genetic diseases to evolution itself.

New Mitochondria Arise from Existing Ones, and They Are Characterized by Unique Functions

Mitochondria have two membranes (inner and outer) that define two concentric separate spaces. The inner space houses hundreds of enzymes, including those involved in the oxidative reactions that supply the energy needed for cell function. The inner mitochondrial membrane contains the energy-conversion apparatus. The majority of the mitochondrial proteins are specified by nuclear genes; they are synthesized on ribosomes in the cytoplasm and then imported into the organelle. A limited set of proteins of the inner mitochondrial membrane, namely, some protein subunits of the oxidative phosphorylation apparatus, are encoded in DNA molecules located within the organelle itself and are synthesized by an organelle-specific translation system. The distinctive structural RNA components of this system—RNAs and transfer RNAs—are also encoded in mitochondrial DNA. Mitochondria do not arise de novo in the cell by self-assembly of their constituent molecules, but are formed by growth and division of existing mitochondria.

The mitochondrial DNA from several organisms, including humans, has been completely sequenced, and much of its informational content has been elucidated. Furthermore, all mitochondrial gene products in humans have been functionally identified. A dramatic outcome of these studies has been the discovery that the mitochondrial genetic system in the organisms studied, except plants, uses a code that differs in several respects from the universal code and, in addition, utilizes for reading the code a novel mechanism, which requires only a restricted set of transfer RNAs.

Excellent Opportunities Exist to Study the Mechanisms of Expression and Replication of Mitochondrial DNA

Studies of the enzymes and ancillary proteins responsible for mitochondrial DNA replication, DNA transcription into RNA, and RNA processing to mature RNA species are making rapid progress, aided by the development of soluble in vitro preparations derived from broken mitochondria and by the use of recombinant DNA technologies. Specific proteins have already been identified and, in some cases, isolated in partially or completely pure form.

As in the case of many nuclear gene transcripts, the coding sequences of several mitochondrial gene transcripts in lower eukaryotes, especially yeast and filamentous fungi, are interrupted by nonfunctional segments, or introns. These introns must be removed to produce the mature RNA. The transcripts of some mitochondrial genes have the remarkable capacity to excise their own introns in vitro in the absence of protein; that is, they function as enzymes acting on themselves. The discovery of mitochondrial introns has opened an active field of research, which is expected to provide deep insights into the mechanisms of RNA splicing in general and into the origin and evolution of introns.

The Formation of New Mitochondria Is Under the Control of Both the Nucleus and Mitochondria

The dual control of mitochondrial formation is most dramatically illustrated by the chimeric structure of nearly all the enzyme complexes of the oxidative phosphorylation system: Each such complex contains some subunits encoded in the nucleus and some encoded in mitochondrial DNA. Because of this dual control, the formation of functional mitochondria requires a quantitative and temporal coordination of expression of the relevant nuclear and mitochondrial genes.

Two main classes of nuclear genes are the object of intensive investigation based on recombinant DNA techniques and on structural and functional approaches: (1) genes coding for subunits of the enzyme complexes of the oxidative phosphorylation system and for mitochondrial carriers used in metabolite transport and (2) genes coding for proteins involved in the expression and replication of the mitochondrial genome. Most of the latter genes are probably distinct from those that code for the homologous components of the nuclear-cytoplasmic apparatus, although interesting exceptions have recently been reported. They concern the possible existence of common nuclear genes for cytoplasmic and mitochondrial components, which could account for at least some of the reported influences of the mitochondrial genome on the remainder of the cell.

Research now under way promises to elucidate the mechanisms and signals involved in the interactions between the nuclear and mitochondrial genomes in the formation of functional mitochondria. Furthermore, research in the area of nuclear-mitochondrial interactions should help us understand how the assembly and function of mitochondria are integrated with those of the rest of the cell and how these processes can be modified in relation to cell respiratory demands, cell differentiation, and senescence.

Proteins Are Imported into Mitochondria

The hundreds of distinct polypeptides that make up a mitochondrion are distributed in a specific way in four compartments: the outer membrane, intermembrane space, the inner membrane, and inner mitochondrial space. After their synthesis on cytoplasmic ribosomes, the nuclear gene-coded polypeptides are imported to their correct location within the mitochondrion. Biochemical studies and the application of recombinant DNA technology have shown that proteins destined for one of the three innermost compartments are usually made from precursors with extensions at the amino-terminal end; these extensions can be as long as 100 amino acids and function as signals directing the proteins to the proper location.

Still unanswered questions include: Which molecules are involved in the import of polypeptides into the mitochondria? What is the role of cytosolic factors in the import process? How do mitochondrial signal sequences perform their function? Why does translocation of proteins across the mitochondrial inner membrane require a gradient of electrical potential across that membrane? Do contact or fusion points between the two membranes function as ports of entry for protein import?

*Crystallographic Studies Should Reveal the Tertiary Structure of the Oxidative
Phosphorylation Apparatus*

The subunit composition of the enzyme complexes of the oxidative phosphorylation system is largely known, as is the location—nuclear or mito-chondrial—of the genes specifying these subunits. From the nucleotide sequence of these genes, the amino acid sequence of the subunits can be inferred. Definitive knowledge about their tertiary structure should eventually come from crystallographic studies now in progress.

This information, together with data derived from other approaches to studies of the spatial relations of the subunits in each enzyme complex, is likely to provide useful models of the three-dimensional structure of each complex and of its topology in the inner mitochondrial membrane. These models will help in interpreting the results of ongoing functional studies on the individual complexes in intact mitochondria and in reconstructed systems.

CELL MOTILITY AND THE CYTOSKELETON

*An Understanding of the Basis of Cellular Motility Is Central to Our
Understanding of the Functioning of All Organisms*

Cell motility is necessary for the survival of virtually all living species. For example, the egg would not be fertilized without a motile sperm, and every cell division that occurs thereafter requires a degree of motility in some cell parts. Without active changes in cell shape and cellular migrations, embryos would not form. Without cellular motility, white blood cells would neither accumulate at sites of inflammation nor ingest invading microorganisms. Without active and rapid movements of organelles in axons and large plant cells, the peripheral parts of these cells would not be nourished.

Cell Structure

*Three Types of Protein Polymers Constitute the Cytoskeleton and Interact with
Force-Producing Enzymes to Cause Cells to Move*

Cells of both animals and plants contain three different types of fibers—actin filaments, intermediate filaments, and microtubules—each of which is formed by the polymerization of distinct protein molecules. Together these fibers provide mechanical support for the cell and thus are considered to be a cytoskeleton. The actin filaments and microtubules also participate in cellular movements, including locomotion of whole cells, cell division, and movement of subcellular components. This combined ability to maintain form against mechanical forces and to produce and transmit force means that this cytoskeletal motility system can determine cell shape and hence the structure of both tissues and whole organisms. A clear understanding of this system will be essential for unlocking the secrets of embryology.

This field is still in an explosive growth phase during which investigators have isolated and started to characterize the major molecular components of these systems. The inventory includes not only the protein subunits of the polymers themselves but also a surprising number of regulatory proteins. For example, in the actin system alone, one cell has already been shown to have almost 20 accessory proteins, which together with the actin constitute 25 percent of the total cell protein. In the developing brain, the microtubule system may include a similarly large proportion of the total cell protein. In skin, the keratin molecules that form the intermediate filaments constitute the major protein in the cells.

The Cytoskeleton Provides Form to Cells and Therefore to Organisms

The polymeric nature and intracellular distributions of the filaments and microtubules suggest that they may mechanically stabilize the cytoplasmic matrix. Recent physical studies on purified cytoskeletal fibers and analysis of the mechanical properties of live cells support this conclusion. Other work has shown that some of the glycolytic enzymes bind to actin filaments and that polyribosomes are associated with isolated cytoskeletons. Thus, beyond imparting mechanical integrity, the cytoskeleton may provide scaffolding for enzymes that participate in cellular metabolism and protein biosynthesis. In this way, the cytoskeleton, like membranes, may be an essential integrator of cellular processes.

It is now possible to describe, in broad outline, how these protein polymers assemble and how some of the steps in the assembly process may be regulated, at least for actin and microtubules. To a large extent, the construction of the system of cytoplasmic fibers can be explained by the process of self-assembly, in which the protein subunits are driven by chemical attraction for each other to polymerize without external direction. This spontaneous process is controlled by a variety of regulatory proteins, some of which must react to signals from the external environment that direct the organization of the cytoskeleton. Cells also contain organelles, such as the centrosome, that help to organize the cytoskeleton. The centrosome is the site where the assembly of microtubules is initiated.

Biochemical and cellular experiments indicate that the mechanisms controlling the assembly and organization of these fibers in the cell are both complex and subtle, as befits a system with such an important influence on cell architecture. To gain a better understanding of how form is determined in biology, considerable new work will be required (1) at the molecular level to elucidate the molecular composition, regulation, and dynamics of the cytoskeleton and (2) at the cellular level to determine how external stimuli affect the assembly of the cytoskeleton.

Cell Movement

Research on the Mechanisms of Cell Movements Is Progressing
on a New Frontier

In parallel with studies on the structural elements of the cytoplasm, work on mechanisms of cell movements has pushed forward rapidly during the past 15 years on two main frontiers; during recent years, a third and possibly a fourth frontier have begun to open. In each case a specific motor protein is responsible for movement.

The first frontier involves the microtubule-dynein system found in cilia and flagella—whiplike organelles (such as sperm tails) that beat rapidly. Cilia are found in groups on the surface of epithelial cells such as those lining the air passages in our lungs, where they are responsible for sweeping mucus and inhaled foreign material out of the lungs. If the cilia are not active, serious infection is inevitable. Flagella form the tails of sperm and propel them toward their meeting with the egg. In cilia and flagella, microtubules interact with a giant enzyme molecule called dynein to convert the chemical energy stored in ATP into a force that bends the cilia and flagella. Since the chemical steps in this process have now been identified, studies on the molecular mechanism that produces the motion can now be pursued vigorously. Dynein is also present in the cytoplasm, outside cilia and flagella where it can move particles along microtubules in the direction from the cell periphery toward the cell center.

The second frontier is the characterization of myosin—the force-producing enzyme long known to cause contraction in highly specialized muscle cells and more recently recognized to exist along with actin in most other cells, even those not specialized for contraction. Superficially most myosins are similar, and it seems likely that all myosins produce force by interacting with actin filaments and ATP in the same fundamental way. The steps in this process have been studied in detail in muscle (an especially favorable test system). Investigations using spectroscopy, x-ray diffraction, electron microscopy, and biochemical methods are also under way to locate the site in the myosin molecule where motion is produced. Myosin and actin are widely believed to be responsible for many forms of cell movements in addition to muscle contraction. For cytokinesis (the final step in cell division), direct experimental evidence validates this hypothesis. Other types of movements required similar experiments in order to explain their molecular basis.

In the past, most cell biologists suspected that either dynein or myosin was responsible for most cell movements, including the ubiquitous rapid movements of cellular organelles in the cytoplasm, but it has recently been discovered that a new type of motor protein called kinesin moves particles along microtubules in the opposite direction from dynein. Together these two motors provide a two-way rapid-transit system for organelles through the viscoelastic cytoplasm. This kinesin-dynein-microtubule system can shuttle a vesicle manufactured in a nerve cell body in the spinal cord to the nerve endings in the big toe (and back) in a few days! Even newer evidence suggests that an unusual form of myosin may pull some organelles along actin filaments. Breakthroughs of this type have raised the hope that we may soon understand how the mitotic apparatus works and how the traffic of organelles is directed to the proper destinations in the cell.

Each of these motile systems operates under exquisite controls that allow cells to respond to internal or external stimuli and to produce a coordinated motile response. In skeletal muscles and heart, the contractile proteins are turned on by calcium, which activates regulatory proteins bound to the actin filaments. In the smooth muscle cells found in internal organs and in nonmuscle cells, the myosin is activated chemically by the attachment of phosphate to the protein. It is not yet understood how these chemical reactions are coordinated in the living cell to produce the complex patterns of movement that are essential for life. The purse-string-like contraction that splits two daughter cells apart at cell division is an example of a movement in response to an internal stimulus arising from the poles of the mitotic spindle. The rapid locomotion of white blood cells to sites of infection and their ingestion of bacteria are examples of complex movements in response to external signals. In these examples, the stimuli and the responses are well documented, but little is known about the mechanisms that convert the stimulus into the response. Regulation of microtubule-based movements presents a similar challenge.

If work on cell motility continues with its current momentum, progress is likely in the following areas.

Molecular Inventory and Characterization. The complete catalog of the molecular components of the cytoskeletal motility system should be completed for a few cell types that are particularly favorable for use as model systems. These include the slime molds *Dictyostelium discoideum* and *Physarum*, the protozoan *Acanthamoeba*, sea urchin eggs, macrophages, platelets, and the intestinal epithe-lial cell. Completion of this molecular inventory may require new functional assays for proteins that have yet to be discovered. The primary structures of the major components need to be determined by sequencing cloned DNA, and the three-dimensional structures of the major components must be determined by x-ray crystallography. The first atomic resolution structure of a cytoskeletal protein (the actin-binding protein called profilin) has been completed, and good progress is being made on actin and myosin.

Cellular Organization and Dynamics. Electron microscopy should lead to more precise localization of the components of these systems inside whole cells. Vastly improved probes consisting of antibodies coupled to colloidal gold, to-gether with better methods to prepare cells, should give us a clearer picture of macromolecular architecture. Furthermore, it should be possible to characterize the dynamics of the cytoskeleton in live cells by using new fluorescence techniques (see Chapter 2). Purified protein molecules can be tagged with fluorescent dyes and then injected into live cells.

Functions and Regulation of the Cytoskeletal Motility System. Perhaps the major challenges in the field will be to determine the functions of the various components inside living cells and to learn how these functions are regulated by the cell. One approach is the use of in vitro assays with purified components. It is remarkable that functions as complex as the contraction of actin and myosin, the movement of an organelle on a microtubule, or the growth of microtubules from the centrosome to the kinetochore of a chromosome can all be reproduced today in vitro. These assays should become more sophisticated, enabling cell biologists to test for the ability of purified components to carry out complex functions outside living cells. Producing mutant cells with defects in single components will also be valuable in demonstrating functions and identifying regulatory mechanisms. This may be a laborious process because there are multiple genes for many components, and even where there is a single gene, the protein itself may be part of a highly redundant system that will function nearly normally without any given component. The microinjection of inhibitory antibodies to inactivate a single component inside a cell and the inactivation of relevant genes are also promising approaches. These and other creative new approaches will be necessary to test current ideas regarding the physiological functions of the cytoskeletal motility system. A long-term challenge will be to characterize the mechanisms by which an external stimulus, such as a chemoattractant, causes a cell to move in a particular direction.

Mechanical Properties. Analysis of the mechanical properties of the cytoskeleton and its component molecules is essential, but has only begun, in part because few cell biologists are familiar with the engineering techniques required for the work. This is an area of potential collaboration between cell biologists and engineers.

CELL MEMBRANE

The Cell Membrane Not Only Forms a Protective Surface But Also Receives Chemical Messages from the Environment

The outer cell membrane is an extremely thin, sheetlike assembly of lipid and protein molecules that provides a boundary to the cell's body. This exquisitely delicate skin, called plasmalemma, is a diffusion barrier for water-soluble substances. In the plasmalemma, the cell assembles all the molecular equipment needed for its exchanges and interactions with the environment.

The Plasma Membrane Shares Many Basic Structural Features with Other Types of Cell Membranes

Membrane structure relies on the use of a continuous bimolecular layer of lipids, the diffusion barrier, which is fluid at the temperature of the environment in which the cell lives. The barrier is traversed by transmembrane proteins that subserve a variety of functions, and it is reinforced by a fibrillar infrastructure made up of other different proteins. Depending on cell type, these infrastructures are built either for imparting tensile strength to a delicate membrane or for controlling the lateral mobility of transmembrane proteins, which if not restrained would move rapidly in the membrane because of the fluidity of the lipid bilayer. The first type of infrastructure has been studied extensively in the red blood cells of humans, and its molecular components and their mode of assembly are well known. Their function is to reinforce the membrane and to give the cell its characteristic shape. More recent work shows that the same or related proteins are used by many other cells to solve similar problems.

Most of the studies on the infrastructure that controls lateral protein mobility have focused on the miniature geodetic cages formed by the protein called clathrin and associated proteins. These clathrins are found on structures called coated pits and coated vesicles associated with the plasmalemma as well as with certain intracellular membrane systems. Coated pits trap functionally important molecules from the environment or from intracellular compartments.

Permeability Modifiers Are Transmembrane Proteins That Facilitate the Transport of Water-Soluble Molecules Across the Lipid Bilayer

Many permeability-modifier proteins transport nutrients such as glucose and amino acids. Others are channels for ions, and still others are energy-driven pumps that move sodium, potassium, hydrogen, or calcium ions in or out of the cells against concentration gradients. Many of these molecular pumps are called ATPases because they obtain the energy needed for their work by splitting ATP. The main function of the pumps is to maintain stable intracellular ionic concentrations at optimal levels for the cell's activities.

During the last few years, many transporters, channels, and pumps have been moved from their previous status as hypothesized physiological entities to that of well-defined protein molecules. Moreover, the amino acid sequence of many of them has been deduced from the nucleotide sequence of the cognate complementary DNAs. Knowledge of the amino acid sequence of these proteins is needed as a first step toward understanding their function and the way they fit into membranes.

Channels and pumps generate differences in molecule and ion concentrations (chemical and electrochemical gradients) as well as electrical charge separations (membrane potentials) across the corresponding membranes. These gradients and potentials are used by cells to propagate signals along the cell surface, as in nerve and muscle cells, and to drive the transport of other molecules and ions across the membrane, as in the cells of the intestine and the kidney.

THE EXTRACELLULAR MATRIX OF ANIMALS

The Cells of Multicellular Animals Are Supported and Organized by a Continuous Extracellular Matrix Composed of Fibrous Proteins and Complex Polysaccharides

In specialized connective tissues, such as bone, cartilage, and tendon, the extracellular matrix is predominant, but even in tissues such as muscle, liver, and brain, each cell is surrounded by a fine matrix. Connective tissues provide the avenues through which blood vessels pass to nourish every organ and serve as homes for the body's defensive cells, including phagocytes and lymphocytes. Consequently, most inflammatory diseases such as arthritis occur in the connective tissues.

The extracellular matrix is produced primarily by cells called fibroblasts, but also by epithelial and muscle cells. For years we have known that collagen (the most abundant protein in our bodies) and elastin form the major fibers in the extracellular matrix. During the past 10 years biochemists have identified more than 10 different types of collagen that are specialized for forming bone, cartilage, and basement membrane (a ruglike structure that all epithelia—lining and covering tissues—stand on). Some collagens form flexible fibers with tensile strength similar to that of steel, while others form three-dimensional networks. The regular arrangement of collagen in tendons has been known for some time, but the elucidation of the molecular organization of less-regular collagen structures such as basement membranes opens a fascinating research opportunity. Rubberlike elastic fibers are responsible for the ability of large blood vessels and skin to recoil when stretched. Elastic elements allow blood vessels to modulate the pulsatile flow produced by beats of the heart. The mysterious loss of elastic fibers during aging has generated a multimillion dollar cosmetic industry to combat wrinkles. The amino acid sequence of elastin in known, but its molecular structure and its association with other molecules in elastic fibers are major research challenges that, when solved, should help explain and perhaps prevent some cardiovascular diseases and effects of aging.

Connective tissues are also rich in a variety of organ-specific complex sugar polymers, some of which are chemically linked to proteins. They are called, as a group, glycosaminoglycans, or GAGs for short. The name derives from their repeating component sugars. Together with collagen fibrils they are responsible for making the cartilage covering most joint surfaces tough and elastic. GAGs also fill the eye and are major components of skin, blood vessels, and other organs.

There is now active research on a variety of proteins that confer biological specificity on the extracellular matrix. For example, an elongated protein called fibronectin binds cells to the matrix. It has binding sites for a receptor protein of the plasma membrane of connective tissue cells, collagen, and GAGs, so it can link them all together. The bond to the cell is relatively weak, so that a cell can gain traction on the matrix but still move through it. Another protein, laminin, binds epithelial cells to the basement membrane. Perhaps the most remarkable feature of these and other adhesive molecules is that they recognize and bind to a very small site (as few as three amino acids—arginine-glycine-aspartic-acid) on collagen and other matrix molecules. The attachments of both fibronectin and laminin can be altered in tumors, and this change is thought to contribute to the ability of tumor cells to invade some tissues—these invasive tumors are the major cause of death from cancer. Future research should reveal ways in which these adhesive interactions can be modified in beneficial ways in tumor therapy. Evidence is also growing that binding to the matrix modulates cellular physiology.

Active work is also being done on specific growth factors that promote the formation of specialized connective tissue such as bone and on other proteins that initiate the formation of the calcium-phosphate crystals that make bone hard. It has long been appreciated that physical forces on bones control their formation and that inactivity of the elderly contributes to the thinning of bones in osteoporosis. Here is a major opportunity for multidisciplinary research by cell biologists, biochemists, and engineers to learn how physical forces are transduced into the cellular activities that maintain normal bone and that fail in osteoporosis. Equally fascinating are the questions of how the information specifying the shape of the skeleton is laid down in the genetic code, how the cells of the connective tissue translate this information, and how matrix molecules influence the development of adjacent tissues.

CELL REGULATION

Cellular Activities Are Regulated by a Combination of Information Provided by the Genes and by Extracellular Signals

The timing of major decisions made by cells, such as whether to grow, to divide, to move to one location or another, or even to die, is determined by genetic programs and also by environmental clues, such as hormones and contacts with other cells. To understand cell regulation, one must study the production and effects of signals coming from the nucleus as well as the mechanisms responsible for transducing extracellular signals into cellular actions.

Cell Division

The Cell Division Cycle Is the Master Program of Cell Regulation That Organizes a Variety of Subroutines

In a very broad sense, to understand cell division, we need to understand more than its component parts: how membranes are assembled and disassembled during mitosis and cytokinesis, how DNA is replicated and organized, and how the mitotic spindle is assembled and disassembled. Beyond this, we need to understand how controls integrate the behavior of the spindle with the replication and segregation of the chromosome, as well as integrate growth and differentiation with cell division. Most importantly, we need to understand how cells "decide" whether or not to leave their normal resting state in tissues and go on to grow, replicate their DNA, and divide. An identification of these controls should lead to a real understanding of several important diseases—most notably proliferative diseases such as cancer and degenerative diseases—some of which are likely to result from a failure of normal proliferation.

Controls that integrate growth with division occur in the first growth phase of the cell cycle, called G1 in yeast and animal cells. Although many of the components of this control system have been identified in yeast, the links between nutrition, protein synthesis, and the apparatus for cell division remain unknown even for this unicellular organism.

After a century of study, the mechanisms that move chromosomes during cell division in both somatic cells (mitosis) and germ cells (meiosis) are finally becoming clear. The main elements are well known: the mitotic spindle composed largely of two arrays of microtubules. One set runs from the centrioles at the spindle poles to the centromeres (kinetochores) of the chromosome. The chromosomes are pulled to the poles as the kinetochore moves along these microtubules toward the poles. Remarkably, this movement seems to be powered by energy stored in the microtubules, whose depolymerization at the kinetochore determines the rate of movement. The second set of microtubules runs from one pole toward the other. These microtubules are slid past each other by an ATP-requiring motor to push the poles apart, which helps to separate the two sets of chromosomes.

A centriole is located at each pole and remains one of the most poorly characterized elements from a molecular standpoint. An understanding of the molecular organization of the centriole will be essential in defining its role in chromosome movement, spindle and aster formation, and its other function as the basal body for cilia and flagella.

Ultimately, the mechanical problems of chromosome movement will need to be placed in the overall context of cell-cycle control, including (before the actual separation of chromatids at mitosis) DNA replication, shutdown of RNA transcription, chromosome condensation, and breakdown of the nuclear envelope; and (after mitosis) nuclear envelope reformation, chromosome condensation, and reinitiation of transcription. Specific protein phosphorylations help drive a cell into mitosis, and the recent development of in vitro systems in which some of these processes occur outside the cell holds promise for a detailed molecular analysis in the near future. Remarkably, some of the central components of the control process are nearly identical in cells as disparate as those of humans and yeast; thus, many of the details of the human cell cycle can be worked on in simpler cells such as yeast, which are readily amenable to a combined molecular and genetic analysis.

Mitosis in somatic cells and meiosis in germ cells resemble each other in such respects as the formation of the spindle apparatus and the general breakdown and reformation of the nucleus. However, details of chromosome behavior differ markedly. Numerous questions need addressing. What causes homologous chromosomes to pair before meiosis? How are the molecular events of crossing over controlled? What causes the unique behavior of the centromeres during meiosis? How is DNA synthesis suppressed before the second meiotic division? These special problems of meiosis remain largely unexplored from a molecular standpoint. Again, important information will come from organisms such as yeast.

Cell-to-Cell Communication

Cells Have Developed Mechanisms for Interacting with Other Cells

Cell-cell interactions are important in simple organisms for such functions as sexual reproduction, colony formation, and attachment to various substrates. In multicellular organisms, cell-cell interactions have become much more complex, since they are essential for the integration of large cell populations into structurally coherent and functionally controlled tissues, organs, and organisms.

Short-range communication depends on direct contact between cells and their neighbors or the surrounding environment. Long-range communication requires the movement of informational molecules (such as hormones and growth factors) from one cell to another through the blood or other extracellular spaces and the binding of these molecules to specific receptors on the surface of the target cell.

Short-Range Communication Requires Plasma Membrane Specializations

The critical importance of cell-cell interactions is illustrated by testing the developmental sequence of multicellular organisms. As the one-celled embryo begins to divide and cells begin to differentiate, mechanisms of short-range cell-cell communications emerge. They consist of gap (or communicating) junctions that link a cell to its neighbors through common transmembrane channels. These junctions create common intracellular environments in cell subpopulations and ensure rapid cell-to-cell propagation of membrane permeability changes and intracellular messengers. Short-range interaction mechanisms also include junctional complexes between cells, which allow the developing organism to build walls of cells, called epithelia, that separate the different parts of its body. In addition, cells in general and epithelial cells in particular participate in short-range interactions with the newly formed extracellular matrix. These interactions are mediated by mutual recognition between cell membrane receptors and specific parts of matrix proteins. Cell membranes have multiple receptors for many matrix proteins, which are large monomeric or polymeric protein molecules with specific sites for binding to the plasmalemma as well as to other matrix proteins. The result is the progressive construction of a mechanically coherent body in which the cells are kept in place by their attachment to one another as well as to structural differentiations (for example, basement membranes and collagen fibers) of the extracellular matrix.

These attachments are effected through rigid plates on the plasmalemma that connect bundles of fibrils from the extracellular matrix to actin filaments or intermediate filaments in the cytoplasm. The rigid plates are maintained in relatively fixed positions by the fibrillar cables, which are under tension because they generally follow the lines of stress propagation within the entire system. This type of construction allows the cells to retain their shape, resist pressure, and recover from deformations.

During embryonic development, the production of matrix proteins follows a sequential program presumably matched by the production of plasmalemmal receptors for specific matrix proteins. Certain cell migrations are controlled by cell-matrix interactions and can be experimentally blocked by antibodies to (or small peptides derived from) relevant matrix proteins. Cell migration is thought to be controlled by a process that activates secretion of matrix proteins and concomitant production of cognate receptors. The cells move along tracks laid down by themselves or by their predecessors.

In the adult organism, gap junctions control the propagation of contraction waves in the heart muscle and in the smooth muscles of the intestine and uterus. Modulations in the construction of junctional complexes also control the permeability of epithelia in the intestine, lung, and kidney as well as the permeability of the vascular endothelium.

Long-Range Communication Requires Messenger Molecules and Receptors

As embryonic development progresses, mechanisms of short-range communication are extended and diversified, but long-range interactions through hormones and growth factors become progressively more important. Long-range communication requires the production of chemical signals, such as hormones and growth factors that react with target cells and modify their activities. The essential elements of any long-range communication system are a chemical messenger molecule, a cellular receptor, and a mechanism that transduces the binding of the messenger molecule to the receptor into a biochemical change in the target cell. The biochemical change then modifies the physiological behavior of the cell.

Messenger Molecules Have a Wide Variety of Chemical Compositions

The chemicals that carry messages from one cell to another are extraordinarily diverse. The classical hormones include derivatives of cholesterol (cortisol, testosterone, estrogens), derivatives of amino acids (thyroid hormone), small peptides (growth hormone), and other types of molecules (epinephrine). More recently, attention has turned to a growing list of protein and polypeptide messengers that regulate the cell cycle of target cells and are grouped together as growth factors.

Growth Factors

Growth Factors Are Ligands That Bind to Receptors and Cause
DNA Synthesis to Begin

Major advances in purification and characterization of the growth factors during the past decade have facilitated an understanding of their modes of action. Until the advent of genetic engineering, studies were limited to those factors available in sufficient quantity from biological sources. Much early work was done with three growth factors: erythropoietin, a glycoprotein from the kidney that stimulates red blood cell production from a common stem cell progenitor of both red and white blood cells; nerve growth factor (NGF), which stimulates the development of neurons; and epidermal growth factor (EGF), which was discovered in part because it induces premature eyelid opening and tooth eruption in neonatal mice by stimulating epidermal cell proliferation. The initial discoveries of these three hormones were made in animal models, which were later used to assay the progress of their purification.

Research on these hormones was accelerated greatly during the past decade through advances in animal cell culture technology. One principal limitation of that technology was a requirement for serum. This was overcome through the realization that serum is a rich source of growth factors, especially those required by mesenchymal cells. That realization led to the identification of platelet-derived growth factor (PDGF), a powerful mitogen for fibroblastic cells that is widespread in nature and that can play multiple roles. Other hormones are present in platelets, including tumor growth factor beta, which is a potent modulator of mesenchymal cell proliferation as well as a powerful growth inhibitor for other cells such as epidermal cells. These observations helped explain the well-documented selectivity for preferred growth of mesenchymal cells in cell culture medium containing serum. Once this major roadblock was overcome, progress in identifying requirements for growing epithelial cells in culture was rapid. Much interest in controlling growth of epithelial cells results from their being the origin of approximately 90 percent of human tumors.

The suspected role of growth factors, growth modulators, and their receptors in cancer has created intense interest in an understanding of their actions and synthesis. Unlike hormones produced by endocrine organs, such as insulin and growth hormone, growth factors are not secreted into the blood. Instead, these paracrine hormones are released at or near their target cells. A functional homologue of EGF, termed tumor growth factor alpha, and growth factors of the insulin-related families are secreted by a variety of cells lines derived from tumors. The normal progenitors of these tumor cells themselves are responsive to these hormones from which the tumor cells have become independent. This class of mechanism, in which a growth factor is produced and utilized by the same cells, has been termed autocrine hormone function. Although uncontrolled autocrine behavior could contribute to tumor progression, it is unlikely that acquisition of a single uncontrolled autocrine mechanism would cause cancer. The stimulation of reproduction of epithelial or mesenchymal cells probably requires the synergistic action of several growth factors acting in a specific temporal manner.

Immunomodulators, such as tumor necrosis factor and interleukins 1 and 2 (IL-1 and IL-2), are substances that influence the expansion of immune cell populations. They are produced and utilized by white blood cells at sites of inflammation, including areas of tumor growth, by a paracrine process. Thus the effective dose of a hormone normally develops only in the area of the cells that release it. A number of paracrine-acting immunoregulatory agents are currently undergoing clinical trials, even though extreme toxicity has frequently been observed at doses sufficiently high to achieve pharmacological effectiveness. Thus the development of drug delivery systems that mimic the local delivery specificity of a physiological paracrine process represents major challenges and new opportunities in current attempts to apply immunomodulators in cancer therapy.

Receptors

There May Be Even More Receptor Proteins Than Messenger Molecules

Receptors are protein molecules that bind specific hormones or growth factors and relay a signal that converts an extracellular message into a biochemical action inside the cell. There may be more kinds of receptors than messenger molecules because the receptors for a single messenger molecule can be different on different types of cells.

The receptors handle their information-transduction functions in a variety of ways. During the past 20 years, our perception of receptors as molecular entities has changed dramatically. Our thinking has evolved from a picture of relatively inert structures able to bind specific ligands that then induce a signaling function to the concept that these proteins contain structural information that enables more diverse functions. We now understand that receptor systems contain structural elements that enable them to bind ligands, participate in signal transduction, and respond to regulation by various cellular mechanisms.

In the simplest case (S1 in Figure 4-3) the hormone (for example, cortisol, sex steroids, vitamin D, or thyroid hormones) is sufficiently soluble in lipids to diffuse across the plasma membrane and act on an intracellular receptor. Alternatively, the receptor protein in the plasma membrane is oriented toward the exterior of the cell, where it can bind the extracellular ligand (S2) and carry its signal across the membrane into the cell (as low-density lipoprotein receptor and trans-

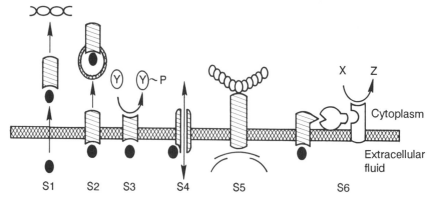

FIGURE 4-3   Strategies for transmembrane signaling. [Adapted from H. R. Bourne and A. L. De-
Franco, in The Oncogenes, R. Weinberg and M. Wigler, eds. (Cold Spring Harbor Press, Cold Spring
Harbor, N.Y., in press), chapter 3]

ferrin). Some receptors are transmembrane proteins with an extracellular portion
that binds a ligand such as insulin, EGF, or PDGF (S3); a transmembrane segment; and an intracellular part with enzyme activity. Binding of ligands to the
extracellular site can stimulate the phosphorylation of protein tyrosine groups by
the kinase activity of the intracellular domain. Still other extracellular signals
(S4), including acetylcholine and γ-aminobutyric acid), act by binding to a transmembrane ion channel; opening of the channel in response to binding of the
ligand allows specific ions to cross the membrane and alter the electrical potential
across the membrane. Finally, a large number of distinct extracellular signals
(S5) are detected by receptors that act through a family of proteins (G proteins)
that bind guanosine triphosphate (GTP) to regulate production of intracellular
mediators of hormone action, known as second messengers [or signaling mechanisms, which include calcium and cyclic adenosine monophosphate (cyclic AMP)].
These second messengers then diffuse through the cell's interior and initiate
various biochemical processes.

   These multiple receptor pathways allow a diverse approach to the control of
receptor function and cell metabolism, permitting rational approaches in the
design of therapeutic drugs to correct altered cell functions.

Transmembrane Signaling

Ion Channels Mediate the Cell's Electrical Communication with Its Environment

   It has been long been known that "excitable" cells in the neuromuscular
system communicate rapidly by means of electric signals generated by the selec-

tive passage of ions through protein channels in their plasma membranes. Now it is clear that virtually all cells use related membrane channels as part of their signal transduction systems. Membrane proteins, called ion channels, form tiny water-filled pores across the plasma membrane that are narrow enough to permit diffusion of one or another of small ions such as sodium, potassium, and calcium between the internal and external solutions. Passage of charged ions through the pore forms an electric current that generates potential changes for signaling. The nervous system generates many kinds of electrical signals by using many types of ion channels, each specific for a few common ions and each opened and closed under the control of specific stimuli, such as membrane potential changes, neurotransmitters, light, or mechanical stimuli.

The study of ion channels has advanced greatly during the past 10 years because of spectacular technical improvements in electrical recording methods and in molecular biology. A new recording method, called the patch clamp, permits routine observations of the opening and closing of single ion channels on a submillisecond time scale. Hardly anywhere else in science has it been possible to monitor conformational changes of one molecule. With the patch clamp, channels have been shown to click abruptly between open and closed conformations. We are now faced with a wealth of recordings of the transitions that need to be understood in kinetic and ultimately molecular terms.

Different channel types can be recognized by their patch-clamp signatures—different ionic preference, different absolute conductance, and different rules for opening and closing. At least 30 types have been recognized in the past 8 years, and new ones are being discovered every month.

The patch clamp reveals that ion channels are present in the plasma membrane of all eukaryotic cells, not only excitable cells like nerves and muscle. Because this ubiquity was not previously suspected, an important task is to determine what roles these signaling molecules play in the housekeeping functions and daily life of cells outside the nervous system.

By far the best-understood ion channel to date is the acetylcholine (ACh) receptor channel that mediates the transmission of signals from nerves to muscles (Figure 4-4). The ACh receptor opens a pore in the muscle membrane when the neurotransmitter ACh is released by the nerve impulse in a nearby nerve terminal. The protein subunits of this pentameric receptor molecule have been sequenced. DNA cloning and sequencing have revealed related ACh receptors in other cells. The cloned messages for various ion channels have been injected into frog oocytes, which then make the proteins and incorporate them into their plasma membranes, where they can be studied by patch clamp-methods. Several groups are selectively mutating these messages to test which parts of the sequence are responsible for each feature of the overall function. The three-dimensional structures of selected channels that should become available in the near future, along with results of ongoing molecular biological and electrophysiological studies, will elucidate the operation of these channels at the molecular level.

RECEPTORS CAN ACT AS SPECIFIC GATES FOR DRUG DELIVERY

The high specificity exhibited by cell-surface receptors for hormones and other ligands, combined with the ability of receptors to be internalized into cells, provides an exciting opportunity for drug delivery to the cell's interior. Such an opportunity is being actively investigated in disease states in which particular cell types need to be eliminated. The strategy links potent toxic molecules covalently to specific receptor ligands, thus promoting specific internalization of toxin with resultant cell death. Experiments with intact cells in vitro have unequivocally demonstrated the effectiveness of introducing toxins such as diphtheria toxin and ricin into cells in this way. When these toxins are covalently linked to peptide ligands such as epidermal growth factor or transferrin, receptors internalize them and rapid cell death results. In order for this technique to be useful in vivo, tissue specificity of the killing action is required. Because most well-characterized receptor systems are expressed in many cell types, the utility of this approach must be perfected so that specific cells or tissues are destroyed.

Recent experiments designed to achieve cell specificity for introduction of cell toxins via receptor-mediated endocytosis are based on differences between cells in respect to receptor expression. Thus tumor cells often express much higher numbers of certain receptors than normal cells from which they were derived. This elevated receptor expression (and the resulting receptor internalization rates) may confer increased sensitivity of such tumor cells to the killing action of the toxin-specific ligand conjugates. Alternatively, some tumor cells may express unique surface antigens. For example, expression of the fetal form of a receptor sometimes occurs in cancer cells. Specific ligands or perhaps specific antibodies against these unique receptors could provide the cell-type specificity required for this therapeutic approach. Thus we have new opportunities to test these hypotheses to develop better strategies for utilizing cell surface receptors to allow targeting of specific cells for drug delivery.

Because ion channels are large, exposed macromolecules having important functions, they have become the target of many classes of natural toxins and of many clinically active drugs. Toxins from cobras, scorpions, cone shells, dinoflagellates, puffer fish, frogs, sea anemones, and many plants act directly on channel molecules. The animal toxins are useful specific labels for the biochemical identification and purification of channel macromolecules. Exploration of such neurotoxic molecules continues to be a rewarding approach to obtaining new reagents for research.

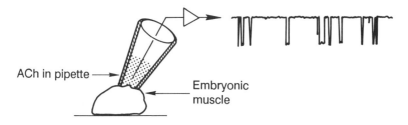

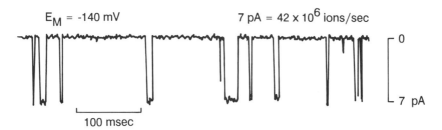

FIGURE 4-4   Acetylcholine opens receptor channels in muscle, a process that can be measured by the patch-clamp technique. [Bertil Hille, University of Washington]

Plant neurotoxins have already been the inspiration for systematic development of standard clinical agents. Thus curare led to neuromuscular blocking agents, cocaine led to local anesthetics and a class of antiarrhythmics, and papaverine led to another class of antiarrhythmics. These agents act on ACh receptor channels, sodium channels, and calcium channels, respectively, by mechanisms that are still only partially understood. Many widely used neuroleptics, tranquilizers, and antipsychotics act directly on channels. There is some hope now that forthcoming knowledge of the three-dimensional structure of channels can lead to the rational design of major new classes of clinical agents with specific actions.

Channel Modulation. All organs of the body are innervated by the two major branches of the autonomic nervous system—the sympathetic and the parasympathetic. In classical terms, signals in the sympathetic nervous system prepare each organ for times of stress—fight or flight—whereas signals in the parasympathetic prepare for more vegetative functions such as digestion. The molecular details of how the body responds to these signals are emerging. The sympathetic neurotransmitter (norepinephrine) and the parasympathetic neurotransmitter (ACh) act on several classes of membrane receptors to produce several intracellular second messengers, which in turn modulate the activities of a variety of ion channels. The receptors, the G proteins activated by the receptors, and effector enzymes and channels are being identified, purified, and sequenced.

Modulation by a cyclic AMP–dependent phosphorylation plays interesting roles in disparate activities. Stimulation of the sympathetic nerves to the heart releases norepinephrine, which speeds the heart rate and strengthens the stroke in each beat—the familar response of the heart to exercise. We believe that in the pacemaker cells of the heart, the rate of depolarization in each cycle is set by the rate of opening of voltage-gated calcium channels and that in the ventricle the force of the pump stroke is set by the number of calcium ions entering per beat. Much of the response to sympathetic stimulation can be attributed to the phosphorylation of voltage-gated calcium channels, which increases the probability of their opening. Each step, from activation of the β-adrenergic receptor by norepinephrine to phosphorylation of the channel, has been carefully documented.

Another example is a learninglike response called sensitization in the sea slug *Aplysia*. Here serotonin released by action of one set of sensory nerve fibers increases the neurotransmitter released by another. The cyclic AMP–dependent phosphorylation of a potassium channel, which here shuts the channel off, seems to account for this sensitization.

A major recent triumph of visual physiology was the discovery of how the light signal modulates the operation of a channel in rods and cones of the retina to initiate the electrical signals leading to vision. The coupling of rhodopsin to transducin to a phosphodiesterase enzyme is described in Chapter 6. The important point here is that the result of a cascade of reactions is the eventual change of the concentration of cyclic guanosine monophosphate (cyclic GMP), which is the direct regulator of the ionic channel. This final stage of transduction was demonstrated with the patch-clamp technique: The channel opened whenever cyclic GMP was applied to the cytoplasmic side of a patch of membrane pulled off the photoreceptor. The possibility that olfactory or taste transduction also requires a cascade of intermediate intracellular messengers offers opportunities for study in the coming years.

G Proteins Are Crucial in Many Kinds of Signal Transduction

Cyclic AMP was discovered as an intracellular second messenger for epinephrine and glucagon more than 25 years ago. Cyclic AMP is synthesized by hormone-sensitive adenyl cyclase. It mediates the cellular effects of a host of polypeptides, biogenic amines, and lipids that regulate mobilization of stored energy (carbohydrates in liver, triglycerides in fat cells), conservation of water by the kidney, homeostasis of calcium ions, contractility of heart muscle, production of adrenal and sex steroids, and many other endocrine and neural functions. Studies indicate that odorant stimuli activate adenyl cyclase in the olfactory epithelium of the nose, suggesting that cyclic AMP is also the intracellular second messenger that mediates the sense of smell. More recently, biochemical and genetic studies of the regulation of cyclic AMP synthesis led to the discovery of G_s, a membrane protein that couples hormone receptors to stimulation of adenyl cyclase.

Although it seemed unlikely that cyclic AMP would be the only intracellular second messenger of hormones, the next system was not identified until very recently. In this system (Figure 4-5), hormone receptors also act through a G protein to stimulate the activity of phospholipase C (PLC), which cleaves phosphatidylinositol bisphosphate (PIP$_2$) to form two distinct second-messenger molecules. One of these is diacylglycerol (DAG), a membrane lipid that binds to and stimulates protein kinase C. The other messenger derived from PIP$_2$ is inositoltrisphosphate (IP$_3$), which acts on receptors in the endoplasmic reticulum to release stored calcium ions and raise their cytoplasmic concentration. The two arms of this signaling cascade, calcium ion (Ca^{2+}) and protein kinase C, act on other cellular components to produce a set of responses that includes contraction of smooth muscle (stimulated by agents such as norepinephrine), modulation of synaptic responses in the central nervous system, a variety of secretory responses, and some, but not all, proliferative responses of cells to growth factors such as PDGF.

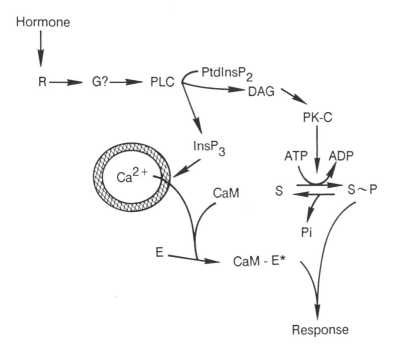

FIGURE 4-5   Calcium ion as a second messenger. [Adapted from H. R. Bourne and A. L. DeFranco, in The Oncogenes, R. Weinberg and M. Wigler, eds. (Cold Spring Harbor Press, Cold Spring Harbor, N.Y., in press), chapter 3]

The discovery of G_s led to the discovery of a whole family of G proteins involved in other kinds of signal transduction. A key feature of the G protein machine is that it allows amplification and regulation of the transduced signal by separating excitation of the receptor from activation of the effector in both time and space. Thus, the encounter of a neuromodulator such as norepinephrine with its receptor may be short-lived. When the encounter generates a GTP-bound G_s molecule, however, the duration of activation of adenyl cyclase depends on the length of time GTP stays bound to the G protein rather than on the tenacity with which the receptor holds onto norepinephrine.

In addition to G_s, at least five other G proteins have been discovered. They include one that mediates hormonal inhibition of adenyl cyclase, two that mediate retinal phototransduction in rods and cones, and others that mediate hormonal regulation of ion channels and production of second messengers derived from phospholipids. Despite their diversity, each of the G proteins is built on the same general plan, comprising three dissimiliar polypeptide chains, called α, β, and γ. The α chains bind and hydrolize GTP and confer on each G protein its specificities for interacting with particular receptors and effectors. The β and γ polypeptides anchor the protein to the membrane. In addition, the highly conserved GTP binding sites of the α chains closely resemble similar sites in the elongation and initiation factors of ribosomal protein synthesis and in the protein products ($p21^{ras}$) of the *ras* oncogenes and proto-oncogenes. Thus, the duplication and divergence of genes in evolution have modified a single molecular machine and made it useful to mediate vision, olfaction, control of cell proliferation, ribosomal protein synthesis, and cellular regulation by many hormones and neurotransmitters.

The apparent parsimony of evolution encourages investigators to extrapolate information obtained from one class of G proteins to another as well as to other proteins with the conserved GTP binding site, such as $p21^{ras}$. Only a few years ago, scientists would have scoffed at the notion that studies of bacterial protein synthesis or retinal phototransduction could reveal anything useful about the cause of cancer.

In addition to explaining the cellular actions of a number of important hormones and growth factors, the cascade of events elicited by these substrates has made two other general contributions to the understanding of signal transduction. First, the discovery of a second system of intracellular messengers encourages the expectation that more such systems will soon be found. The enormous array of chemically distinct membrane phospholipids provides an almost inexhaustible source of potential precursor molecules for generating second messengers.

Second, understanding this system has enriched our understanding of the interplay between different second messenger systems in individual cells. For example, epinephrine can stimulate the release of glucose from the liver via two different pathways using different receptors and different G proteins to generate both cyclic AMP and IP_3-DAG in the same cell type. The two second-messenger

G PROTEINS AND HUMAN DISEASE

Growing knowledge of G proteins has already contributed significantly to molecular understanding of human disease. The pathogenic toxins of Vibrio cholerae (cholera) and Bordetella pertussis (whooping cough) are now known to act as enzymes that attach adenosine diphosphate–ribose (ADP-ribose) to the α polypeptide chains of two specific G proteins, G_s and G_i, respectively. The modification of G_s by cholera toxin stabilizes the G protein in its GTP-bound form and thereby increases activation of adenyl cyclase (AC); the resulting cyclic AMP rise in mucosal cells of the intestine causes the profuse secretion of salt and water responsible for the often fatal diarrhea of cholera. In contrast, attachment by pertussis toxin of ADP-ribose to the a chain of G_i produces a G protein that is uncoupled from interaction with hormone receptors; as a result, pertussis toxin blocks hormonal inhibition of AC. In other cell types, pertussis toxin's action on G_i or a G_i-like molecule block ligand-stimulated phospholipid metabolism and elevation of cytoplasmic calcium (thereby inhibiting, for example, release of allergic mediators from mast cells).

In the other direction, the ability of these toxins to introduce radioactively labeled ADP-ribose into the G protein α chains has made them valuable tools for studying G proteins. Indeed, the G_i molecule was discovered by investigators primarily interested in the mechanisms of action of pertussis toxin rather than in hormone action.

The pivotal biological roles of the G proteins suggest that mutations in the corresponding genes should produce inherited disorders of signal transduction. One of these possible diseases has been identified: inheritance of a mutant allele of the gene that encodes the α chain of G_s produces an inherited human disorder called pseudohypoparathyroidism, type I (PHP-I). The hypocalcemia, mental retardation, convulsions, and other clinical manifestations of PHP-I result from partial resistance to hormones that utilize cyclic AMP as a second messenger. Many endocrine responses mediated by cyclic AMP are only mildly affected in PHP-I; resistance to parathyroid hormone, the master hormone of calcium ion homeostasis, is a prominent feature of PHP-I, suggesting that the amount or activity of G_s is normally more rate-limiting for actions of this hormone than of others.

systems act in parallel to stimulate the breakdown of glycogen in the liver cell. In contrast, the IP_3-DAG and cyclic AMP work in opposite directions in platelets. By imagining only one or two more second-messenger systems in addition to cyclic AMP and IP_3-DAG, we could account for almost any degree of complexity in cellular responses to extracellular signals. Regulation of cell proliferation involves just such a complex interplay of transduction pathways; these pathways are just beginning to be understood.

Cytoplasmic Free Calcium Acts as a Second Messenger

Three mechanisms for raising the free calcium ion concentration in cytoplasm have been found: the voltage-gated calcium channel of the plasma membrane, release of calcium ions from the endoplasmic reticulum of muscle when the surface membrane is depolarized in response to a nerve impulse, and release of calcium ions from the endoplasmic reticulum in response to hormonal signals acting on plasma membrane receptors. We understand how IP_3 is produced, but not what it does to the endoplasmic reticulum to release calcium. Because of the central role of calcium ions in signaling, the mechanisms of excitation-contraction coupling in muscle and that of calcium ion mobilization by IP_3 in many cells needs vigorous investigation.

Two processes regulated by intracellular calcium also need to be studied. One is the release of vesicles of packaged neurotransmitters, hormones, or enzymes. Although we know that calcium entry is required, we do not understand any of the subsequent steps that, for example, cause the transmitter ACh to be released within 0.1 millisecond of the excitation of a nerve terminal. Both botulism and tetanus are bacterial diseases whose lethal consequences are due to toxins that interfere with this calcium-dependent secretion by nerve cells. Another calcium-dependent process is cell motility, including muscle contraction. The calcium-sensitive component molecules of the contractile machinery have been characterized in some detail, but except for muscle the physiological mechanisms coordinating motility are poorly understood.

Progress is now possible in the analysis of calcium-stimulated events with the advent of new optical methods to measure intracellular calcium ion concentrations. Newly synthesized calcium-indicator dyes can be injected or allowed to permeate into cells. The time course of changes within a single cell can now be monitored reliably. Computer-aided spectroscopic techniques permit making accurate maps of the changing free calcium in each pixel of a video image of the cell observed under a high-power light microscope. The technique has already been used to observe the fertilization of eggs and the stimulation of liver cells by peptide hormones.

Transduction Mechanisms Use a Variety of Strategies to Alter Cellular Chemistry in Response to Messenger Molecules

More intracellular second messengers remain to be discovered, and the means by which they and the "established" second messenger systems integrate control of cell function are still to be elucidated. Thus, a large number of hormones and other chemical signals act on receptors whose transduction mechanisms are unknown. These include growth hormone, many of the lymphokines, nerve growth factor, and tumor necrosis factor. Understanding the actions of these ligands may uncover new molecular mechanisms of signal transduction. Moreover, each transduction system can alter its sensitivity to extracellular stimuli. Determining the molecular mechanisms that regulate sensitivity represents important new opportunities in research.

Oncogenes

A New Class of Genes Can Transform Host Cells

An oncogene (the *src* gene) was first identified in the transforming Rous sarcoma retrovirus. When suitably expressed, this gene alone was capable of transforming avian cells. Normal cellular homologues of oncogenes, or cancer genes, were found subsequently in virtually every organism analyzed, which suggested that transposed DNA from normal hosts had been incorporated into the retrovirus genome, under control of viral promotors, and was capable of initiating and maintaining transformation of infected cells. Subsequently, more than 25 oncogenes have been identified along with the normal cellular counterparts, the cellular proto-oncogenes. Oncogenes may be resolved into families that include elements of many of the cell-regulatory systems reviewed above: (1) the protein products of the *src* gene family appear to be tyrosine-specific protein kinases (and thus similar to growth-factor receptors) that are located at the plasma membrane or in cytoplasm; (2) other oncogenes are related to growth factors or to their transmembrane receptors; (3) the *ras* gene proteins bind and hydrolyze GTP and seem to act as signal transducers at the plasma membrane and cytoplasm; and (4) nuclear oncogenes probably act as transcription factors or DNA-binding proteins.

Research opportunities include the identification and functional characterization of additional proteins encoded by oncogenes. Once defined, the pathways by which these proteins influence cell growth and those that regulate proto-oncogenes should provide insight into the cell cycle, cell proliferation, and control by pharmacological means. Research into oncogenes should help us better understand normal cellular proliferation, the abnormal proliferation characteristic of the cancer cell, and pathways of significance to cellular differentiation. Other research using genetically engineered molecular hybrids should provide understanding at the molecular level of the function of each of the protein domains of these gene products responsible for transformation.

GENERAL PLANT CELL BIOLOGY

Plant Cells Differ from Animal Cells in Many Ways

Cells of plants and animals have many features in common, such as nuclei, endoplasmic reticula, Golgi complexes, mitochondria, plasmalemmas, coated vesicles, microfilaments, and microtubules; but in some features and processes, cells of these two kingdoms are distinct from one another. For example, plant, but not animal, cells contain plastids (the generic term that includes chloroplasts and modified chloroplasts such as amyloplasts that produce and store starch), large vacuoles that serve as the lytic compartment (lysosome-like) and contain most of the plant cell's water; they also have relatively rigid cell walls composed of strands of cellulose embedded in amorphous carbohydrate polymers together with a small amount of protein.

At higher levels of organization, plant cells are connected to one another by strands of cytoplasm (plasmodesmata), rather than by junctional elements of the type found in animal cells. The tissues and organs of plants and animals are not organized along similar functional lines. Leaves, stems, roots, and buds have no conspicuous counterparts in hearts, livers, skins, or lungs.

At many fundamental levels—biosynthetic pathways and structures of proteins and nucleic acids—plant and animal cells are often, but not always, the same. Thus, differences, when they occur in similar processes, often reveal much about the basic nature of a process. It is often profitable to determine whether newly discovered features of plants occur in animals and vice versa.

The features of plant cells and tissues that differ distinctively from those of animals each require direct investigation. For example, in plants, mitochondria perform the same respiratory functions (and perhaps some other, as yet undiscovered functions) as they do in animals, but they generally contain much more DNA. They also seem to use the standard genetic code, whereas the mitochondria of animals and yeast use a nonstandard one.

In addition to photosynthesis, a number of phenomena are unique to plants, and some are targets of active research. For example, each plant cell contains three rather than two gene-containing compartments (the nucleus, mitochondria, and plastids), thus greatly complicating protein targeting and the integration of genome expression. Other research is focused on the plant's rigid cell walls, which are constituted in different ways at different stages of development and for various specific functions of different cell types. The composition of the cell walls is reasonably well understood, but many questions remain unanswered.

Self-incompatibility, the inability of some plants to fertilize themselves, is an important mechanism for regulating outcrossing. The expression of plant genes can be affected by symbiotic or pathogenic microorganisms through complex interactions. In this area, important advances have been made, especially in studies of symbiotic nitrogen fixation, but the total effort is still very small. Some transposable genetic elements have been characterized molecularly, but how they are integrated into and excised from the genome remains to be understood.

Chloroplasts

Photosynthesis Is the Biological Process That Connects Life on Earth to the Sun

Through photosynthesis, light energy is converted to chemical-bond energy stored in sugar molecules. In higher plants and algae, photosynthesis is carried out in chloroplasts—organelles containing vesicles bounded by energy-transducing membranes in which the chlorophyll is localized. In primitive, noncompartmentalized cells—prokaryotes—these vesicles lie free in the cytoplasm.

Progress in understanding photosynthesis has been intertwined with advances in chemistry, photophysics, and biology. The path followed by carbon during photosynthesis—from carbon dioxide to sugar—is now known in detail, and most of the enzymes have been identified. The enzymes themselves are being studied, and, surprisingly, a number of them are found to be regulated (in ways yet to be understood) by light and by certain of the small molecules that are intermediates in the biosynthetic chains of photosynthetic carbohydrate production. The regulatory mechanisms and how each enzyme interacts with other proteins and with its substrates are questions currently being addressed.

Chloroplast Genes Are Being Mapped and Sequenced at a Rapid Rate

The chloroplast genomes of tobacco and liverwort have been fully sequenced, and restriction maps of the chloroplast genome have been completed for at least 40 to 50 species of plants. Most gene mapping is limited to genes that were located and initially mapped and sequenced in maize, spinach, tobacco, and the green algae *Chlamydomonas* and *Euglena.*

As more and more chloroplast DNA is sequenced, interest will grow in identifying the gene products of unrecognized proteins; their identification should move us toward a better understanding of photosynthesis and plastid metabolism. This DNA sequencing will also reveal features of plastid genes. At least one type of promoter sequence—resembling the prokaryotic type—has been recognized. The existence of other kinds of control sequences remains to be established. The identification of these and other controlling elements of plastid or nuclear-cytoplasmic origin that constitute the parts of the machinery for control of differential gene expression may well be the most interesting and outstanding problem in this research area. Its resolution is likely to illuminate the mechanisms underlying the transcriptional control of chloroplast gene expression and to reveal research approaches leading toward an understanding of intergenomic integration.

*Gene Expression in Chloroplasts Is Both Developmentally and
Functionally Regulated*

Like mitochondria, plastids contain genetic material, but not all of their components are products of these genes. Many proteins of the chloroplast are imported from the cytoplasm, and these are encoded in nuclear genes. For example, the larger of the two subunits of ribulose bisphosphate carboxylase/oxygenase (rubisco) is the product of a chloroplast gene, whereas the smaller subunit is the product of a nuclear gene.

How the photosynthetic apparatus of chloroplasts is produced, that is, its biogenesis, is a question fundamental to understanding the life of a plant or alga cell. A number of related questions are under active investigation: What are the special characteristics of chloroplast genes? What are the enzymes and mechanisms for their replication? What are the mechanisms for controlling the expression of sets of developmentally regulated chloroplast genes? How is the expression of these genes integrated with the expression of nuclear genes for chloroplast components? How are nuclear gene-coded, cytoplasmically synthesized proteins targeted to plastids, and what are the mechanisms for their uptake and integration into the life of the plastid? How does the machinery for chloroplast gene expression interact with the machinery of the nuclear-cytoplasmic compartment? Among the most interesting aspects of the plant eukaryotic cell is the integration of the activities of its multiple compartmentalized genomes—of nuclei, plastids, and mitochondria. The nature of the integrating mechanisms can now be investigated.

The Origin of Plastids Is Still a Mystery

One of the great puzzles of modern biology is how eukaryotic cells originated. A unique characteristic of eukaryotes is the presence of multiple compartmentalized genomes. In plants, these include the nucleus, mitochondria, and plastids. The question of how the expression of these genomes is integrated has already been raised. An older unanswered question is, How did the multiple genomes come into existence? There are two obvious possibilities. One is that membranous compartments formed in the structural equivalent of a modern prokaryotic cell, and some genes then became sequestered in each compartment. A second possibility, favored by most available evidence, is that the major organelles characteristic of eukaryotic cells—mitochondria (which are found in the cells of all but a very few eukaryotes) and chloroplasts—are the descendents of symbiotic bacteria that entered early eukaryotic cells. To account for the genetic organization of contemporary eukaryotic cells it is necessary to assume the movement of genes or gene functions from the symbiont to the host genome. Many researchers have interpreted the available evidence as suggesting multiple origins, involving different groups of bacteria, for chloroplasts and perhaps for mitochondria also. These hypotheses require further analysis.

What evolutionary pressures led to the existence of the different information-processing systems now found in the nuclear-cytoplasmic and organelle compartments? Are the different systems relics of the independent evolution of two progenitor cell types that subsequently joined to become the ancestral form of the modern eukaryotic cell? Alternatively, did two or three information storage and processing systems evolve within a single cell? What we learn about the molecular biology of the organelle and the nuclear-cytoplasmic systems may lead to a better understanding of the origin and evolution of eukaryotic cells as well as of the forces and mechanisms underlying the shifts of genes among genomes.

Evolutionary relations among genomes are much better understood now than they were half a dozen years ago because of the accumulation of nucleic acid sequence data. However, the forces that have led to the segregation of genes in nuclei, mitochondria, and plastids are unknown. Gene-transfer methods are beginning to be used to explore these questions from the focus of the nuclear genome. Efficient organelle gene transfer (transformation) methods that would greatly aid such investigations remain to be developed.

The Plant Cell Wall

Research on the Extracellular Matrix (Cell Wall) of Plants Is Crucial to Understanding How Plants Grow

Lacking a skeletal structure and subjected to a fluctuating osmotic environment, plant cells rely on rigid cell walls to serve as cementing substances between cells, to provide mechanical strength, and to support high internal osmotic pressures. For many years, it has been known that the ultimate shape and strength of such walls are largely determined by the pattern and extent of deposition of extended fibrils composed of cellulose (glucose molecules joined in β-1,4 linkage). In recent years, the concept has evolved that a second framework, composed of cross-linked extended rods of a hydroxyproline-rich glycoprotein called extensin, also contributes to strength and shape in some plant cell types. Interspersed within these frameworks exist a variety of matrix polysaccharides, some having complex structures. One example is the recently discovered, highly branched rhamnogalacturonan II polymer that contains a number of variously linked sugars including a novel monosaccharide called aceric acid, which was previously unknown in nature. Progress in determining the structure of these polysaccharides has advanced considerably in recent years, in part as a result of vastly improved techniques for the analysis of complex carbohydrates. These techniques require the use of expensive instrumentation such as mass and nuclear magnetic resonance spectrometers. Increased access to such instrumentation would hasten progress in the study of extracellular matrices of both plants and animals.

Understanding plant cell-wall structure is crucial for the ultimate understanding of how plants grow. Plants increase in size by expanding their rigid cell walls, but our understanding of the mechanism by which this occurs is still limited. The process is known to be under hormonal control; wall loosening is believed to occur by processes of breakage and reformation of crucial linkages and by turnover of some wall polymers, but the specific processes remain elusive. Recent structural analyses have provided evidence for intra- and interchain linkages between extensin molecules through isodityrosine residues, presumably formed in the wall by a peroxidase-mediated reaction. Similar enzymes may also be responsible for the formation of cross-links between phenolic components attached in ester-linkage to matrix polysaccharides. Other recent studies have implicated hormonally regulated degradative enzymes involved in the turnover of some wall polysaccharides, such as xyloglucan, in the process of wall extension.

On the basis of current knowledge of wall structure, one can estimate that there must be more than a hundred different enzymes required for wall assembly. Not one of these enzymes has yet been purified and characterized in detail, although a number have been detected and assayed in crude membrane preparations isolated from plants. Similarly, no gene responsible for coding for these enzymes has been identified, mapped, or cloned. Exciting progress has been made recently for the cell-wall protein extensin, since a gene for this protein has recently been cloned and characterized. Coupled with previous structural information on the protein, data from the gene sequence now give the entire amino acid sequence of the protein so that we can identify sites of glycosylation, cross-linking, and possible areas of interaction of this polymer with other wall components.

We now recognize that the cell wall is a dynamic structure. Not only do changes occur during normal growth and development, but the wall also responds quickly to external perturbations such as mechanical damage, water, or temperature stress and upon interaction with symbionts, pathogens, or parasites. Recent work has defined some of these "wound" responses in some detail. Specific changes identified are a cessation of cellulose synthesis coupled with induction of synthesis of a related beta-glucan called callose, which seals off areas of assault; induction of lignin or suberin synthesis; and elevation of synthesis of the soluble precursor to cell-wall extensin. This latter compound in some cases seems to serve as an agglutinin for invading pathogens. Most exciting is the recent work implicating fragments of wall polysaccharides as regulatory molecules. One example is an oligogalacturonide released from the wall by invading pathogens; this oligosaccharide serves as a specific inducer of the synthesis of phytoalexins—plant-made antibiotics that are toxic to invading microorganisms. Other examples include small phenolic compounds that may be released from the wall and that signal interactions crucial for the establishment of plant parasites or symbiotic associations with bacteria (see Chapter 11).

Recent advances in our ability to analyze the structure of wall components indicate that the time is rapidly approaching when data from structural studies can be integrated with growth physiology studies to achieve an overall understanding of plant growth. Recent development of specific dyes and monoclonal antibodies that interact with specific linkages in the wall should lead to a much better understanding of the localization and interaction of the various polymers in the wall. Cloning of the extensin genes will now allow a study of the regulation of various members of this family of genes; the patterns of expression of these genes should help clarify the various functions of this unusual polymer. The exciting discovery that fragments of wall components serve as regulatory signals will undoubtedly open a whole new area of study on ways plants communicate and interact with other organisms. Finally, recent advances in the biochemistry of membrane proteins may lead to breakthroughs in the identification and isolation of enzymes and their corresponding genes involved in wall synthesis. Since plant cell walls are a major sink for biomass, much of which is only poorly digestible, it is hoped that our ultimate ability to modify wall structure by modifying the expression of genes controlling wall assembly will lead to a greater understanding of just how flexible such wall structure can be, and perhaps also to the development of plants with improved agronomic or nutritional value.

5

Development

*The Molecular Mechanisms by Which Organisms Develop Can
Now Be Investigated*

Since the time of Aristotle, a major preoccupation of biologists has been the description of how an organism develops from an embryo to its adult form. By the beginning of this century, the elaboration of the cell theory, the discovery of the details of fertilization, and the development of improved histological techniques had led to an accurate description of the anatomical and cellular details of development in many kinds of organisms. As they learned more about the elaborate processes of gastrulation, neurulation, and pattern formation, however, biologists yearned for mechanistic explanations. Emerging theories emphasized the importance of the structure of the egg, the lineage of cell divisions, the accurate timing of these divisions, the importance of cell-cell interactions, and the role of inductions. More recently, with the advent of new techniques for dealing with the genetics and molecular biology of developing systems, this essentially anatomical description has been extended to a detailed analysis of selective gene expression and to the discovery of genes that regulate developmental decisions.

Despite this progress, a molecular explanation for the processes of development remains an elusive goal. In many branches of biology, phenomenological accounts of life processes have been replaced by detailed chemical descriptions; many complex processes, such as DNA replication and virus assembly, have been reconstituted from purified components in vitro. Comparable results have been difficult to achieve in developmental biology, despite their fundamental importance for advances in the field.

Giant strides may at last be possible in developmental biology, however, because advances in cell and molecular biology, as well as in the precise study of developmental systems themselves, have greatly improved understanding of the properties of eukaryotic cells. In the field of cell biology, cellular components such as microtubules and actin are now reasonably well understood. Differential cell adhesion can now be described in terms of specific molecules, receptors, and known elements of the extracellular matrix. Communication between cells, hitherto mysterious, can be explained in terms of such components as soluble growth factors, second messengers, cell receptors, and cell junctions.

Molecular biology has likewise contributed much to the improved prospects for an understanding of development. For example, cell differentiation can now be understood primarily as differential gene expression, coupled with the modification and turnover of macromolecules. In addition to providing sensitive probes for following developmental events—essential for biochemical work with individual embryos—molecular biology has begun to provide us with an understanding of how gene expression is controlled during development. Our understanding of the links that connect cell morphology, the extracellular environment, and gene expression is still incomplete, but we have already learned much about how metazoan developmental systems function.

Armed with these new techniques, many scientists have begun to address the classical problems of developmental biology with renewed vigor. A number of different systems have been investigated, each yielding important contributions to our understanding of the overall problems involved. In *Drosophila*, for example, the combination of genetics with molecular biology has led to the discovery of important regulatory genes. Other organisms have been studied because of the regularity of their cleavage process. At the same time, the classical objects of developmental studies, such as sea urchins and frogs, have continued to reveal important facts about the physiology of fertilization, the regulation of gene expression, morphogenetic movements, induction, and cell-cycle regulation. Our expanding knowledge has made it clear that the basic cell biological processes of development are common to all organisms, so that the combination of these results has proved especially fruitful.

This chapter is designed to provide a selective tour through the sequence of fundamental developmental mechanisms, emphasizing the interplay among them. Differentiation of two highly specialized cells, the egg and the sperm, contains the instructions for the earliest steps of development. The tightly controlled interaction between these gametes, which is called fertilization, breaks the developmental arrest that is characteristic of germ cells and initiates other controls of cell growth and cell division. As the single fertilized cell divides, the daughter cells begin to differentiate along separate pathways by expressing different subsets of genes from their identical genomes. The unfolding of the genetic program of each cell is directed by both internal and external cues. Internal cues include the poorly understood "determinants," which are inhomogeneously distributed in the oocyte and distributed unequally to certain cells as the embryonic cells divide. External cues are derived from interactions with other cells and with extracellular matrices laid down by other cells. The external information is enriched by very specific cell movements that allow different kinds of cell-cell interactions during different developmental periods.

The emphasis of this chapter will shift among different organisms and different techniques, but is designed to explicate our understanding of the major developmental mechanisms in cell biological and molecular terms. The prospects for advance in this field are extraordinary, with its major questions approachable today in ways that they could not have been conceived of even a few years ago. These new investigations will not only begin to answer the intellectual questions that have preoccupied scientists and philosophers since at least the time of Aristotle, but they hold the promise of considerable practical impact on our understanding of the entire field of biology, especially on medicine and agriculture. An understanding of the way an organism develops is of fundamental importance to comprehending and utilizing the properties of that organism.

DEVELOPMENT BEGINS WITH GAMETOGENESIS

Understanding the Differentiation of Germ Cells Will Give a Key to the Initial Steps of Development

In one sense, development begins with the productive encounter of the sperm and egg at the time of fertilization. However, fertilization actually has its basis much earlier, during the long and complicated process of the growth and development of that sperm and egg. Sperm and eggs develop from a specific group of cells in the embryo called germ cells. In some organisms, the precursors of the germ cells can be identified and followed throughout the course of an embryo's development. Eliminating the precursors from a fertilized egg renders the resulting animal sterile. It is likely that specific molecules in the egg, the so-called determinants, are responsible in such cases for the development of the germ cells. The molecular nature of these determinants and the mechanisms of their action are unknown; however, transplantation of cytoplasm containing these determinants can induce the ectopic development of germ cells. In other kinds of organisms, the presence of specific determinants has not been detected; in these cases, it is unclear whether determinants, comparable to those in the animals in which they are known, are actually absent or simply not detectable by available methods. In mammals, germ cells can be identified at a certain point in development; the origin of these cells is uncertain, however, and the mechanisms by which they appear are unknown.

In most animals, the germ cells differentiate along two different pathways, one of which results in the production of a large, immobile gamete called the egg, while the other leads to the production of a small, mobile gamete called the sperm. Mature gametes contain half the amount of DNA (half the number of chromosomes) present in somatic cells; their union restores the diploid number of chromosomes. In most animal species, eggs and sperm usually reside in separate female and male individuals, respectively. In each sex, the differentiation of germ cells into mature gametes is a process of key importance since it represents the basis for understanding the initiation of development; such understanding is likewise of great practical importance.

The Egg Contains Both Nutritive Materials and Positional Information Needed for the Early Stages of Development

The process of oogenesis, or egg production, results in cells that contain sufficient stored material to support at least the first stages of development. In addition to nutritive and structural materials, the egg also contains information necessary for directing subsequent development. This information is produced and appropriately distributed according to instructions included in the egg genome and is also influenced by the contributions of other maternal cells surrounding the egg. Information is stored in different kinds of molecules, mainly proteins and RNAs, some of which are nonrandomly distributed in the egg cytoplasm. Other molecules, less well known, probably also contribute to the information pool. In addition, the plasma membrane of the egg contains information in molecules unequally distributed on the surface. Furthermore, the development and maturation of the egg is controlled by numerous outside factors (mostly hormones), which are essential for normal oogenesis, although they may not contribute directly to the complexity of the egg structure. By synthesizing the appropriate receptors, egg cells can regulate, to a degree, their responsiveness to these outside factors.

Oogenesis is a unique process, combining cytoplasmic diversification with significant growth. In nonmammalian species, fully grown oocytes contain all of the raw material needed to support embryogenesis. Even in mammals, in which embryonic growth is largely supported by materials from the mother, the egg is several hundredfold larger than somatic cells. Although the egg contains a tremendous amount of information, this information alone is not sufficiently complex to direct every detail of development. Instead, subsequent development is a series of interacting processes that call upon the genome and the existing maternal materials to generate further complexity. It is the proper understanding of this series of reactions, whereby the crude information of the egg is transformed into the detailed information of the organism, that constitutes the main goal of developmental biology.

The enormous growth of the oocyte occurs while the cell is arrested in meiotic prophase. In order to continue development, the cell must be released from its state of arrest to complete the meiotic divisions. In mammals this process is under the control of gonadotropin hormones; this control is likely mediated by the follicle cells surrounding the oocyte. How they do this remains one of the most important questions in reproductive biology. The recent development of in vitro techniques that enable at least partial reproduction of oocyte growth and meiotic maturation should contribute significantly to the understanding of the biochemical basis of this process. In addition, these techniques will significantly enhance our ability to control the reproductive capacity of mammalian species; their clinical, agricultural, and ecological applications can be readily visualized.

Some progress has been made in identifying the genes that are active and necessary for oogenesis. For example, in *Drosophila*, many of these genes direct the first stages of the program of morphogenesis.

Sperm-Specific Changes in Chromatin Structure Seem to Control Gene Expression

The spermatozoon contributes half of the genetic material to the developing individual and provides the necessary stimulus for the activation of development in the egg. During spermatogenesis, the chromatin undergoes unique changes as the protamines replace histones; this process is reversed after fertilization. Since the changes in chromatin structure are probably associated with changes in transcriptional activity, one can speculate that gene expression in the sperm genome is specifically controlled. Unique methylation changes occur during spermatogenesis in many species, and these changes might cause specific information storage in the sperm DNA.

The morphology of the adult sperm is species-specific and varied. Sperm morphology, controlled in part by the sperm genome and in part by supporting cells, is probably another example of complex interactions between the process of development and particular sets of genes. Although it is unclear whether sperm provide any nongenetic information crucial for development, the fact that the eggs of many animal species can undergo parthenogenetic development—development into individuals of normal appearance in the absence of fertilization—argues against this possibility. The absence of parthenogenesis in mammals, however, might suggest that the contributions of the sperm and egg are different and that both are essential for normal development. Recent results indicate that during mammalian oogenesis and spermatogenesis, the genome of the gamete undergoes different imprinting that modifies its activities. The mechanism and extent of this imprinting are unknown and should be studied since they represent an important mode of gene control.

The Interactions Between Sperm and Egg Are Tightly Regulated at Several Levels

During the course of evolution, mechanisms have originated that ensure the specificity of the sperm-egg interaction and prevent the fusion of more than one pronucleus from each parent in the formation of the zygote. After being released from the male genital tract, sperm undergo a final maturation step (capacitation),

---

DO THE EGG AND SPERM CONTRIBUTE EQUIVALENT GENETIC INFORMATION TO THE ANIMAL EMBRYO?

In mammals the egg and sperm do not contribute equivalent sets of genetic information, even beyond the obvious differences contributed by the sex chromosomes. Micromanipulation techniques have made it possible to replace the male pronucleus—the sperm nucleus before fusion—in a mouse zygote with the female pronucleus from another zygote and vice versa. In this way, it has been possible to study the preimplantation development of constructed embryos containing two sets of female (biparental gynogenomes) or two sets of male (biparental androgenomes) genomes. After being implanted, gynogenetic embryos fail to develop the extraembryonic components—the trophoblast and yolk sac—whereas androgenetic embryos fail to develop the embryo proper. Both classes of embryos eventually abort, indicating that male and female genomes must be present and that each performs a different but essential role in development. Genetic experiments have localized functional differences between male and female genomes to several chromosomes or subchromosomal regions. It should now be possible to identify those gene sets that are expressed differently in the chromosomes contributed by the male and female. Knowing what make these identical genetic elements function differently depending on their parental derivation is a matter of great theoretical and practical importance.

---

which renders them competent to fertilize the egg. Sperm-egg interactions are probably regulated by specific recognition signals, and several molecules involved in this process have now been characterized. Both capacitation and sperm-egg interaction are important targets for contraceptive intervention and must therefore be studied further.

Great progress has been made over the past decade in our understanding of the physiology of fertilization, largely as a result of improvements in techniques to measure small changes in ion and lipid concentrations. The process of fertilization represents a case in which two cell types that have spent some time in a quiescent storage state must rapidly resume a high level of metabolic activity and fuse with each other. This change in state has been termed activation. In both eggs and sperm, activation is accompanied by a dramatic change in intracellular ion concentration, particularly that of Ca^{2+} and H^+ (more commonly referred to as pH_i). These changes must be important steps in activation because altering either pH_i or $[Ca^{2+}]_i$ tends to stimulate the parthenogenetic development of invertebrate eggs. The Ca^{2+} ionophore A23187 is a general activator of eggs, apparently because of its ability to elevate $[Ca^{2+}]_i$. In contrast, ammonia and other weak bases, which elevate pH_i, activate some but not all of the events of early development, including DNA synthesis, chromosome condensation, and increased rates of protein synthesis in the sea urchin egg.

Changes in pH_i and $[Ca^{2+}]_i$ are potent effectors of metabolism and development in numerous species, in different cell types, and at various developmental stages. Such changes accompany normal fertilization in many eggs, but the mechanisms by which the changes are achieved are poorly known. In the case of $[Ca^{2+}]_i$, a specific type of lipid in the plasma membrane of the eggs of both invertebrates and vertebrates (phosphatidylinositol-bisphosphate or PIP_2) rapidly cleaves after activation, releasing inositol trisphosphate (IP_3) into the cytoplasm, while diacylglycerol (DAG) remains in the plasma membrane. Both of these molecules are important second messengers since IP_3 triggers Ca^{2+} release from intracellular stores and DAG activates an enzyme called protein kinase C, which turns on other important cellular proteins by phosphorylating them. This chain of events leads to an increase in $[Ca^{2+}]_i$. Perhaps the most interesting aspect of this story is that this same process of inositol lipid cleavage occurs in a wide variety of other cell types that respond to the binding of a variety of external agents, including hormones, growth factors, neurotransmitters, and chemotactic molecules at their plasma membrane. Thus studies of fertilization may help to explain signal transduction in other systems and serves as a model system for understanding the role of second messengers in other kinds of systems.

CELL DIVISION, GROWTH, AND DEVELOPMENTAL TIMING

Cell Division and Cell Growth Must Be Exquisitely Regulated for Normal Development

Regulating cell division and growth together is a requirement for homeostasis. If growth outpaces division, the cell will become larger and larger. If division is faster than growth, the cell will become smaller and smaller. All eukaryotic cells exhibit a chromosome cycle and a cytoplasmic cycle, which are well correlated with one another. The chromosome cycle includes a period of DNA replication (S phase), during which each gene is duplicated, and a period of mitosis (M phase), during which the genes are segregated to the daughter cells. In the cytoplasm, growth is more or less continuous and the bulk content overall is duplicated; only the centrosomes undergo a discrete duplication (usually in S) and a segregation (in M). In most cells, the S phase is separated from the M phase by periods during which neither replication nor segregation occurs. The period between M and S (when a eukaryotic cell is typically diploid) is called G1, whereas the period between S and M (when a cell is typically tetraploid), is called G2.

Many important problems in biology, medicine, and agriculture concern how this simple cell cycle functions. In the embryo and in renewable tissues, such as bone marrow, cells are continually entering this cycle. Most differentiated cells, however, are not dividing and are resting in G1. Some cells, such as mature nerve cells of the central nervous system, will rest in G1 for the life of the organism. Cells in some tissues are only contingently arrested in G1. For example, cells such as the fibroblasts in the skin or parenchymal cells in the liver are capable of entering the cell cycle if the organ is injured. Under such circumstances, these normally quiescent cells initiate DNA replication and divide until sufficient cell replacement has occurred. The inability of some cells, such as nerve and muscle cells, to reenter the mitotic cycle limits the capacity of their tissues to recover from injury. At the other extreme are tumor cells, which reenter the mitotic cycle too easily and do not respond to the normal signals that arrest division. The specific events of the cell cycle are also critical points for the life of the cell. Errors in DNA replication cause mutation, and errors in meiosis or mitosis cause chromosome abnormalities.

In embryonic development, growth is precisely regulated. For an animal to develop, specific cells must divide at a specific time and cease to divide at a specific time. Some cells are even programmed to die. Recent studies in the development of the embryo of a nematode, a member of a simple group of worms, give a dramatic example of the regularity of the growth process. Each individual is almost identical to the next. The cells divide on schedule and with a reproducible orientation. Certain cells keep dividing while others cease dividing, and some die on cue. In human development, although the regularity is less extreme (presumably other factors ensure the successful production of the embryo since our pride would never allow us to admit that a nematode is put together more exquisitely than a human), it is likely that the same principles underlie the regulation of cell division in both organisms.

Not until recently has our understanding of cell division progressed appreciably beyond the descriptive stage, and we are just beginning to understand the underlying biochemical mechanisms. Simpler systems such as yeast are being studied to enhance our knowledge of the principles involved, but embryonic systems have also proved useful for this purpose. The egg and early embryo of many animals are essentially nongrowing systems, dividing their cytoplasm into increasingly smaller cells. In such systems, only the pace of the cell cycle, and not the accumulation of a certain minimum cell mass, determines when division occurs. In such embryonic systems, intracellular factors that regulate the cell cycle have been described for the first time.

Improved methods for protein separation, immunological characterization, and cloning have recently increased our understanding of the extracellular growth factors that may cause quiescent cells to begin to divide. We now know a considerable amount about what these factors are, but still have little knowledge of how they act after reaching the plasma membranes of their target cells.

Many fundamental questions about the way in which cell growth is regulated during the course of embryology are under active investigation. For example, we would like to know, What are the factors that time cell division and other events? How are they segregated equally to daughter cells? What causes specialized cells such as neuroblasts to proliferate, and what causes them to cease proliferating? What prevents cell death in one daughter cell and yet causes its sibling to die? How is cell-cycle timing related to differentiation? Are other developmental events tied to the cell-cycle timer in the way our morning alarm is tied to a clock? These old questions are currently benefiting from new investigations in which results from different organisms are being elegantly tied together to produce new results of generality. As we shall see, the story of the current investigations into the cell cycle carries us from tumor viruses to nematodes and from yeast genetics to frog embryology. It is a search for new molecules and a continuing search for the ways old molecules function.

Yeast Provide a Good Model System for Studying Molecular Mechanisms of Cell Division

Yeast are unicellular fungi, simple in some respects, but with many of the same features that characterize mammalian and all other eukaryotic cells. Like all eukaryotes, yeast cells have a nucleus, chromosomes, mitotic activity, and a cell cycle divisible into G1, S, G2, and M. The growth of yeast cells, like that of other eukaryotic cells, can be arrested in G1 if they are deprived of nutrients; similarly, growth can be arrested in the same stage by factors that specifically inhibit division, such as the mating pheromone. Yeast cells can easily be manipulated genetically, and many mutants that cause cells to arrest at specific points in the cell cycle have been identified. In recent years the study of the cell cycle in baker's yeast, which divide by budding, and fission yeast, which divide like normal mammalian or plant cells, has yielded important genes that are involved in control of the cell cycle in all higher organisms. Starting with the striking observation that a human gene can replace a yeast regulatory gene controlling cell division there has been a steady stream of reports showing homologies between human and yeast structure and function. Recently these studies in yeast have been combined with those in frog, sea urchin, and human to reveal some of the basic workings of the control mechanisms for the cell cycle. The yeast cell may be of particular utility in demonstrating feedback control of the cell cycle that occurs when there is damage to DNA or when mitosis is inhibited. The future will see many opportunities to exploit the genetic advantages in both budding and fission yeast with the biochemical advantages of other systems. It will be of particular interest to combine studies of cell signaling through the mating pheromones with the cell-cycle arrest that ensues. This may turn out to be a good analogy with the many examples of control of the mammalian cell cycle by growth factors or other extracellular regulators. However, to exploit the potential fully will require the development of in vitro systems for yeast.

Study of Embryonic Cell Division Is Leading to a Better Understanding of the Cell Cycle in All Cells

A toad embryo does not pause to grow. Instead it divides at an incredibly rapid rate, with a cell cycle 25 times as fast as normal somatic cells in culture. The *Xenopus* egg, for example, does not pause between M and S or between S and M during its first 12 divisions; yet before fertilization the cell cycle has been shut down completely. The large amount of cytoplasm that these eggs possess makes them ideal subjects in which to distinguish the contribution of the cytoplasm from that of the nucleus. Experiments on these eggs demonstrate that the regulating machinery of division is in the cytoplasm; the nucleus is merely a responding element.

Several years ago, a factor was identified from frog eggs that induced meiosis when injected into frog oocytes. This factor, called maturation-promoting factor, was subsequently shown to be present at mitosis of all eukaryotic cells. In vitro experiments have been carried out in which nuclei have been reconstituted from DNA and soluble components and induced to break down and undergo mitosis by the addition of maturation-promoting factor. The complex steps of nuclear assembly and disassembly may soon be reducible to pathways similar to those which have been demonstrated for virus assembly. In the case of the cell cycle, however, the assembly process is carefully regulated. We know that the final reaction regulating the assembly and disassembly of the nuclear envelope is phosphorylation of a specific protein.

From a different direction, other embryonic studies have provided another piece to the solution of the puzzle of the cell cycle. Although embryonic cells are endowed with a generous supply of all known structural proteins needed for cell division, they must still synthesize proteins in order to pass through the cell cycle. This has suggested to some that protein synthesis serves a regulatory function, a view that was reinforced when the amounts of specific proteins were found to oscillate during the embryonic cell cycle in clam and sea urchin eggs. These proteins, named cyclins, induce cell-cycle transitions in frog eggs, which tie them to the mitotic factors described above. In the past two years, work on maturation-promoting factor from frog eggs, cyclins in sea urchins, and regulatory genes in yeast have come together in a spectacular way. Maturation-promoting factor has been purified and found to contain as its principal component the homologue of the yeast regulatory gene. Cyclin activates maturation-promoting factor from a stockpile of inactive material. The basic workings of the cell cycle have been reconstituted in extracts in which the intimate biochemical relationships between cyclin and the yeast gene can be studied. All of these basic systems have been immediately applicable to all organisms. The use of advanced molecular biological techniques, like the polymerase chain reaction, makes it possible in a matter of weeks to identify genes from one organism to another across hundreds of millions of years of evolutionary history. It is thus no longer necessary to find the optimum system for study; one can use eggs for biochemical analysis, yeast for genetics, and *Drosophila* for developmental studies. In the future, this approach will become more prevalent. Progress on control of the cell cycle will be accelerated by assuming that everything basically works the same. Differences will emerge, but the power of assuming the basic conservation of fundamental biological processes like cell division will be more and more obvious.

Growth Factors and Their Receptors Play Many Roles in Shaping the Organism

Extracellular factors that stimulate the entry into a proliferating stage of the cell cycle have been much better characterized than intracellular factors. Some, like epidermal growth factor (EGF), have been purified for many years, and their receptors in the cell membrane have been well characterized. Recently, the connections between oncogenes and growth factors and oncogenes and growth-factor receptors have become clearer. For example, several oncogenes and growth-factor receptors are tyrosine kinases. Since most oncogenes have a cellular homologue, their study gives us new approaches to normal cell-cycle regulation. At present the number of growth factors and receptors is unknown. Some are clearly tissue specific; others, such as EGF and insulinlike growth factors, are widely distributed.

Our knowledge of growth-factor receptors and soluble factors has two important gaps. First, even when the activity of a receptor has been identified (for example, that it is a tyrosine kinase), we do not know the substrates for this activity; even in those cases in which we can identify some of the substrates we do not know how they induce cell proliferation. The second major gap is our knowledge of the role of the growth factors in normal development. These factors modify other properties of the cell in addition to proliferation. Some, for example tumor growth factor, control the production of extracellular matrix. Others, such as fibroblast growth factor, induce the proliferation of blood vessels. The classical embryology literature is replete with descriptions of factors, often obtained from heterologous sources, that will induce the formation of early tissue types, such as mesoderm or neural tube. It seems likely that some of these factors will ultimately prove to be known growth factors or proteins encoded by proto-oncogenes.

Although the well-studied growth factors have been soluble proteins, embryological studies have provided evidence that the extracellular matrix can be an important inducer of differentiation events and can control cell proliferation. Recently, several extracellular matrix proteins have been shown to have sequences related to EGF. The potential importance of interactions of these kinds in normal development has been highlighted by the discovery of a neuronal defect mutation in *Drosophila*. This mutation is in a gene encoding a large secreted protein with a region homologous to EGF. Another discovery has been a lineage-defect mutation (a mutation causing a particular set of cells to assume the wrong identity) in a nematode. This mutation is in a gene encoding a large protein also having an EGF-like sequence.

A fundamental question is how complex the signals are for induction. Are the inducers influencing simple properties such as the decision to proliferate, or are they specifying detailed positional information? At this point, we can say only that there appear to be several classes of signaling molecules: trophic factors, which prevent cell death; growth factors, which induce proliferation; transforming growth factors, which change expression; and differentiation factors, which may determine cell type. It seems unlikely that these categories can be so neatly maintained; more probably, different factors will function differently in different circumstances. Most likely we will need to understand the cellular processes of growth and differentiation to understand fully what the growth factors are doing in any specific example. In this respect, the history of the cells may also be important. Classical embryological studies have suggested that before the retina induces the epidermis to form a lens, the epidermis has received several sequential inductive signals. Each signal may be from a growth factor or an extracellular matrix element, and after each induction the cell has presumably been functioning and changing. The ultimate response will probably be a product of the intermediate responses as well as specific past inductions. Growth factors and their actions during embryogenesis, along with cellular responses to them, need to be described more completely.

Several Different Time-Keeping Mechanisms Are Used During Development

Adult organisms have a well-elaborated nervous system and brain, which enable them to respond rapidly to events, and also a well-developed circulatory system and endocrine organs, which allow them to respond on a longer time scale. In adults, normal feedback loops control the estrous cycle or reflex and control rapid signals and responses. On the longer time scale of circadian or even annual rhythms, our knowledge is less secure.

In contrast to adults, embryos need to do more things, with a high degree of control, and without the benefit of highly elaborated communication systems, such as the nervous system. This control is shown in the course of development of any embryo. A fundamental question about the spatial control of embryonic development has been whether form is predetermined (prelocalization and mosaic development) or develops progressively (epigenetic development). The timing of embryonic development might also be epigenetic (a series of sequential events, each leading to the next) or prelocalized (the result of timing mechanisms in place in the oocyte that would signal events at the appropriate time). Timing can be either autonomous within an individual cell or influenced by cell-cell interactions. Biochemists are most familiar with simple cause and effect, such as gene repression by an end product; simple causal relationships, such as those mediated by growth factors, are important in embryonic development also. Recently, however, long-term timing mechanisms have been described in developmental systems. The early embryonic cell cycle in most species is rapid and often synchronous or at least metachronous. Studies in the frog and the sea urchin have shown that this synchrony is not due to pacemaker cells, but to a clock that exists in each blastomere.

In the early embryonic cleavages, the cell cycle becomes variable in close correspondence with the specific developmental fates of particular cells and cell lines. By experimentally shifting cytoplasm from one blastomere to another in the nematode, one shifts the cell-cycle time as well as the developmental fate of the cells. Yet recent experiments in the nematode show that cell communication rather than prelocalized information determines the behavior of some of the cells and their corresponding division patterns. In *Drosophila*, the synchronous pattern of nuclear division stops after 12 divisions. In the next two divisions the asynchrony is not random, but locally patterned. Is this pattern of cell division merely a reflection of the underlying developmental events? Or are the division patterns in some way linked to the expression of a specific fate?

Mechanisms for controlling developmental time have only recently been investigated. In *Xenopus*, as in *Drosophila*, the early synchronous cleavages end after the twelfth division; this cell-cycle transition is correlated with a major increase in transcription, and in *Xenopus* with the onset of cell motility as well. It is becoming clear that this transition is caused by the depletion of some maternal component by the rapidly proliferating nuclei. At a certain threshold, cell division stalls and new developmental events are initiated. Yet in the frog, not all early developmental events are timed by the arrival of the nuclei at a certain threshold. The onset of gastrulation movements in the cortex and the translational control of certain proteins seem to be timed by a clock set off at fertilization and operating independently of nuclear mass.

In *Drosophila* recent studies have combined our knowledge of the biochemical events in the cell cycle with studies of the cell-cycle transition after 12 divisions. A yeast gene that acts in the regulatory pathway for cell-cycle control was also found in *Drosophila* to control this important developmental transition.

The overall results on this cell cycle show that we must consider the cell cycle in terms of external cell signals as well as developmental decisions. The progression through the cell cycle depends on the successful completion of events, such as DNA replication and mitosis. Yet the cell cycle is not merely the sum of external and internal signals. We now know that maturation-promoting factor and cyclins regulate the basic cell-cycle progression and that other factors influence the activity of these molecules. The exact linkage between the intrinsic regulators and extrinsic signals is important for study in both embryology and in cell biology.

DIFFERENTIAL GENE EXPRESSION

Developmental Control of Gene Expression Is Precise with Respect to Tissue, Time, and Position

With few exceptions, every cell in an individual organism contains the same complement of genes; yet the multiple cell types of on individual result from the expression of significantly different subsets of these genes. The determination of the molecular mechanisms of this control has been a major goal of developmental biologists. Although recent progress has been impressive, the problem is difficult and will probably not be completely resolved in the near future.

In bacteria, elegant models of gene expression, controlled at the level of transcription either by gene-specific repressors or by novel subunits of RNA polymerase that recognized specific classes of gene promoters, were developed early. The many recombinant DNA techniques now allow equally specific analyses of gene expression in the cells of eukaryotic organisms. These analyses have demonstrated many additional ways in which gene expression can be controlled. The control can be at any of several levels, including the transcription, processing, and translation of RNA, as well as RNA stability. In a few instances, specific changes in DNA, such as the amplification of particular sequences or gene rearrangement, are associated with transcription. Examples of each type of gene control are now being studied; to date, perhaps most has been learned about transcriptional control. For many genes, specific parts of the sequence have been shown to be necessary for tissue- or temporal-specific transcription. An exciting finding has been the identification of enhancers, short sequences that can act from a distance of thousands of nucleotide pairs to increase transcription of nearby genes; since they act in either orientation, they apparently accomplish their task by affecting chromatin structure. Gene rearrangements, such as those that initiate transcription of immunoglobulin genes, can act by bringing the coding sequence into the vicinity of a tissue-specific enhancer. The identification of transcriptional control sequences offers a way to construct genes for expression at specific times or locations.

As emphasized throughout this chapter, the developmental program is not laid down in its entirety at fertilization. Instead, the program unfolds through a series of interactions as development proceeds. Some of these interactions occur between members of sets of genes that direct particular aspects of development. Two sets of genes that are being studied with particular success at this time are those that direct early patterning and segmentation and those that direct sexual differentiation. Studies of these two systems can be expected to provide paradigms for the analysis of the interplay of gene functions in the development of other organ systems. These systems are also providing new evidence of mechanisms for control of gene expression. For instance, studies in *Drosophila* have shown three genes in the pathway for sex determination that are transcribed in both males and females, but spliced differently in the two sexes.

The Genetic Program of Each Organism Must Carry Instructions Not Only for How Cell Types Will Differentiate, but Also for Where These Cell Types Will Be Located

Although the bodies of different kinds of animals contain essentially the same cell types, the body shapes often differ greatly. Understanding how the three dimensions of an organism's shape are encoded in the linear nucleotide sequences of DNA appears to be one of the most difficult problems of developmental biology. Recently however, many laboratories studying *Drosophila* have made exciting progress. *Drosophila* is ideally suited for genetic analyses of developmental processes. The wealth of information that is available from years of genetic studies is now being combined with molecular analyses of these genes and their products.

Geneticists have isolated mutations in a number of genes that affect the dorsal-ventral or anterior-posterior axes of the animals. These genes, therefore, convert the more-or-less apolar oogonial cell into an egg cell that can give rise to an individual with a head and a tail, a front and a back. Other identified mutants indicate the existence of genes that establish the segmental pattern of the individual. The segmentation provides the basic framework for the regional specialization of head, thoracic, and abdominal segments.

Contemporary genetic studies have shown that the genes that determine embryonic axes and segmentation form an interacting hierarchy. The genes that act first in the establishment of the primary axes are maternal genes, which perform their functions during the growth of the egg cell. Some of these genes have prelocalized messenger (mRNA) transcripts and proteins, whose spatial distribution is essential for regulating developmental processes. Immediately after the egg is fertilized, embryonic nuclei multiply rapidly and become distributed throughout the cytoplasm of the large egg cell. Because of the action of the maternal genes, the egg cytoplasm is already inhomogeneous in the distribution of the developmental determinants. As the embryonic genome becomes activated, embryonic genes begin to direct additional steps in the development of axes and segmentation. When membranes form, isolating embryonic nuclei into individual cells, those cells are already partially instructed about their role in morphogenesis.

As embryonic development proceeds past the blastoderm stage, the pattern of cells with detectable amounts of product from any given segmentation gene changes. The mechanisms underlying the changing patterns of RNA distribution may vary with cell type and with the particular gene considered. In some cases, the presence or absence of an RNA may depend on the regulation of transcription of its coding gene. In other cases, changes in the rate of degradation of the RNA may be specific to cell type. Other levels of cell-type-specific regulation of segmentation genes may be imposed on the translation and further metabolism of their protein products; however, studies of the proteins involved have not progressed as far as the RNA studies have.

The net result of the changing patterns of the distribution of segmentation gene products is that the body of the developing embryo is divided into a series of segments. The identification of bands of cells as segmental boundaries or interior regions seems to depend on the combinational expression of particular sets of the segmentation genes during early embryonic periods. These studies are moving rapidly. New interactions between the known segmentation genes are being defined, and further members of the set are being identified. Within the next few years, we should have a detailed picture of the molecular basis of the program that divides this embryo into repeating segments. Since segmentation is an early step in the development of many animals, the results of these *Drosophila* studies should have broad implications.

The third level of the hierarchical network that directs the body plan of *Drosophila* contains the homeotic genes, which specify the identity of each segment; these genes differentiate the segments into head, thorax, and abdominal parts. A mutation in a homeotic gene often causes a segment to form structures normally found in another segment. For example, the *Antennapedia* gene is not normally expressed in the head, but regulatory mutants causing inappropriate expression of the gene lead to the development of legs rather than antennae there. This result had led to the concept that homeotic genes are "selector" genes, capable of directing a complete switch from one developmental pathway to another. Each of the major known homeotic loci is expressed in a distinct, relatively nonoverlapping pattern along the anterior-posterior axis of young embryos. The expression of each homeotic gene is controlled by the segmentation genes and by interactions with other homeotic genes.

Many of the genes identified by these genetic analyses have now been cloned, and sequences of the encoded proteins have been determined for some of them. Although these steps are significant, finding a molecular function for a protein, even when its sequence is known, can be extremely difficult. Fortunately, the sequences of many of these proteins have been informative, at least to the extent that they suggest testable hypotheses for the functions of the proteins. Major results include the finding of the "homeobox," a 60-amino-acid domain common to proteins encoded by all of the major homeotic genes, as well as by several genes involved in the determination of segmentation or polarity. This homeodomain has structural homology to DNA-binding proteins from bacteria and yeast, suggesting that proteins containing the homeobox may act as transcriptional regulators, controlling the expression of other genes. Another protein sequence homology has been found between the early-acting segmentation genes and transcription factor TFIIIA of *Xenopus*. TFIIIA binds both DNA and RNA. All of these proteins show a nuclear localization consistent with the obvious hypotheses for their functions. Their actual functions in vivo, however, remain to be demonstrated, and the meaning of the DNA binding needs to be discovered. Other members of the regulatory gene sets may act in different ways.

SOME GENES CAN MAKE DRAMATIC CHANGES
IN THE SHAPE OF AN ANIMAL

Most animals, from flies to fishes, are made up of essentially the same structural proteins; nevertheless, they come in an amazing variety of forms, which we know to be genetically determined. Thus there must be genes that control the pattern of expression of structural proteins in any species. By studying mutant individuals, geneticists are beginning to identify such controlling genes. Mutations in these controlling genes offer possible explanations for ways in which different animal species evolved. For example, *Drosophila* is an insect with a single pair of wings, although some insects, such as dragonflies, have two pair of wings. A number of genes have been identified that affect the placement of wings on individuals of *Drosophila*. By combining mutations in three particular genes (*abx*, *pbx*, and *bx*) it is possible to produce *Drosophila* with two pairs of wings (Figure 5-1). Although the individuals resemble *Drosophila* fully in almost every respect, in this one characteristic they recall the distant ancestors of the flies, which existed hundreds of millions of years ago and which certainly had two pair of fully formed wings.

A breakthrough in the study of development came from the observation that vertebrate genomes also have sequences homologous to the homeobox. For example, the mouse has at least 16 genes containing the homeobox. Transcripts of several of these genes have been detected in embryos, but there is not yet enough information to determine whether they play a role in the control of developmental processes in vertebrates. If the strong evolutionary conservation of the homeobox sequence reflects a conservation of functions in developmental control, the study of development will accelerate since there are no other obvious means of identifying genes that control early vertebrate embryogenesis. The genetic approaches used with *Drosophila* are not feasible with mammals. Even after 50 years of intensive study, only a handful of mutations that affect early mouse embryogenesis are known to exist; virtually nothing is known about the products of the genes identified by this genetic analysis.

Studies on all classes of these morphogenetic genes are progressing rapidly. For *Drosophila*, a rough extrapolation from the available data suggests that at most a couple of hundred genes (of the estimated 5,000 to 10,000 genes in the total genome), will be found to act as central components of the regulatory hierarchy. Although we think we understand the rough outlines of that hierarchy,

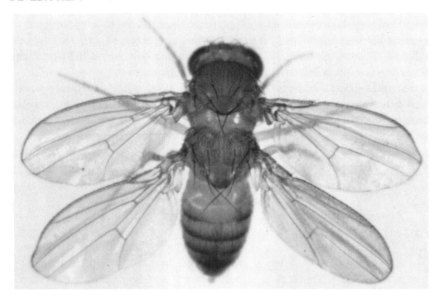

FIGURE 5-1   Double-winged *Drosophila*. [Photo: Edward B. Lewis, California Institute of Technology]

gaps are still to be filled and almost certainly a few more surprises will be found. There is a growing appreciation that a given gene can have a key role in more than one developmental process; schemes that take this relationship into account in identifying network elements are finding genes that had not been detected earlier.

A key attribute of the segmentation and homeotic genes, one that distinguishes them from nearly all other kinds of genes, is the positional specificity of their expression. The patterns of expression may vary with developmental stage or tissue type, but in every case the expression also depends on the positions of the cells in the organism. Understanding the molecular mechanisms of the positional control of these genes is an important goal. At least some of these elements are likely to be related to the control elements of vertebrates, if not in sequence, at least in mechanistic principles; thus, the insect models will provide hypotheses to be tested in other organisms. The potential of using homeotic genes as tools for genetic engineering is considerable. If they can be used to direct the position-specific activation of other genes, they could also be useful medically.

Another important research goal is the identification of the genes controlled by the homeotic genes. How do the proteins containing the homeobox (or combinations thereof) instruct a cell to follow a particular pathway of morphogenesis? This area is virtually unexplored, in part because the experimental material—defined cell types of the appropriate developmental potential—is limiting.

In addition, regulatory elements, by their nature, are expected to be relatively rare; attempts to overproduce them may well have unexpected effects, such as cell lethality. Techniques for the introduction of specifically engineered genes into organisms now provide a basis for imaginative approaches to the problem.

Studies on homeobox-containing genes in vertebrates must determine whether these genes have a role in the control of developmental processes. For technical reasons, this problem will be even more complex to study than it has been in the invertebrates.

Technical Advances Are Needed for Detailed Study of Gene Activity in Development

The need to study gene function in an intact organism is especially obvious in the case of genes that regulate functions such as morphogenesis and sex differentiation. Recently developed techniques that allow the stable integration of specific gene sequences into chromosomes provide a basis for powerful experimental approaches to developmental questions; they are already yielding important results. For example, it is possible to link genes to heterologous regulatory elements, to dissect these elements, and to generate efficient levels of expression, different patterns of expression, or modulation of protein products. As we learn more about protein structure and function, we will be able to devise even more critical experiments.

Many interesting experiments are possible only if the introduced gene sequence can be kept silent until the time of testing. To do this, we need better inducible gene promoters. Using promoters that can be induced by heat shock or heavy metals has yielded some success. Unfortunately, heat shock and probably heavy-metal stress significantly change cellular metabolism, which may cloud the results. Detecting other inducible promoters or better understanding the biology of the existing ones will expedite these studies.

In addition, techniques for efficiently eliminating specific gene function in vertebrates would lead to rapid progress. The ideal solution would be an equivalent of "homologous recombination," in which an altered form of the gene displaces the resident gene, which has been so effective for yeast genetics. Gene displacement is not yet routinely possible in vertebrates. A possible alternative way to eliminate gene activity is by binding the mRNA to complementary, or antisense RNA. (The mRNA that codes for a specific protein is called "sense," while the complementary strand of RNA that does not code for protein is called "antisense.") Antisense RNA is naturally used for regulating some prokaryotic genes. More work will be required to make this technique reliable and useful for eukaryotic cells. For example, the early amphibian embryo has an activity that specifically unwinds RNA-RNA hybrids. This activity may be important in activating stored maternal information, but it may also prohibit the use of antisense RNA as a tool in these cells.

CONTROL OF THE TRANSLATION OF A SINGLE GENE PRODUCT BLOCKS UNSCHEDULED EMBRYONIC DEVELOPMENT

The mature oocyte is loaded with materials necessary for the early steps of development, yet it remains quiescent until fertilized by the sperm. The holding pattern of the oocyte results from the action of a number of control mechanisms that we are just beginning to understand.

Immediately after fertilization, the embryo begins a series of rapid nuclear replications. Studies of clam embryos have shown that the oocyte does not have a store of deoxyribonucleotides, the immediate precursors for nuclear DNA synthesis. It is, however, richly supplied with ribonucleotides and with the large subunit of ribonucleotide reductase, the enzyme that converts ribonucleotides to deoxyribonucleotides. The small subunit of ribonucleotide reductase is missing, although untranslated RNA coding for this subunit is present. Immediately after fertilization, translation of the small enzyme subunit begins, allowing the enzyme to start synthesizing deoxyribonucleotides. Thus the specific translational inhibition of this particular messenger RNA acts as a rapidly reversible block to DNA synthesis before fertilization.

CELL MOVEMENT AND CELL ADHESION

We Can Now Begin to Unravel the Mechanisms That Control Cell Movement and Cell Adhesion

As we have seen, animals begin development as a single cell (the fertilized egg), which cleaves into a single-layered ball of cells (the blastula). How this monolayer of cells is transformed into a multilayered embryo (the gastrula), and how the cells in these layers interact with one another to form the complex patterns of tissues and organs that are characteristic of a mature animal, has puzzled biologists for centuries.

Over the past several decades, developmental biologists have learned a great deal about how the cellular phenomena of cell motility and cell adhesion help shape the development of the overall body plan, a process called morphogenesis. From the earliest stages of animal development, cells change shape, move relative to one another, and ultimately interact and coalesce in intimate contact with other cells. Understanding the mechanism by which cells correctly associate with others, however, continues to be a major problem in animal developmental biology.

Cell movement plays a major role in morphogenesis. The cells often move as sheets. During gastrulation, for example, the epithelial sheet of the blastula expands around itself, or invaginates into itself, or both, so that by the end of gastrulation, several sheets are formed, one inside another. These concentric sheets (often a tube within a tube), known as the germ layers, interact with each other to form the tissues and organs. Similar events occur during many other morphogenetic events, including the formation of the neural tube in vertebrates. After induction from the underlying mesoderm, part of the surface ectoderm thickens into the neural plate; it then folds inward to form a tube, which ultimately develops into the central nervous system. What forces drive these dynamic morphogenetic processes?

Cells can also move as individuals. In vertebrates, the epithelium at the site of closure of the neural tube gives rise to the important group of migratory cells known as neural crest cells. They break loose from the epithelium and migrate along specificpathways throughout the embryo, giving rise to such tissues as peripheral sensory and autonomic neurons, Schwann cells, pigment cells, gland cells, and, in the head, various connective tissues. What decides the pathway of migration for a particular neural crest cell? What decides where cells stop migrating, and what differentiated cell type they will become?

Finally, cells can elongate as one part of a given cell moves relative to another part. The best example is in the elaboration of axons and dendrites by neurons. An ameboid process, called the growth cone, extends along specific pathways toward its appropriate target, leaving behind an axon whose shape records the history of the growth cone's choices. Ultimately, the growth cone must recognize its target and finally halt, leading to the formation of synaptic connections. What factors guide neuronal growth cones along their highly specific pathways? And what factors control the events of cell-cell recognition as neuronal growth cones recognize specific target cells?

In each of these examples of cell movement, the events of morphogenesis are controlled in part by the ability of cells as individuals or as collectives to move selectively and to adhere to one another and to their extracellular environment.

All of these cell movements are characterized by repetitive cycles of extension, adhesion, and contraction at the leading edge of the cell. The extended processes of the cell must initially adhere to something solid, such as another cell's surface or a basement membrane, so that during the contractile phase the cell can be pulled toward that surface. Ultimately, however, the adhesive bond must be broken if the cell is to continue on its journey. Recent experimental data suggest that proteases or other secreted enzymes may be involved in regulating how rapidly the adhesive bonds are broken during cell migrations. However, much remains to be learned about the molecular mechanisms that control how cells begin to move, why certain cells move and others do not, and what causes cells to stop moving.

Over the past several years, great strides have been made in our ability to unravel the mechanisms controlling cell movement and cell recognition by identifying some of the cell surface and extracellular matrix molecules that mediate and modulate cell adhesion. Rather than try to catalogue all the molecules and all their interactions, we will instead try to highlight a few examples of the most recent discoveries and a few of the insights they have provided for understanding developmental biology.

Over the past decade, immunological and biochemical approaches have revealed the distribution of adhesion molecules at various stages and in various tissues of the embryo. Such studies have revealed that in general these adhesion molecules are neither tissue- nor cell-specific, but rather that many are displayed on different cells in different tissues, appear and disappear or are modified at specific times and places, and in many cases are restricted to particular domains of a cell's surface. Thus much of morphogenesis is specified by the temporally and spatially controlled display of a moderate number of molecules that mediate and modulate adhesion. Some of these molecules are embedded in cell membranes, whereas others are secreted into the extracellular matrices to which cells bind and on which they often move.

A Variety of Cell-Adhesion Molecules Have Been Discovered, and Several Have Been Purified and Characterized

Cell-adhesion molecules, or CAMs, were first definitively identified by means of immunologically based adhesion assays in which specific antibodies capable of blocking cell adhesion were used to purify cell-surface molecules. Over the past few years, a variety of CAMs have been discovered; several have been purified and characterized chemically. All of these CAMs are large cell-surface glycoproteins, several of which come in different protein forms; most appear to be intrinsic membrane proteins with extracellular, transmembrane, and cytoplasmic domains. The genes encoding most of these CAMs have been cloned and the sequences of the proteins deduced. Some of the CAMs appear to be structurally related, whereas others are not. The emerging picture is one of a variety of different CAMs, many of which are not specific for a single cell type or tissue. Rather, many of the different CAMs seem to be expressed at different times and places throughout development and to assist in the direction of the events of morphogenesis.

The best characterized of the CAMs is neural CAM (N-CAM). Initially discovered in neural tissue, N-CAM was subsequently found in a much wider and highly dynamic range of tissue. The mechanism of N-CAM binding is homophilic; that is, N-CAM on one cell surface binds to N-CAM on another. A particularly striking modification related to the function of N-CAM is the unusually high amount of a particular type of carbohydrate, sialic acid, present on N-CAM molecules isolated from embryonic brain. In contrast, N-CAM molecules in adult brain contain only one-third the sialic acid present per unit in embryonic brain. This decrease is correlated with a threefold increase in the rate of binding of the molecule: the more sialic acid, the less sticky the molecule. This change in the chemical properties of N-CAM, coupled with changes in the amount and localization of the molecule during development, help contribute to the diversity of functions of just this one adhesion molecule.

Other CAMs with interesting specificities have been discovered in the immune system, and some of these are also expressed in the brain. Still other adhesion systems seem to utilize carbohydrates and either soluble or cell-surface carbohydrate-binding proteins (called lectins). Two questions remain: How many different types of adhesion molecules exist, and with what specificity are they expressed?

These are particularly important questions for the developing nervous system, with its enormous diversity of cell types and specificity of cellular interactions and connections. Much current interest is being focused on the search for other, more specific neuronal adhesion and recognition molecules by means of which neurons find and recognize one another as growth cones are guided and synapses are formed. Recent discoveries have revealed a number of relatively rare glycoproteins that are transiently expressed on the surface of different subsets of neuronal processes during the events of growth-cone guidance. In addition, several groups of related carbohydrate structures have been discovered that label specific subsets of neurons and their targets in the developing spinal cord. Whether these specific proteins and carbohydrates actually function as neuronal recognition molecules, and whether in so doing they either directly mediate or modulate specific cell adhesion, awaits future studies.

In addition to the more transient forms of cell adhesion involved in cell movement and cell recognition, a variety of more specialized and long-term cell-surface contacts, called cell junctions, provide more stable adhesion and communication. These junctions also play an important role in morphogenesis. Such junctions, or specialized regions of the opposing cell membranes, are classified as adhering junctions, which mechanically hold cells together; impermeable junctions, which not only hold cells together but seal them in such a way that molecules cannot leak in between them; and communicating junctions, which mediate the passage of small molecules from one interacting cell to another.

All of these different types of stable junctions between cells allow them to hold onto one another, communicate with one another, and alter one another's shape and state as a collective during development in a way that helps provide proper form and shape of tissues and organs. In recent years, many of the molecular components of these junctions have been characterized and cloned. With this new molecular analysis of their structure and function, we are beginning to get a more detailed understanding of how these junctions are assembled and how they work.

Studies of the Extracellular Matrix Are Important for Understanding
Morphogenesis

The extracellular matrix (ECM) is an insoluble macromolecular meshwork that surrounds most cells of the body; it consists primarily of fibrous proteins embedded in a hydrated polysaccharide gel. This meshwork provides tissues with tensile strength, compartmentalizes tissues, and anchors cells. It also plays a major role during morphogenesis. The composition of the ECM varies from tissue to tissue, but all matrices contain the same types of molecules: collagens, glycoproteins, proteoglycans, and hyaluronic acid.

The synthesis and secretion of ECM components is tightly controlled during development. Many of the matrix components can self-assemble into insoluble structures; various components also bind to one another. In addition, cells help to mold the matrix into its proper geometry. Once formed, the matrix becomes an insoluble, semistable structure somewhat independent of the cells that synthesized it. Cells that come into contact with it can now use it for anchorage. Cells may also receive growth and differentiation signals from the ECM. Perhaps most important for morphogenesis, cells use many ECM components as adhesive substrates on which they move during cell migrations. A delicate balance probably exists between attachment and detachment of cells to the ECM, determining whether a cell will remain stationary or migrate through tissues.

Of the ECM glycoproteins that promote cell adhesion, fibronectin and laminin are the best understood. Both are large, complex glycoprotein complexes made up of several subunits. Sequence analysis of the fibronectin genes and its mRNAs reveals that alternative splicing products can generate more than 200 different dimeric forms of fibronectin, each potentially expressed in a cell- or tissue-specific manner. The fibronectin molecule has many different binding sites for other ECM molecules and for cell-surface receptors. The diversity of alternatively spliced fibronectins suggests that this molecule is expressed with different combinations of binding sites, thus promoting different combinations of interactions.

Fibronectin appears to play a major role in the invagination of a sheet of cells during gastrulation and in the migration of primordial germ cells and cranial neural crest cells during development.

A distinctive feature of laminin is that it has an impressive stimulatory effect on the outgrowth of neuronal growth cones in tissue culture. It will be important to elucidate the structure of the laminin site that mediates the binding of neuronal growth cones and to understand the role that laminin plays in promoting neurite outgrowth in the developing embryo.

The glycoproteins in the ECM exert their effects on cell movement and cell adhesion through the interaction of individual matrix components with cell-surface receptors. The binding site for many of these receptors includes the simple tripeptide arginine-glycine-aspartic acid. Recent results show the existence of a whole family of receptors related to these tripeptide receptors.

Many questions remain for the future. But with the advent of monoclonal antibody and recombinant DNA technologies, the events of cell movement, cell adhesion, and cell recognition during development can finally be studied at the level of molecular mechanisms. With the discovery of a number of major cell and substrate adhesion molecules and their receptors, and with the recent progress in uncovering additional adhesion and recognition systems, we can expect great advances over the next decade in our understanding of what mechanisms control these basic events of morphogenesis and how these events help control the development of tissues and organs throughout the body.

POSITIONAL INFORMATION

Species Differ in the Patterns in Which the Cell Types Are Arranged, Not in Their Cell Types

What distinguishes one group of organisms from another, and indeed one part of an individual organism's body from another, is the way in which cell types are arranged with respect to one another. The mechanisms that operate during development to ensure the correct spatial arrangement of cells, tissues, and organs are included in the term *pattern formation*. Although no unified view of pattern formation yet exists, an understanding of the behavior of embryos and their constituent cells is recognized as necessary to deduce probable mechanisms. Probable mechanisms can then be tested experimentally; if they survive such tests, they can be used to guide the formulation of questions about the molecular nature of patterning mechanisms.

Although the problem of pattern formation lies at the heart of not only developmental biology but also evolutionary biology, it is a late-bloomer compared with other problems in development. The current status of the field has been likened to that of genetics at about the time of Mendel. However, considerable progress has been made in recent years, and advances in understanding pattern formation promise to occupy center stage in developmental biology in the next decade.

Diverse model systems have proven advantageous for investigating the various aspects of pattern formation. There are four major episodes of patterning events.

1. The placement of the cytoplasmic constituents important to subsequent development during oogenesis, most likely in a spatially organized pattern. In response to fertilization, this early pattern is extensively and precisely reorganized in many species.

2. The establishment of the main axes of the body in the multicellular embryo, which has emerged as a consequence of repeated divisions of the egg. In accomplishing this feat, cells meet different fates along both the anterior-posterior and the dorsal-ventral axes of the body. During this process, cells initiate exten-

sive movements to bring previously separated regions of the body into proximity, and patterning information is transferred (induced) between newly opposed sheets of cells.

3. The development of appendages, such as legs and wings, at particular positions along the main body axis.

4. The development of patterned structures at the body surface, such as scales, hairs, and feathers.

Does Cell Lineage Completely Determine the Fate of a Cell?

Perhaps the simplest of the views of animal development is one that considers the egg to have highly detailed information sufficient to specify the features of the adult. This view has found support in such well-studied embryos as those of leeches and nematodes, in which cell lineage is normally invariant during development. Further experimental analyses, however, have cast doubt on the validity of this simple view of things. For example, in *Xenopus* embryos, the pattern of cell lineage is precise enough to enable a detailed fate map of the major body parts to be constructed from early cleavage stages. Nevertheless, experimentally produced variations in this cleavage pattern, while altering the "standard" fate map, have no consequence for the emergence of the final form. In fish embryos, early cleavages seem to yield reproducible patterns of cell lineage, but these bear no relation to the final pattern of the body; during early gastrulation, prior to the establishment of the main body axis, cells from different lineages migrate individually and mix randomly. Even in nematodes and leeches, in which cell lineage under undisturbed conditions is invariant, examples are accumulating that suggest that cell lineage and the determination of cell fate are not obligatorily coupled.

These studies suggest that the environment outside a cell, whether this is other cells or molecules, is important in the emergence of pattern even in situations in which cell lineage predicts cell fate. But only when it is possible to follow the fates of cells isolated from early cleavage stages, either alone or after transplantation to a new site, can the extent to which the environment affects early development be assessed. However, evidence from a variety of embryos suggests the presence in eggs of determinants that become segregated into certain lineages. The best-studied example of such a determinant is the one that specifies the germ cell lineage in a variety of animals.

The emerging view is that localized determinants exist, but that specify key positions within the embryo are fewer than would be required to specify the entire body pattern. Patterning of the cells that lie between the specialized key positions (for example, between the extreme ends of the body) involves interactions, either short- or long-range, between cells. We anticipate that in the next few years, the molecular identity of at least some determinants will be uncovered and will lead in turn to an understanding of how cytoplasmic determinants interact with the genes to determine cell fate. In addition, the existence of precise information about normal cell lineage, made possible by a battery of new techniques for cell marking, will be of great assistance in unraveling the patterning events that do not depend on the inheritance of determinants.

How Can the Environment of the Developing Cell Play a Role in Determining Its Fate?

Development proceeds with increased spatial complexity in the embryo. Most recent studies have concluded that, although there may be some prelocalized information such as RNAs in the egg, the spatial complexity of the egg is fundamentally simple, being confined to localization of information in the anterior and posterior poles of eggs such as frog and flies. As cells divide, the embryo becomes more complex and most of this complexity arises through cell-cell interactions. Evidences that cells influence and transform the fate of their neighbors goes back to early experiments on the induction of the axial organization of frogs. Such inductive interactions have been found in all organisms.

The search for the inducers was a history of frustration until recently. It was long known that dead or heterologous tissues had potent capacity to induce new tissue types. Only recently have known growth factors been tested in the appropriate assays and been shown to be extremely effective in eliciting induction. With sensitive molecular biological techniques it has been possible to examine embryonic tissues, where it has been found that embryos in the earliest stages of development contain both the mRNAs for growth factors and the molecules themselves. In the frog the mRNA for a relative of a known growth factor involved in wound healing has been found localized in the egg in a region where the earliest inductive signals are generated.

Most of the signaling molecules act locally, which is consistent with the behavior of embryos in classic transplantation experiments. In some cases there is good evidence that molecules exist, called morphogens, that act over a long range and provide positional information for tissue organization.

The best candidate thus far for a morphogen in vertebrate systems is retinoic acid. This lipid-soluble compound, derived from vitamin A, can have dramatic effects on cells. For example, low concentrations of retinoic acid cause teratocarcinoma cells in vitro to differentiate into heart muscle cells, whereas high concentrations favor the differentiation of neurons. Most importantly, in a number of different systems, exogenously applied retinoic acid and its analogues seem to affect patterning dramatically, and research is currently directed at determining whether or not retinoic acid acts as a morphogen during normal development. Even if retinoic acid is not an in vivo morphogen, it will be very useful as a probe because once we understand how it alters patterns, we will know more about the patterning process itself.

Local Cell-Cell Interactions May Play a Large Role in Pattern Formation

Patterns may emerge as a result of local cell-cell interactions by a process of intercalation rather than as a result of long-range signaling by morphogens. Intercalation occurs when cells that are normally not adjacent come into contact, either as a result of rearrangements during development or wound healing or as a result of grafting. This contact stimulates cell division, which continues within the system until all the intervening structures are replaced by the proper pattern. In insects, the epithelial cells of the imaginal disk carry out intercalation. In amphibians, connective tissue fibroblasts play this role, as is most clearly seen in the regeneration of lost limbs. After removal of part of the appendage, wound healing brings normally nonadjacent cells into contact, producing a discontinuity in the normally smooth gradation of positional values, which stimulates cell division and intercalation to reduce the discontinuity. The studies on amphibians raise the intriguing possibility that mammals may one day be stimulated to regenerate their limbs if a way can be found to reactivate the developmental programs used for forming limbs in the embryos.

Most of the experimental evidence on pattern formation comes from regenerating systems; it is not yet clear to what extent intercalation may establish and regulate the primary body pattern of animals. Future research is needed to specifically address the issue of cell-cell interactions and intercalations during development of the early embryo.

Research on pattern formation is aimed at understanding how cells, as the units of development, interact with one another and their environment in producing the characteristic patterns of organisms and their parts. Answering this question will require not only understanding which genes are active at what times, but also appreciating what activities the cells are engaged in and how gene activity relates to this. For example, unequivocally identifying a morphogen in the near future would not by itself explain pattern formation, just as knowledge about insulin and its structure have not explained its mode of action. Complete understanding will come from knowledge of the molecules that act as signals in conjunction with knowledge about the responses of cells to these signals.

DEVELOPMENT IS FOR ADULT ANIMALS TOO

The processes of development do not cease with the hatching of an egg or birth of an animal. Metamorphosis in insects and amphibians, limb regeneration, and even the attainment of sexual maturity by mammals during adolescence, are illustrations of development as a process that continues throughout life; development could even be considered to include the controlled phenomena of death.

How Is Cell Number Controlled in Different Tissues of the Adult?

A number of materials that control the activity of partially differentiated stem cells have been discovered. For example, the recently cloned colony-stimulating factors stimulate the production of macrophages and granulocytes, and erythropoietin stimulates the production of red blood cells. Such systems cannot simply be maintained by inductive signals, however, since each type of cell has a characteristic lifetime and concentration in the body. Homeostatic mechanisms must measure such numbers and control the totals of each type of cell present. Neutrophils, for example, have a half-life of about 5 hours. Our bodies contain constant numbers of these cells, averaging about 4×10^{10} per individual. The numbers rise in response to bacterial infection or shock and eventually return to their steady-state levels upon recovery. Something in the individual must be monitoring and controlling these numbers, yet at present we understand little about the process.

Other cells of the hematopoietic system have different lifetimes and inductive signals; they are presumably monitored by other systems. Likewise, nearly every part of the body that is subject to renewal in adult life must have some kind of homeostatic monitoring system, from the cell lining of the intestine, which turns over rapidly, to the liver.

In adult rats, liver cells (called hepatocytes) turn over relatively slowly: they have half-lives of about 7 days. If 90 percent of the liver is removed surgically, however, hepatocyte division is rapidly induced, and the organ is restored to its original size within a few days. The regenerating liver does not significantly overshoot or undershoot its size goal when reconstituting itself; this indicates the existence of accurate mechanisms for measuring the proper size of an adult rat liver and inhibiting hepatocyte proliferation as this size is reached.

Again, little is known about how this mechanism might work. It might monitor blood levels of metabolites handled or produced by the liver. Or, it might measure the size of the liver by means of specialized junctions between cells, called gap junctions. Gap junctions are small channels in cell membranes that connect neighboring cells in various tissues, including the liver. These junctions allow the free diffusion between cells of small molecules; theoretically, the concentrations of such molecules within the liver cell could serve as a measure of the total size of the organ.

What Genetic and Physiological Mechanisms Determine the Life-Span of Cells and Organisms?

A chapter on growth and development would be incomplete without mentioning aging and death. Accidents aside, biological processes control these events, both at the level of the whole organism and at the level of the individual cell. Not only do different species live for different lengths of time, but even within a given species, such as the laboratory mouse, different strains have different life expectancies. Obviously there must be genetic influences on life-span. Some mouse strains are susceptible to diabetes, autoimmune disease, neuromuscular problems, or particular types of cancer; these diseases contribute to shortened lifetimes in these strains. Such differences, however, cannot account for the fact that mice consistently live for shorter times than some related species of rodents, such as rats.

Even within a given organism different cells have vastly different life-spans. Certain nerve cells exist for as long as the individual itself. Other kinds of cells, such as neutrophils and intestinal cells in vertebrates, turn over rapidly. Even individual cells of a particular type may survive for different lengths of time depending on other events. For example, mammalian T lymphocytes are formed in the thymus. Within the thymus, 95 percent of such cells die rapidly, within two or three days of their formation, unless they are selected by the thymus because of a particular specificity of their receptor for antigen. Successfully selected cells are released by the thymus and migrate to other parts of the body, where they become part of the large pool of T lymphocytes in the animal, responsible for fighting off infections. Even there, these cells have a relatively short half-life—less than a week—unless they encounter an antigen to which their receptors can bind. If this happens, the T lymphocyte divides and produces various hormonelike factors (lymphokines), which stimulate the B cells (antibody-producing cells) to divide and help rid the animal of the invading antigen. Once the invader is destroyed some of the progeny of the once-dividing T lymphocyte become "memory cells"; they stop dividing and producing lymphokines, but they survive in the animal more or less indefinitely, with a life-span approximately that of the individual itself. By this means the immune system builds up a pool of long-lived T lymphocytes, which are useful in fighting off the types of infections that its host will encounter during life. A single encounter with an antigen, and burst of cell division, changes the life expectancy of the human T cell, without any further cell division, from less than a week to more than 10 years.

SPECIAL PROBLEMS IN PLANT DEVELOPMENT

Plants and Animals Share Many Structures and Developmental Mechanisms, But There Are Some Major Differences

The first and most obvious difference between plant and animal development is that plant development is usually repetitive in nature and indefinitely long. A tip of a maple twig will put out pairs of leaves all season long and then again the following year. Roots are less periodic in structure, but they too grow and branch indefinitely. The plant body is basically the accumulation of the products of its past developmental activity. Many of the cells remain alive. By design, however, many do not, resulting in the accumulation of wood and bark.

Occupying ever more volume and intercepting ever more light is obviously a key strategy of high adaptive significance for plants. This strategy is generally implemented by having the tip regions elongate in length indefinitely, a process called primary growth. Subterminal regions stop elongating and, in perennials, increase in girth at their periphery, while producing dead but functional wood cells toward the interior of the stem or root. The well-known rings in cross-sections of woody stems that grow in areas with a seasonal climate are the result of this secondary growth. It is not too far off the mark to characterize plant development as continued, or repeated, embryology.

A second special feature of plants is that the cells do not move relative to one another. Even the male gamete lacks flagella in most groups of contemporary plants and is borne to the vicinity of the egg within a pollen tube. In plant embryology, there is no phenomenon comparable to the sudden contact of a group of cells with a new cellular environment, as occurs in animal gastrulation. Nonetheless, plants exhibit diverse cell types and complex morphogenesis.

A third major difference between development in plants and animals is that in plants the germline is not distinct. Cells in many different parts of the plant—as in flowers on many branches—may undergo meiosis. The products of meiosis are not gametes, as they are in animals, but rather haploid spores, which divide mitotically, forming a haploid phase in the life cycle called the gametophyte. In many ferns and bryophytes, the gametophytes are green, photosynthetic, and free-living, whereas in seed plants—flowering plants and gymnosperms—they are highly reduced, enclosed within and completely dependent nutritionally on the sporophytes on which they are borne.

Describing the Sequential Details of Development and Experimentally
Modifying This Sequence Are the Major Approaches to Studying
Plant Development

Important data on plant development have been obtained by using clonal analysis. Here x-rays induce visible heritable changes in individual cells, tagging them and all their progeny through further development. These clones are ideal for cell lineage studies, and they can reveal how many cells are involved in the formation of a given plant organ, such as a leaf. The intriguing result is that the number is never 1. The initiation of an organ is a "group donation" of 10 to 20 cells. Tissue character can be independent of cell lineage: epidermal cells normally divide as a coherent surface sheet; however, when an epidermal cell occasionally divides and contributes a cell to the interior of the leaf, the cell differentiates as an interior cell. Other studies have shown how the cytoskeleton changes in relation to cell differentiation and to the initiation of organs.

Through in situ hybridization of nucleic acids it is possible to find out which cells make specific transcripts and when. Spatially and temporally defined patterns of transcription have been found, among other places, in the interaction between pollen and stigma surfaces that precedes pollen-tube growth and fertilization in the flowering plants. This phenomenon is discussed in more detail in Chapter 11.

Developmental mutations offer the prospect of revealing key steps in the developmental chain, where the presence or absence of a single kind of protein determines a major change in the course of development. Mutations of this sort include those which convert the normally complex, bilateral form of a snapdragon flower to a radially symmetrical one, like that of a morning glory. Another set of well-defined mutations converts a compound pea leaf into a set of tendrils or makes the leaf into a set of round stalks. Thus single genes can profoundly affect organ character.

Other mutations put the right organ in the wrong place. Certain mutants in *Arabidopsis* have extra sets of stamens, have petals in place of stamens, or exhibit other deviations from the normal condition. These are homeotic mutants, with well-known equivalents in *Drosophila* and other invertebrates. Other mutants disrupt the timing of developmental events.

Light effects on plants, beyond photosynthetic effects, have been well known since the time of Darwin. Such effects, along with the hormone activity that underlies them, are discussed in Chapter 11. Plant scientists have succeeded in identifying many key control points along the causal chain from genome to a full-grown plant. In general, the nature of the agent with an effect—for example, mutation, light absorption, hormone structure—is well understood; the nature of the responding system, on which the agents act, is not. Some of the central research opportunities in the field of plant development center on supplying this missing information.

Plant Cell Growth

The Plant Cell Stands as a Key Intermediate Unit in the Sequence from Gene to Phenotype

On the one hand, the genome produces a cell with a repertoire of physiological and developmental activities; on the other, it is the integration of these activities over time and space that ultimately produces the roots, shoots, and flowers that constitute the mature, reproductive plant. The remaining parts of this section will concentrate on the development of the plant cell. This subject has a strong biophysical component, because for the cell to grow, the cell wall, a strong structure, must yield to high pressure—approximately six times that in a pressure cooker. Most of the controls mentioned above have their ultimate effects through some modification of the biophysics of the plant cell wall. Thus, a portrayal of this subject is a convenient format for the illustration of the unique features of plant development.

Plant Growth and Morphogenesis Are Dominated by the Plant Cell Wall

Plant cells cannot move appreciably relative to each other; to cover distance they must grow across it. This they do in impressive fashion: more than 100 meters in the height of a redwood tree, scores of kilometers of root length in a typical prairie grass. Much of plant development, and thus many issues in agriculture and forestry, center on how fast and in what direction this growth occurs. A blade of grass, an orchid flower, or a ponderosa pine tree all achieve their configurations by growth and division of walled cells.

The two features of major significance in plant cell growth are rate and direction. These two features are controlled, for the most part, by the two main structural components of the cell wall. A strong fibrous component is made of cellulose microfibrils, the orientation of which controls direction. The fibrils are embedded in the second component, a gellike matrix of other carbohydrate and some protein; the properties of this matrix control rate. The structure is thus like that of the hull of a fiberglass boat. The fibers provide great strength; the matrix distributes stresses equitably so that they do not concentrate at any point to break the fibers. Capitalizing on this principle, plants have come to physically dominate much of the earth's surface.

Control of Growth Rate Is Influenced by Hormones

It has been known for more than 50 years that a hormone, made in a localized region of a plant, can promote elongation in other parts. The compounds auxin and gibberellic acid play major roles in this process, and their structures, synthetic pathways, and modes of degradation are relatively well known. The investigation of the precise ways in which they affect growth, however, is one of the central problems of plant development.

Several major features of this process are already clear. The wall extends in response to turgor pressure inside the cell. The cell is able to use osmosis to inflate itself to high pressure with water. The three potential physical controls on the rate of expansion are the turgor pressure itself, the yielding properties of the wall, and the ability of water to enter the cell rapidly. Two of the three possibilities have largely been eliminated. There is no support for the idea that plant growth hormones increase turgor directly, and the rate of entry of water is clearly not limiting. In view of these relations, the hormones must be able to cause the wall to stretch or yield.

How do the hormones "soften" the wall? The simplest conception of the process views the wall matrix as viscous, like tar or taffy; in such a model, the hormone would simply act by reducing wall viscosity. Such a simple physical model cannot apply in any direct way, however, because the expansion process requires continuous metabolism: For example, inhibitors of oxidative metabolism stop the growth process as soon as they arrive at the growing site. The conclusion is that plant growth hormone action is complex, either coupled to the synthesis of new compounds or to the activation of special metabolism.

Plant growth, which depends on the growth of cells, is a self-stabilizing and therefore complex process. If the wall-softening processes get out of hand, the cell will burst and die. If synthesis, which must in the long run compensate for the stretching of the wall, is excessive, the wall will be too thick ever to elongate again. One can thus look for a control circuit with both physical and biochemical components. The full clarification of this circuit constitutes one of the major research opportunities in plant growth and development, with important implications for progress in agriculture and forestry.

New methodology promises to make this easier. A remarkable "pressure probe" device enables one to measure turgor pressure continuously in the growing cell. This allows the intimate study of metabolic action on the wall's physical properties because changes in wall properties bring about changes in pressure. Understanding the detailed mechanism by which hormones stimulate rate of growth and the way certain wavelengths of light inhibit it are major prerequisites to the understanding of plant growth.

Control of Direction of Growth Depends on the Direction of Cellulose Synthesis

Control of the shape of plant cells can be achieved by the highly localized control of growth rate, as is the case for some unicellular structures such as root hairs and pollen tubes. The cells in multicellular plant organs, such as roots, stems, and leaves, however, grow throughout their length and hence change shape by a different method. These cells are directionally reinforced, the cellulose microfibrils in their walls lying transverse to the cylindrical axis of the cell. The cell resembles a barrel made of hoops, with the body of the cell extending through the center of the hoops. Without such directional reinforcement, turgor pressure would swell the cells into spheres. Such a pattern can be explained only by cellulose synthesis in a particular direction (as opposed to random synthesis), and this directionality in turn determines the shape of a plant. The shoot attains its height and the root its depth through the directed extension and periodic division of transversely reinforced cells.

The control of direction of cellulose synthesis is achieved by means of cytoplasmic microtubules that lie just inside the plasma membrane of the cells. Any depolymerization or disruption of the alignment of these microtubules will randomize cellulose alignment. Understanding this relation has made possible an important distinction between cellulose synthesis as such, which continues during disruptive treatments, and the control of the direction of this synthesis, which is attributable to alignments in the cytoskeleton of the cell around which the synthesis is taking place. Progress has also been made in visualizing the synthesis of cellulose. Parallel strands of cellulose polymer seem to coalesce into microfibrils as they emerge from rosettes of protein molecules in the plasma membrane. The involvement of such complex structures at the site of production may explain why such cellulose production has not yet been achieved in vitro. A current major research opportunity is to understand the nature of the coupling between microtubule direction and the control of the direction of the spinning out of new microfibrils. This connection is important in understanding the geometry of plant bodies.

Microtubule orientation, which is central to cell wall formation, is also involved in certain plant hormone responses. For example, ethylene, a gas that helps fruit ripen and acts as a plant hormone, causes many organs to swell. Such swelling involves a rotation of cellulose synthesis in cell walls from the transverse to the longitudinal, thus bringing on a corresponding rotation in the direction of growth of the cells. Whereas ethylene causes cells to become less markedly cylindrical, gibberellin, enhances their cylindrical form, apparently by improving cellulose microfibril alignment. When gibberellin is active, the microtubules are more numerous and better aligned. The connection between hormone presence and microtubule influence on the pattern of cell-wall growth is a key factor in understanding plant development and one that has come into prominence only recently.

The influence of hormones on microtubule orientation and hence on cell form and their subsequent influence on growth rate are of considerable practical importance for agriculture and forestry. For example, one class of weed killers contains growth hormone analogues that disrupt relative growth rates so that undesirable plants will die; another class of weed killers destroys weed seedlings by interfering with their directional growth. In the future, further mastery of these processes, combined with the techniques of genetic engineering, will have the potential to produce crops of improved form, thus enhancing productivity directly.

The Development of Plant Organs

Plant Meristems May Be Viewed as Developmental Engines

As the study of the extension of the cell wall dominates studies of plant development at the cellular level, so the study of cyclic or continuous morphogenesis in well-defined meristems dominates the study of whole-plant development. Meristems are zones of continuous cell division located at the tips of stems and roots in plants. For example, the shoot tip returns repeatedly to the same configuration while continuously producing leaves and additional stem. Such a tip differs from all mechanical analogues since it continually incorporates new material in its products by forming new cells. Three characteristics—persistent activity, the production of new organ structure, and ever-renewed cell composition—must be combined in any coherent theory of meristem development.

The shift to flowering occurs when the vegetative meristem (producing a consistent leaf pattern) of a plant becomes an embryonic inflorescence. Within these embryonic flowers, the internodes are greatly compressed. Unlike the consistency seen with leaf production, the floral organs change character in successive rounds of meristematic activity. From a central mound of tissue come sepals, petals, stamens, and carpels—the members of the four whorls that make up a complete flower. The ways in which the interaction of light and hormones bring about this conversion are explored in Chapter 11; their cellular details remain poorly understood. For example, why do stamens arise at certain places on the meristem? The application of various techniques that have recently become available promises to shed light on these processes in the near future.

6

The Nervous System and Behavior

The Objective of Modern Neuroscience Is to Understand How the Nerve Cells of the Brain Direct Behavior

Many central issues with which neurosciences is concerned, such as how we perceive the world around us, how we learn from experience, how we remember, how we direct our movements, and how we communicate with each other, have commanded the attention of thoughtful men and women for centuries. But it was not until after World War II that neuroscience began to emerge as a separate and increasingly vigorous scientific discipline that has as its ultimate objective providing a satisfactory account of animal (including human) behavior in biological terms. This ambitious goal has as its basis the central realization that all behavior is, in the last analysis, a reflection of the function of the nervous system. It is the organized and coordinated activity of the nervous system that ultimately manifests itself in the behavior of the organism. The challenge to neuroscience then, is to explain, in physical and chemical terms, how the nervous system marshalls its signaling units to direct behavior.

The real magnitude of this challenge can perhaps be best judged by considering the structural and functional complexity of the human brain and the bewildering complexity of human behavior. The human brain is thought to be composed of about a hundred billion (10^{11}) nerve cells and about 10 to 50 times that number of supporting elements or glial cells. Some nerve cells have relatively few connections with other neurons or with such effector organs as muscles or glands, but the great majority receive connections from thousands of other cells and may themselves connect with several hundred other neurons. This means that at a fairly conservative estimate the total number of functional connections (known as synapses) within the human brain is on the order of a hundred trillion (10^{14}). But what is most important is that these connections are not random or indiscriminate:

They constitute the essential "wiring" of the nervous system on which the extraordinarily precise functioning of the brain depends. We owe to the great neuroanatomists of the last century, and especially to Ramón y Cajal, the brilliant insight that cells with basically similar properties are able to produce very different actions because they are connected to each other and to the sensory receptors and effector organs of the body in different ways. One major objective of modern neuroscience is therefore to unravel the patterns of connections within the nervous system—in a word, to map the brain.

A second, related objective is to identify the differences that exist between nerve cells. For although nerve cells have a number of properties in common—especially their abilities to respond to signals from other cells and to conduct signals along their processes—on morphological grounds alone, thousands of different classes of nerve cells are evident. The morphological differences were the first to be recognized once techniques had been developed that reveal the form of individual neurons. Some cells were found to have only a single process, others just two processes, and still others—including the overwhelming majority of neurons in the brains of vertebrates—have several, often scores, of processes. In most cases we can recognize a single process, the axon, that serves to conduct information—usually in the form of all-or-none signals known as action potentials or nerve impulses—to other cells. Variable numbers of receptor processes or dendrites receive information from other cells, integrate it, and relay it to the nerve cell body and beyond it to the axon.

But it is not only in the morphology of the processes that nerve cells differ. We now know that dozens of different classes of neurons can be recognized on the basis of the chemical messengers or neurotransmitters that they use to communicate with other cells. The discovery in the early 1950s that almost all nerve cells communicate with each other through the release of chemical neurotransmitters at specialized sites along the course and at the ends of their axons was one of the major events that marked the beginning of modern neuroscience. Only in the past decade, however, have we come to realize that there may be not just a handful of chemical transmitters as was once thought, but perhaps a hundred or more, and that it is the subtle and distinctive actions of these transmitters that account for much of the functional complexity of the nervous system. We have also come to realize that neurotransmitters can act upon other cells only if the cells have the necessary receptors to selectively bind the neurotransmitter. The interaction of the neurotransmitter with its appropriate receptor is what initiates the response of the target cell. Again, it is only in the past few years that we have come to appreciate that the target cells can respond in several different ways depending on the nature of the transmitter, the types of receptors involved, and the mechanisms that the transmitter-receptor interaction activates. In some cases the response of the target cells is rapid and transitory, with a time course of just a few thousandths of a second; in other cases the cell responds over a fairly long period—perhaps many seconds; and in certain situations the behavior of the target cell may be modified for many hours, or even days and weeks. A third major task of neuroscience, therefore, is to understand how nerve cells generate signals, often over long distances, and how these signals change the various target cells with which the neurons are in functional contact.

The cellular and molecular mechanisms involved in nerve signaling and synaptic transmission are currently among the most intensively studied and best understood aspects of neuroscience. Less well understood, but no less important, are the longer term changes in nerve cells that must underlie the acquisition and storage of information that we commonly refer to as learning and memory. Although there is a vast body of literature on human learning and memory and on the effects of damage to various parts of the brain on its ability to acquire, store, and retrieve information, it is only relatively recently that the longer term effects of synaptic activity that must be involved in these processes have begun to be studied at the cellular and molecular levels. The first insights that we have gained into these processes suggest that a wide variety of behavioral phenomena may well prove to be explicable on the basis of just a few general mechanisms such as the covalent modification of particular molecules involved in nerve signaling or the activation of specific genes and the synthesis of new proteins.

A fourth major objective of neuroscience is to account for the unusual cell biology of neurons. Although nerve cells share many properties with other cells, their special roles in the transduction of sensory information, in the transmission of signals over considerable distances, in being able to respond to signals from other cells, and, in turn, in being able to modify the activity of their target cells imposes on neurons a number of highly specialized functions. These considerations raise a number of intriguing questions including (1) how the enormous phenotypic diversity seen among nerve cells is generated, (2) how different parts of each neuron become specialized to either receive or transmit signals, (3) how nerve cells are able to maintain such lengthy processes given that the genetic information is confined to the cell nucleus and most of the synthetic machinery is confined to the relatively small cell body, (4) how communication is maintained between the nerve cell body and its various processes, and (5) what changes occur in the cells in response to "experience" and aging. The fact that most neurons have to survive and continue to function effectively throughout the life of the organism—for 70 or more years in the case of neurons in the human brain—is one of their most impressive characteristics. Recent developments in molecular and cell biology are beginning to influence the study of these phenomena, and there is every reason to be confident that they will soon be as well understood as the mechanisms involved in impulse conduction and synaptic transmission.

Undoubtedly the greatest challenge to contemporary and future neuroscience is to understand what might be referred to as the "information-processing" capacity of the brain—to determine how the various systems within the brain are organized and function to direct and mediate such behavioral phenomena as sensory perception, language function, motor actions, emotion, cognition, and thought. Again, although we know from clinical neurology and pathology that the destruction of certain areas of the brain seriously impairs or effectively abolishes these capacities, how these higher functions are normally carried out remains largely unknown. That it has been possible to produce machines that can duplicate some aspects of these higher brain functions has suggested that developments in computer science and especially artificial intelligence may inform our understanding of how the brain functions in much the same way as molecular and cellular neuroscience have been informed by concurrent developments in molecular and cell biology. But this remains to be seen, and for the present the single greatest challenge to neuroscience is to elucidate how the brain works.

It Is Important to Use Both Reductionist and Synthetic Strategies for Studying the Nervous System

The strategies that neuroscientists have adopted for studying the nervous system have varied over the years as new techniques and methods have been developed. But from the beginning they have all been based on a few general premises. Among the more important of these have been the following: (1) Animal behavior reflects the activity of the nervous system; (2) no matter how simple or complex the nervous system or the behavior, the essential units—the neurons—are alike in most significant respects; and (3) because of the structural and functional similarity in neurons, it is important to select the neuronal system or the neurons that are most advantageous for study, regardless of where they are found. Before the introduction of micropipettes as recording electrodes, the giant axons that control the mantle musculature of the squid were the objects of choice in the study of the physico-chemical mechanism of nerve signaling: They were several centimeters long and up to 0.5 millimeters in diameter, they could be easily dissected out, they remained viable in a recording chamber for many hours, and long, insulated wire electrodes could be inserted for some distance down their length so the potential difference across the axonal membrane (axolemma) at rest and during the conduction of an impulse could be measured.

Later when it became important to study the electrical events that occur in both the pre- and postsynaptic elements during synaptic transmission, an unusual "giant" synapse in the squid nervous system proved to be invaluable. Over the years the most useful preparation for studies of synaptic transmission in vertebrates has been the neuromuscular junction, because it is readily identifiable and easy to handle, and because it is relatively easy to record from the postsynaptic cells. Similarly, the modified muscle cells that compose the electric organs of certain fish have provided the richest source of the receptor for the neurotransmitter released at nerve-muscle junctions. The availability of the acetylcholine (ACh) receptor in such large amounts made possible, first the isolation and biochemical purification and characterization of the four subunits of which the receptor is composed, and later the cloning of the genes for each subunit.

The earliest studies of behavior were largely descriptive, but in the hands of the neurophysiologists and ethologists they provided a wealth of information about reflex (including conditioned reflex) and "instinctive" behavior, some of which has since been analyzed electrophysiologically. Because of the difficulty of analyzing behavior in most complex organisms, considerable attention has been paid to the analysis of more simple, rigorously definable behaviors in simple organisms. For example, much of what we know about the way in which motor activity is programmed has come from the analysis of locomotor behavior in crayfish and leeches. More recently, scientists have taken advantage of certain large, readily identifiable neurons in the sea snail *Aplysia* to analyze the short- and long-term changes that occur in neuronal structural and function in certain well-defined behaviors including an especially useful model for nonassociative and associative learning. The *Aplysia* nervous system has also proved useful for studying gene expression in specific neurons and for determining the neuropeptides that mediate even complex behavior such as egg-laying.

In an attempt to work out the complete organization of a nervous system in an animal which is readily amenable to genetic manipulation, the nematode *Caenorhabditis elegans*, which has just over 300 neurons, has been studied. The lineage of each of these neurons has been determined (by direct inspection), specific lineage mutants have been identified, and the complete wiring pattern of the nervous system has been reconstructed from serial electron micrographs.

Of even greater use for genetic studies is the nervous system of *Drosophila*. Mutations that affect different parts of the nervous system have been known for many years, but it is only in the past 15 years that a major effort has been made to identify behavioral mutants, to analyze the development of the nervous system (including the eye), and to clone the genes for a number of interesting neuronal proteins such as rhodopsin and one of the potassium (K^+) channels. The power of this approach, which has already yielded so much in other systems, holds great promise, especially for understanding the molecular mechanisms involved in neural development.

While we cannot emphasize too strongly the importance of this search for simple systems in which to analyze specific aspects of neuronal functions, in the last analysis the greatest interest lies in understanding the functions of the human brain. Until recently the opportunities for doing this were limited. Clinical neurologists led the way by analyzing brain function resulting from localized or more general brain pathologies. Much of what we know about the organization of the human brain has come from studies of this kind, and when patients with specific brain lesions (such as an interruption of the corpus callosum that unites the two cerebral hemispheres, or localized damage to the speech areas or to those concerned with memory processing) have been carefully studied by neuropsychologists, they have revealed aspects of brain function that could not have been determined in any other way.

In the past decade a number of noninvasive methods have been developed for studying the human brain. These include computerized axial tomography (CAT scanning), positron emission tomography (PET scanning), magnetic resonance imaging (MRI), event-related potential recordings from the scalp with computerized averaging techniques, and magnetoencephalography. The full impact of these technologies has yet to be felt, but it is already clear that they will enable us to study many aspects of brain function in the intact human brain that hitherto could be analyzed only in the brains of experimental animals. For the study of those distinctive aspects of human behavior, such as speech, they should prove invaluable.

Useful as these noninvasive methods are proving to be, at present they suffer from severe limitations in spatial or temporal resolution (or both). They usually provide information only about the summed activity of larger numbers of neurons rather than about the functions of individual neurons. For studies of this kind, we must still turn to experiments on animals. The need for such experiments will continue, for, despite efforts to find alternatives, the only hope we have in the forseeable future of understanding the organized activity of the brain is by directly studying the brain itself. Simulations and computer modeling of brain functions are no substitute for direct observation.

One of the most promising developments in this regard has been the perfecting of techniques for recording the activity of individual neurons in conscious, behaving primates. This approach, first introduced about 15 years ago, has become increasingly popular for studies of sensory perception and motor control. It has been possible in several instances to train the experimental animal to carry out a psychophysical task for which comparable human performance can be measured; investigators are now beginning to collect substantial data on the activities of neurons in parts of the brain (including the so-called association areas, which had hitherto defied analysis) that are likely to be directly applicable to human brain function. An extension of this approach that permits simultaneous multiple recordings from many neurons is one of the promising recent developments in neuroscience.

Nerve Cells Are the Signaling Units of the Brain

As we have seen, almost all nerve cells have at least three or four main parts: (1) a cell body that contains the nucleus and most of the cell's biosynthetic machinery; (2) a number of relatively short processes, called dendrites, which extend from the cell body and provide the largest receptive surface for inputs to the cells; (3) an axon, which usually extends for some distance from the cell body and is used for long-range signaling; and (4) specialized regions, commonly at the end of axons, called synaptic boutons or synaptic endings, where communication with other nerve cells or special effector tissues (such as gland or muscle cells) is carried out.

The best way to understand how these various components of a neuron work is to consider them in the context of a simple behavior—for example, the reflex withdrawal of a hand that touches a very hot object. Contact with the hot object activates a group of sensory receptors in the skin that respond to heat and causes them to fire a burst of all-or-none signals called action potentials. These action potentials propagate along the length of the sensory neurons, past the cell bodies, and to the axons, which extend into the spinal cord. At the ends of the axons the action potentials cause a chemical transmitter to be released. The chemical transmitter released at the ends of several axons interacts with receptors on the surfaces of the dendrites of certain spinal cord neurons giving rise to an activating signal called an excitatory postsynaptic potential. If the excitatory potentials elicited by impulses in the sensory axons are of sufficient amplitude, they trigger a nerve impulse, or a group of impulses, in the spinal cord cells. These, in turn, through their axons, activate a group of motor or effector nerve cells. The axons of the motor cells extend out from the spinal cord to the muscles in the forearm and hand, where again a chemical transmitter is released at the nerve-muscle junction. The binding of the transmitter to the appropriate receptor in the muscle causes a brief change in the surface membrane of the muscle cells that leads the muscles to contract and the hand to withdraw. Concurrent with the excitation of neurons in the spinal cord that activates motor neurons, some of the branches of the sensory axons contact yet other spinal neurons that, when activated, inhibit the activity of the motoneurons that normally cause the forearm and hand to extend. The activation of yet other neurons in the spinal cord leads to the propagation of information about the sensory stimulus (its location, nature, and intensity) to higher levels within the nervous system. These lead, among other things, to the conscious perception that the hand has been in touch with a hot object, and if the stimulus is severe enough, a generalized arousal of the individual that focuses attention on the stimulus and its behavioral significance. If the hand is jerked back with sufficient vigor, there may also be a number of reflex adjustments within the spinal cord to maintain the subject's balance and posture.

This simple example serves to make several general points about the nervous system and its role in behavior. (1) Most behavior occurs in response to an external sensory stimulus of some kind; (2) sensory signals must be transduced into nerve signals; (3) nerve impulses travel along specific pathways to defined areas of the central nervous system; (4) nerve cells communicate with each other through specialized junctional zones known as synapses; (5) synaptic transmission can be either excitatory or inhibitory depending on the chemical neurotransmitter involved; (6) most behavior manifests itself in the form of overt motor actions; and (7) many sensory stimuli are also consciously perceived as a result of the transmission of information to higher brain centers including the central cortex, and this perception may result in conscious arousal and the focusing of attention on the stimulus and its behavioral consequences.

Considerable progress has been made in recent years in our understanding of all of these steps: sensory transduction, the nature of the nerve impulse, synaptic transmission, the anatomical pathways involved in a variety of sensory and motor mechanisms, and how these pathways are assembled during development. In the sections that follow we shall try to summarize what is known about these issues and to point out the directions in which future work seems to be headed.

NERVE CELL COMMUNICATION

Nerve Cells Communicate by Electrical and Chemical Signals

To understand how neurons and synapses work, we need to understand how a nerve impulse or action potential in a presynaptic neuron causes the release of a chemical neurotransmitter at the synapses formed by its axon. But first we must focus on the ionic currents that produce the action potential in the presynaptic neuron and on the way these currents interact with the structures in the terminal parts of the axon to bring about the release of the transmitter.

The use of certain naturally occurring neurotoxins that bind specifically to the sodium channel has made possible a preliminary molecular characterization of the sodium channels in muscle and brain. The channel is a large glycoprotein with a molecular weight of 270,000 whose amino acid sequence has been determined from the corresponding complementary DNAs. The amino acid sequence has, in turn, suggested several ideas about the function and evolution of the different segments of the channel protein. For example, the molecule contains four similar sequences (homologous internal repeats), each about 150 amino acid residues long: These repeat sequences suggest that the channel may have evolved from a single ancestral DNA segment that was duplicated within the gene three times. By examining the distribution of specific amino acids within the entire peptide, it is possible to identify candidate domains concerned with various functional properties. In particular, each internal repeat has five long hydrophobic areas that probably represent the transmembrane domain and a charged segment that is thought to serve in the gating process (that is, the opening and closing of the channel) and in the selectivity of the channel for sodium. The channel probably responds to changes in membrane potential by undergoing a conformational change and a masking of the positive charges that bound its pore. The techniques of site-directed mutagenesis to modify specific sites in the channel protein should make it possible to test these ideas directly.

The analysis of the voltage-gated sodium channel has brought to light two features we shall encounter again when we consider the acetylcholine receptor. First, several stretches of hydrophobic amino acids seem to correspond to transmembrane alpha helices. Second, the channel is symmetrical, consisting of similar subunits arranged in the plane of the membrane around a central aqueous pore. These early findings encourage us to think that all membrane channels are

---

RESTING AND ACTION POTENTIALS

At rest, all nerve cells have an electrical potential across their plasma membranes of about 50 millivolts (mV), the inside of the cell being negative with respect to the outside (Figure 6-1). An action potential is an all-or-none signal that not only reduces but actually reverses the membrane potential, moving it a total of about 100 mV from its resting value of -50 mV to +50 mV. The action potential is generated by the movement of ions through two types of intrinsic membrane proteins called "voltage-gated" channels: those for sodium and those for potassium. Each of these channels is closed at rest and opens in an all-or-none fashion as a result of a reduction in the membrane potential below a critical value. When open, each channel permits the rapid movement of ions through the membrane, giving rise to pulses of current of variable duration but constant amplitude. Sodium, being a positively charged ion which, at rest, is at higher concentration outside the neuron, moves rapidly into the cell from the outside, bringing in positive charge; this movement causes the rising or depolarizing phase of the action potential. In turn, potassium ions carry positive charge out of the cell; this movement is responsible for the declining or repolarizing phase of the action potential.

---

likely to share certain common structural features. Once these are understood, it may turn out that all membrane channels work in much the same way. But at present, the conceptual gap between the primary structure of the channel proteins and their function is too large to allow us to make this prediction with any degree of confidence.

Synaptic Transmission: The Nerve-Muscle Junction as a Prototypical Example

The nerve impulse is essentially a form of electrical signaling, with the wave of ionic currents sweeping down the surface of the axon at speeds in the range of 1 to 100 meters per second. Communication between different neurons and between neurons and other cells is chemical in most instances—the release from the nerve endings of a small amount of a specialized neurotransmitter that diffuses across the space separating the two cells. The binding of the transmitter to receptor molecules in the membrane of the postsynaptic cell gives rise, in turn, to a new class of signals called synaptic potentials. Thus, whereas the action potential is a purely electrical signal, the synaptic potential is an electrical signal initiated by a chemical one. In the past two decades a large number of such

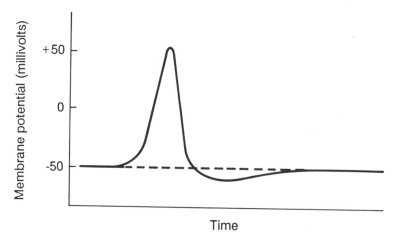

FIGURE 6-1   Resting and action potentials.

chemical transmitters haves been identified: They range from small molecules such as ACh, glutamate, noradrenaline, and serotonin to much larger molecules including a rapidly growing number of peptides.

Over the past 30 years, each of the steps involved in synaptic transmission has been characterized in considerable detail, primarily through the use of intracellular microelectrodes and thin-section electron microscopy. The pace of research on synaptic biology has increased rapidly in recent years since the introduction of rapid freeze-fracture electron microscopy, patch-clamp microelectrode techniques, and the application of modern methods of protein chemistry, recombinant DNA probes, and monoclonal antibodies to the isolation and characterization of the molecular mechanisms involved.

Because of its ease of access and because it was the first site at which chemical transmission was identified, the neuromuscular junction—between the motor axon terminals and muscle cells—has been the most intensively studied synapse and illustrates the major features of synaptic transmission.

When an action potential invades a motor nerve terminal, it releases the transmitter ACh after an irreducible delay of about 0.5 to 1.0 milliseconds. The transmitter then diffuses across the 50-nanometer synaptic cleft between the nerve terminal and the muscle cell in a matter of about 200 msec before binding to ACh receptors in the junctional region of the muscle membrane. The interaction of the transmitter with the receptor leads to a conformational change in the receptor and the opening of its channel. This is followed by an influx of sodium ions that depolarizes the postsynaptic membrane. In contrast to the all-or-none nature of action potentials, the depolarization of the muscle membrane—the end-plate potential—is a local response proportional to the amount of ACh released. Under normal ionic conditions the end-plate potential is usually large enough to trigger an action potential that is then conducted away from the end-plate region along the surface of the muscle cell. The action of the synaptically released ACh is brief: The end-plate current decays within 1.0 to 2.0 msec. The duration of action of the released ACh is limited by its breakdown by an extremely active enzyme, acetylcholinesterase, that is concentrated in the synaptic cleft.

The discovery in the early 1960s of α-bungarotoxin, an 8,000-dalton peptide that binds specifically and with high affinity to ACh receptors in skeletal muscle, provided a crucial tool for biochemical studies of ACh receptors. The receptor, first purified from electric organs of the rays *Torpedo california* and *Torpedo marmorata* (uniquely rich sources that are embryonically related to muscle cells) is a 275,000-dalton glycoprotein made up of four subunits. Complementary DNA cloning techniques have revealed that each subunit contains four hydrophobic regions that presumably span the lipid bilayer. A fifth, amphipathic domain, located between the third and fourth hydrophobic regions, may also span the membrane.

The ACh receptor has led the way because of the relative ease of its purification from receptor-rich electric organs, but complementary DNAs that code for other receptors or ion channels (ligand-gated and voltage-gated) have now been cloned and sequenced. It has been known since 1914 that ACh activates two broad classes of receptors. Nicotinic ACh receptors of the sort we have so far considered are present at motor end plates, at synapses within autonomic ganglia, and at a few synapses within the central nervous system (CNS). Muscarinic ACh receptors are found in various autonomic effectors including smooth muscle, cardiac muscle, and exocrine glands. Most ACh receptors within the brain are muscarinic. This receptor has now been cloned, as has the β-adrenergic receptor.

Relatively Little Is Known About the Molecular Details of Transmitter Release at the Neuromuscular Junction

In addition to the two major currents involved in the action potential (the sodium and potassium currents), a third, minor current is particularly important at the presynaptic terminals of the synapse: the calcium current. The calcium current is small, only about 1/100 of the sodium or the potassium current, and therefore it does not usually contribute importantly to the action potential per se. Rather, it serves as a messenger carrying into the cell information that is necessary for release of the chemical transmitter.

The function of calcium in this context can best be understood if we shift our attention from the receptors on the postsynaptic cell—the muscle—to the presynaptic terminals of the motor neuron. The chemical transmitter is released from the axon terminals, not as isolated molecules, but in packets containing about 5,000 molecules of ACh. Enclosed in small subcellular organelles called synaptic vesicles, these packets of ACh are released from the expanded terminals of the axon by an exocytotic process, in which vesicles fuse with the external membrane of the synaptic terminal and extrude their contents into the synaptic cleft. Exocytosis occurs only at certain points, called active zones, within the axon terminal. At these active zones, a gridlike array of "dense projections" provides a cluster of docking sites for the vesicles. Calcium is thought to be essential for fusion of the synaptic vesicles with the external membrane of the axon, and hence its entry into the axon terminal is a necessary prerequisite for the exocytotic release of the transmitter.

One of the best-characterized proteins associated with synaptic vesicles is synapsin I, a phosphoprotein with a globular head and an elongated, collagenlike tail. Synapsin I seems to be involved in transmitter release, although it remains to be determined whether it regulates vesicle-membrane fusion (directly or indirectly) or if it acts on another process such as vesicle mobilization.

There Are a Variety of Neurotransmitters

The number of putative neurotransmitters and neuromodulators of synaptic function has increased tremendously in the past decade, and many more probably remain to be identified. In addition to ACh, catecholamines (norepinephrine, dopamine), indolamines (serotonin), and amino acids (glutamate, γ-aminobutyric acid or GABA, and glycine), attention has been directed recently to purines and to a wide variety of peptides, several of which were first identified and characterized in other tissues as possible local or more distally acting hormones. Although many of these peptides have been shown to excite or inhibit the firing of nearby neurons at relatively low concentrations, and while their selective distribution within the nervous system argues strongly that they may function as neurotransmitters, at present the great majority of neuroactive peptides must still be regarded as transmitter candidates. Neuropeptides tend to alter the membrane properties of target neurons on a slower time scale, causing them to respond differently to other ongoing synaptic drives; for this reason the term neuromodulator may be more appropriate than neurotransmitter.

Yet another distinction between peptides and more conventional transmitters is the manner in which they are synthesized. Most conventional transmitters are synthesized by enzymes located in nerve terminals; peptides, on the other hand, are synthesized as parts of larger polyproteins on polyribosomes in the cell body. The polyprotein may contain several active peptides or multiple copies of the same peptide: To cite just one example, pro-opiomelanocortin gives rise to adrenocorticotropic hormone (ACTH), β-endorphin, melanocyte-stimulating hormone, and another small peptide, all of which are cleaved from the parent protein by appropriate enzymes. Undoubtedly one of the greatest challenges to neurobiology is to determine the role of the more than 50 neuropeptides that have been identified and of the 100 or more that we believe are yet to be discovered.

Some Synaptic Receptors Activate Ion Channels by Means of
Second Messengers

Ionophoric receptors and those for ACh function as ion channels and mediate fast synaptic transmission. Depending on which ions pass through the membrane, the effect on the postsynaptic cell will be excitatory or inhibitory. Whatever the sign, the effect is brief (usually just a few milliseconds) because it depends on a change in the conformation of the receptor protein molecules. In contrast to ionophoric receptors, other receptors are coupled to ion channels by guanosine triphosphate proteins that bind second messengers, such as cyclic adenosine monophosphate (cyclic AMP), cyclic guanosine monophosphate (cylic GMP), calcium, diacylglycerol, or arachidonic acid. Second messengers typically diffuse within the cell to deliver the information from the receptor to various target proteins, modifying these proteins' activities and thereby modulating the physiological responses of that cell.

The information that a second messenger delivers can be amplified to alter both the magnitude and the duration of the signal. At their briefest, these changes persist for as long as the store of receptor-stimulated second messengers lasts— usually minutes to an hour. However, the activation of one second-messenger pathway may lead to changes in others, and cumulatively these changes can be prolonged for hours. If the cascade of receptor-stimulated reactions alters gene expression (for example, inducing the synthesis of new protein molecules within the stimulated cell), the effects may endure for days, weeks, or even months. All second messengers studied so far operate by modifying the activity of proteins within the stimulated cells. Although most of the modifications studied are produced ultimately through the activation of one or another enzyme, some second messengers (like cyclic GMP in photoreceptor cells of the retina) may act directly on their target molecules without the intervention of other enzymes. To date, the best-studied neuronal second messengers are calcium, cyclic AMP, and the products of phosphoinositol turnover that are released from the plasma membrane by receptor-mediated activation of the enzyme phospholipase C.

THE CYTOSKELETON AND THE TRANSPORT OF MATERIALS WITHIN NERVE CELLS

Among animal cells, neurons are distinguished by the number, length, and variety of their processes. Since some of the processes may be extremely long (the axons of motor cells in the spinal cord that innervate muscles in the human foot extend a meter or more), and since all the genetic and most of the biosynthetic machinery of the cell is confined to the cell body, an elaborate transport system has been evolved to support and maintain neuronal processes. Although the same transport system delivers material to the dendrites, it is usually referred to as axonal transport.

The materials transported along neuronal processes move with different velocities over the range of about 0.5 to 500 millimeters per day. The more slowly moving materials are associated with the transport of cytoskeletal components—microtubules, intermediate filaments, and so forth—while the more rapidly moving are membrane and synaptic components.

Among the key issues involved in axonal transport are: (1) What materials are transported at each rate? (2) What is the nature of the transport mechanism? (3) What role does axonal transport play during the growth of neural processes and during normal neuronal function?

Slow Axonal Transport

Several Components of Slow Axonal Transport Have Been Defined, and It Is Likely That They Will Be Understood in Terms of Molecular and Physical Mechanisms Within the Next 5 Years

Slow axonal transport moves materials at a rate of between 0.5 and 5 mm/day and consists primarily of cytoskeletal proteins transported from the cell body to the end of neural processes. In mature neurons, slow axonal transport maintains the nerve cell processes, whereas in developing axons the rate of process extension corresponds fairly closely to the rate of slow axonal transport. The finding in axons treated with imino-dipropionitrile that the intermediate filaments separate from the microtubules to form a peripheral cylinder suggests the presence of at least two motile complexes, which is consistent with an earlier characterization of axoplasmic transport. In order to directly measure the microtubule movement in slow axonal transport, fluorescently labeled tubulin has been followed as it migrates down a process after being microinjected into the perikaryon. Similarly, specific inhibitors of axonal transport can be injected into the cell, and their distribution, as well as their possible modes of inhibition, can be determined directly.

The newer technologies of microinjection, quantitative fluorescence video microscopy, and light and electron microscopic labeling with antibody probes have to a large extent replaced earlier experimental approaches that depended on monitoring bulk movement of radioactivity down axons or the damming back of materials after nerve ligation. The recent cloning of many genes for the cytoskeletal proteins opens the possibility of manipulating the molecular architecture of specific proteins and determining the subsequent alteration in function in vivo. Few examples exist at present, but improvements in expression systems should stimulate many such studies. Most of the clones have been used to follow the amounts of messenger RNA (mRNA) present during development or differentiation. For example, nerve growth factor (NGF) stimulation of certain responsive cells sets in motion the synthesis of new mRNA for the *tau* proteins that bind to microtubules and stabilize them. In this case switching cells from an undifferentiated state to a characteristic neuronal phenotype involves the coordinate modulation of the levels of a number of different proteins within those cells.

Rapid Axonal Transport

Recent Studies of Rapid Axonal Transport Exemplify How Current Technologies Can Be Used to Understand the Cell Biology of Neurons

A broad spectrum of materials are transported along neuronal processes with velocities in the range of 20 to 500 mm/day. Included in these materials are several synaptic components including some of the synaptic vesicle membrane proteins and a number of transmitter-synthesizing enzymes.

The general problem is how materials move from one point to another within cells and, in the case of neurons, how material can be rapidly moved from the perikaryon to the distal ends of processes. The application of modern video technologies to light microscopy has allowed investigators to follow the movements of transport vesicles within axons, and by modifying the buffer conditions, in dissociated axoplasm. Single filaments were seen to support the bidirectional transport of vesicles. Sequential light- and electron-microscopic examination of the transport filaments revealed that they were single microtubules; single microtubules could also move on glass in the presence of a soluble protein fraction from axoplasm. Through the use of microtubule movement as an assay of the "motor protein," a new microtubule-based molecule, kinesin, was purified. Considerable evidence now exists that the motor does indeed drive vesicle movements from the cell body to the synapse in neurons. A cytoplasmic dynein (an enzyme that converts chemical energy into movement in association with microtubules) powers transport in the opposite direction, from the synapse to the cell body. The major questions of how vesicles are programmed to move in the retrograde or anterograde direction will also be addressed.

Although several rapidly transported proteins have been identified and characterized, much remains to be done before we can completely account for the to-and-fro movements of the materials within nerve cells. However, it is already clear from studies of NGF that the retrograde movement of trophic molecules provides an essential link between events occurring in the vicinity of axon terminals and the regulation of genomic function in the cell nucleus.

DEVELOPMENT OF THE NERVOUS SYSTEM

Despite a Century of Research on Development, Many Important Questions Remain Unanswered

As developmental biology itself matured, the development of the nervous system became a subject of interest to a generation of embryologists who were fascinated by the complex sequence of morphogenetic events that give rise to the brain and spinal cord and lead progressively to its elaboration and refinement. More recently, interest has been directed toward elucidating some of the developmental anomalies that affect the nervous system, including a wide spectrum of genetic disorders whose consequences are frequently devastating to the affected individual.

The sorts of questions that have been addressed and that continue to attract the most attention are, What embryonic tissue gives rise to the nervous system? Where are the billions of nerve cells and tens of millions of glial cells generated? How do the cells reach their definitive locations, and how do they selectively aggregate with other cells of like kind? What molecular events underlie the differentiation of neurons and confer upon them their distinctive morphological, biochemical, and physiological features? How do axons find their way through the developing nervous system and finally identify the appropriate target structures with which to establish connections? To what extent is the development of the nervous system genetically determined, to what extent are epigenetic factors involved, and to what degree is the immature nervous system capable of responding to environmental factors?

Considerable progress towards answering these questions has come from the identification of suitable model systems and from the recognition that, despite some differences in their patterns of development, vertebrates and invertebrates are remarkably similar. The main advantage of the invertebrate nervous systems that have been studied is that they tend to be rather stereotyped and often contain relatively small numbers of neurons. And, it is often possible to follow the fate of individual cells throughout their life history, and in some cases (*Drosophila* is a good example) to perform a variety of genetic manipulations that affect the nervous system. Against these advantages is the fact that development in most invertebrates seems to be rigidly programmed and shows little of the plasticity so characteristic of vertebrate development. Needless to say, among vertebrates, the greatest interest centers on understanding the development of the human brain, but as yet we have hardly progressed beyond the descriptive level.

The development of the vertebrate nervous system consists of a number of interrelated steps beginning with the phenomenon of neural induction.

Factors Affecting Neural Induction Need to Be Determined

The emerging nervous system of vertebrates first appears as a thickening of the ectoderm (the outermost layer of the embryo) in the dorsal midline. This thickened region, the neural plate, arises in response to the inductive influence of the underlying notochord and mesoderm, which during the process of gastrulation have invaginated from a region called Hensen's node or the dorsal lip of the blastopore and extend forward toward the future head-end of the embryo. Despite considerable effort, mainly in the period between the two World Wars, the nature of this inductive influence is still poorly understood. In large part this is because the amounts of tissue available for study are severely limited and because, until relatively recently, our knowledge of gene activation (which must underlie this process) was rudimentary. The availability of modern molecular genetic techniques should make it possible in the near future to identify the factors involved in neural induction and to isolate and characterize them.

Coincident with the induction of the neural plate, the tissue along its margins is induced to form the presumptive neural crest. The neural crest itself is a transitory structure that is first recognizable as a longitudinal band on the dorsal surface of the neural tube. Almost immediately, the cells of the crest become widely dispersed, migrating along predetermined pathways to the skin, gut, head, and so forth, where they give rise to a remarkable number of different tissues. The enormous phenotypic diversity of the derivatives of the neural crest has made it a subject of special interest in recent years. Among other tissues it is known to give rise to nearly all pigmented cells, much of the mesenchyme and the skeletal components of the head and face, and certain of the endocrine glands, as well as to most peripheral sensory neurons, the neurons and supporting cells of the autonomic ganglia, and the Schwann cells of peripheral nerves. From a variety of ingenious experiments in amphibian and chick embryos it has become clear that the precursor cells in the neural crest are pluripotent, and the fate of their progeny is largely determined by the environment through which the the cells migrate and the regions in which they finally come to reside. At least some of the cells can change their phenotype relatively late in life given the right conditions. For example, cultured sympathetic neurons that normally synthesize only the neurotransmitter noradrenalin can, in time, begin to synthesize and release acetylcholine.

Proliferation of Both Neuronal and Glial-Cell Precursors Occurs in a Highly Programmed Manner

Neuronal proliferation in the CNS occurs for the most part within the ventricular lining of the original neural tube or its later derivatives in distinct spatiotemporal patterns. In a few regions, secondary proliferative foci are set up in specialized areas referred to collectively as the subventricular zone. Unfortunately, despite a good deal of effort, we know comparatively little about the factors that regulate the patterned cell divisions in the nervous system, but it is these complex patterns that ultimately determine the numbers of neurons and glia found in different regions of the brain and establishes the initial size of each neuronal population.

The technique of autoradiography with tritium-labeled thymidine has made it possible to establish, for a large number of structures, the times at which their constituent neurons lose their capacity for DNA synthesis and the orderly sequence in which different neuronal types appear.

One of the major limitations for our understanding of the factors regulating cell proliferation in the CNS is the absence of suitable markers for the stem cells of different neuronal lineages. Attempts to generate such markers were until recently largely unsuccessful, but in the past four or five years a number of monoclonal and polyclonal antibodies against the major classes of central and peripheral glial cells have become available. The introduction of identifiable genomic sequences into transgenic or chimeric animals should rapidly transform this area and lead to the sorts of insights that have been so valuable in the study of neuronal lineages in the invertebrate nervous system (especially in the nematode *C. elegans*).

The Overwhelming Majority of Nerve Cells Must Undergo at Least One Major Phase of Migration in Order to Reach Their Definitive Locations

The withdrawal of neurons from the cell cycle appears to be the trigger for their outward migration from the ventricular or subventricular zones. In most parts of the CNS the initial migration of neurons is more or less radial with respect to the ventricular lining of the neural tube, and it seems to occur mainly along the surfaces of radially oriented processes of glial cells whose bodies lie within the ventricular zone.

Perhaps, not unexpectedly, in the course of migration some neurons become misdirected and end up in ectopic loci. If the ectopic cells are able to make the appropriate connections they are able to survive; if they do not they are usually eliminated by cell death.

Selective Cell Aggregation

Neuronal Aggregation Involves Specific Cell-Adhesion Molecules

It has been known for almost 50 years that dissociated embryonic cells, if artificially mixed together, can sort themselves out in a tissue-specific manner, ectodermal cells associating with ectodermal cells, mesodermal with mesodermal, and so on. Recently, developmental neurobiologists have taken advantage of this phenomenon to explore the molecular basis for the selective associations of neurons during development. A number of cell-surface molecules that seem to mediate such cell-cell interactions have also been identified and the genes that encode them have been cloned.

One such cell-adhesion molecule (or CAM) referred to as N-CAM because it is predominantly expressed in neural tissue, shows homophilic binding (two molecules on different cells stick to each other) has been referred to in Chapter 5. It is widely distributed on the surfaces of all neurons and occurs on certain nonneuronal cells as well. During development it undergoes a characteristic embryonic-to-adult modification with an increase in binding affinity. A second, termed NG-CAM because of a presumed role in neuron-glial interactions exhibits heterophilic binding. Recent immunocytochemical studies have established that NG-CAM is preferentially expressed on growing axons. Antibodies to N-CAM selectively prevent nerve cell aggregation in vitro and also perturb normal neurite fasciculation. The widespread distribution of both molecules at key stages in the development of the brain and spinal cord suggests that they may each play a critical role in several of the morphogenetic events being considered here. However, there is at present no conclusive evidence that they are specifically involved in the selective aggregation of neurons to form the various nuclear groups and cortical layers that characterize the CNS or the various ganglia of the peripheral nervous system. However, rapid progress has been made in the past decade on these and several other related cell-cell and cell-substrate adhesion molecules, and this continues to be an area of promise for future investigation.

Most Neural Cells Seem to Undergo Their Major Differentiation Only After Reaching Their Final Locations

When cells migrate away from the ventricular zone, they display many of the differentiated features characteristic of neurons or glia, but their major phase of differentiation usually occurs only after they have reached their final destinations. Formally, one can recognize three aspects to this phase of neuronal differentiation.

First, the cells acquire a distinctive morphology, usually characterized by the development of several dendrites and a single axon. Second, the cells acquire a number of distinctive membrane properties. These properties generally do not all emerge simultaneously, but appear over a period of time.

The third aspect of neuronal differentiation is associated with adoption of a particular mode of synaptic transmission. In most neurons, the cells synthesize one or more neurotransmitters or neuromodulators and generate all the necessary cellular machinery for their transport to the axon terminals and for their exocytotic release. Simultaneously, the cells express a variety of receptor molecules that become inserted into the appropriate postsynaptic sites on their own surfaces.

Recently, researchers have focused on the cloning of the genes for (1) peptide transmitters, (2) enzymes involved in the synthesis of the more conventional neurotransmitters, and (3) receptor molecules involved in synaptic transmission. This work holds great promise that the regulation of these molecular aspects of neuronal differentiation will soon be well understood.

Although Much Work Has Been Done on Axonal Outgrowth, the Matter Is Still Far from Being Resolved

The central issue in developmental neurobiology is, "How are the highly specific patterns of connections that characterize the mature nervous system generated?" The general question encompasses three separate issues. The first concerns how the cells acquire individual "addresses" that define their position in the three-dimensional neuronal complex within which they lie. The second concerns the expression of this acquired "positional information" in the outgrowth of the cells' axons and the identification of their appropriate targets. The final question is how the cells identify both the region in which they should terminate and the appropriate subset of neurons on which to form synapses.

Several lines of evidence suggest that most cells acquire their address either at the time they are first generated or when they first assemble with their fellows to constitute the primordium of the definitive neuronal population. The nature of the addressing mechanism, however, is unknown. By contrast, the mechanism of axonal outgrowth is reasonably well understood. Axons are extensions of the cell that grow by the addition of new materials at their expanded ends, referred to as growth cones. These growth cones are highly motile structures that bear large numbers of delicate fingerlike processes known as filopodia; the filopodia are thought to play a key role both in the recognition of the axon's appropriate course and in the identification of its appropriate target. Most axons do not grow in isolation but in association with other axons from the same neuronal population, and it is becoming clear that they can use a variety of strategies to find their way. These include selective axon-axon interactions, selective axon-substrate adhesivities, the identification of key landmark structures, chemical tropisms, and even simple mechanical factors. What is most impressive is that even radical experimental perturbations of a group of axons (such as deliberately forcing them to grow into an abnormal pathway) rarely succeed in preventing them from reaching their predestined targets.

A different set of factors seems to be involved in the identification of the desired target. This issue has proved to be the most difficult to study. The most widely accepted view (the chemoaffinity hypothesis) was first put forward to account for the uncanny ability of regenerating axons in the mammalian peripheral nervous system and in the CNS of fish and amphibians to "home in" on their targets and to reestablish orderly connections with their original targets. According to this hypothesis, each small group of neighboring neurons requires a distinctive cytochemical label that is also expressed on the surfaces of the growing axons; the presence of matching or complementary labels on the target cells enables the axons to recognize and form synapses with their appropriate partners. A good deal of experimental evidence is consonant with this hypothesis (and none yet contradicts it), but the nature of the proposed molecular labels has so far escaped identification.

The final event in the establishment of the initial pattern of connections is the formation of synaptic contacts between the related populations of cells. To date, this problem has been most carefully studied at the neuromuscular junction. Here functional contacts can be established extremely quickly—within minutes, in fact, of the axon's contacting the muscle cell. The assembly of the entire complement of presynaptic and postsynaptic components does not occur for some time, however, and must involve a complex set of inductive interactions between the axon terminal and its target cell.

Because growing axons have to make their way through a veritable jungle of other neuronal and glial processes, it is perhaps not surprising that some of them enter an incorrect pathway or grow to an inappropriate target. Most of these aberrant or erroneous connections are eliminated during the next two phases of neurogenesis, which are concerned with the progressive refinement of the initial pattern of connections.

Nerve Cell Death During Neural Development Allows for the Fine-Tuning of the Nervous System

It has been known for more than 80 years that some nerve cells die during normal neural development, but it was not until the late 1940s that the full significance of such cell deaths came to be appreciated. We now know that, in almost every part of the nervous system, neurons are initially overproduced; at some later period, between 15 and 85 percent of the initial population degenerate. In a few situations it has been possible to establish that the phase of cell death is temporaily related to the period during which connections are being established within the target fields. And the finding that the number of neurons that finally survives is closely related to the size of the target field has led to the suggestion that the axons of the cells compete with each other for some entity (probably a trophic agent) that is normally available in the target area in only limited amounts. The cells that are successful in this competition survive; those that are unsuccessful die.

To date the only well-characterized trophic agent is NGF, which is essential for the survival of many sensory neurons, sympathetic ganglion cells, and central cholinergic neurons. If an excess of NGF is made available to the axons of embryonic sensory ganglion neurons, the normally occurring death of 40 to 50 percent of the cells can be completely prevented. A vigorous search for other neuronal growth factors is under way; a number have been identified and partially characterized, and at least two (fibroblast growth factor and epidermal growth factor) are known to be present in the CNS and capable of maintaining dissociated neurons in culture.

Naturally occurring cell death seems to serve at least three functions. (1) It matches the sizes of individual neuronal populations to each other and to the functional requirements of their targets. (2) It allows for the elimination of developmental errors, especially errors in connectivity. (3) It defines the limits of particular cellular lineages.

A Second Regressive Phenomenon During Neural Development Selectively Eliminates Excess Connections

Just as more neurons are generated than seem to be needed, there is generally an initial excess in the number of connections formed by each neuron; over a period of some days or weeks these excessive connections are progressively eliminated until the mature number is reached. For example, muscle cells and neurons that normally receive only one or a few more inputs frequently receive many synapses early in development and progressively lose the supernumerary contacts over a period of three or four weeks. In other cases, longer collateral pathways are selectively eliminated, and whole fiber systems come to be reorganized. For example, all areas of the cerebral cortex initially send axon collaterals to the opposite hemisphere through the corpus collosum. Later, the callosal projection becomes restricted to certain functionally defined zones, not through the death of the cells that initially project to the contralateral side, but rather through the selective loss of callosal collaterals.

The mechanism underlying the selective elimination of particular axonal branches is far from clear, but may well also involve trophic factors. Certainly in the case of NGF, both the responsive cells and each of their processes require a continuous supply of the factor for their maintenance. Electrical activity also seems critical for the maintenance of individual axonal branches. Blocking activity with a drug such as tetrodotoxin (which blocks sodium channels) completely prevents the refinement of connections in several developing (and regenerating) neuronal systems. This is of considerable importance, for the early excess of neuronal connectivity and the critical dependence of connections on the maintenance of appropriate patterns of activity must provide the substratum for much of the plasticity of the immature nervous system.

A Rapidly Expanding Body of Literature Indicates how Widespread Plasticity Is in the Nervous System

To the age-old question of whether the development of the brain is shaped more by nature or by nurture, the answer is now clear: Both are important at different times and in different ways. The early development of the nervous system—including all the events that lead to the establishment of the initial patterns of connections—seems to be determined largely by genetic and locally acting epigenetic factors. But once the initial neural framework has been laid down, environmental factors become increasingly important. In several instances we now recognize what are referred to as critical periods during which the relevant neural systems seem to be particularly susceptible to external environmental influences. The final form of the brain and its functional capacities are shaped by the interplay of intrinsic (genetic and epigenetic) and extrinsic (environmental) influences.

Consider, for example, the importance of hormonal influences on the development of the brain and, in particular, the sexual determination of those parts of the hypothalamus that control the pattern of release of the gonadotrophic hormones from the anterior pituitary gland. Although an animal's sex is genetically determined, the regulation of hypothalamic function is determined largely by the levels of circulating sex hormones during a critical period in development. If these hormonal levels are significantly perturbed, the pattern of gonadotrophic hormone release can be completely reversed from the male to the female pattern, or vice versa. Animals experimentally subjected to this type of sex reversal display all the behavioral characteristics normally associated with the opposite sex.

Even more attention has been focused on the capacity of the major sensory systems to respond to normal or altered sensory experience. The most striking findings have come from work on the visual system, in which, during a critical period, any marked abnormality in the animals's visual experience can result in an irrevocable alteration in its later visual behavior.

For example, if the upper and lower eyelids of one eye are sutured closed throughout this critical period and then the eye is reopened, the animal subsequently behaves as if it were blind in the deprived eye. When the visual cortex is explored neurophysiologically, the cells that would normally be activated by stimulating that eye are found to be either silent or dominated largely by inputs from the nondeprived eye. And when the cortex is examined anatomically, the regions in which the inputs from the deprived eye terminate are found to be substantially reduced in size, whereas those related to the nondeprived eye are correspondingly expanded.

Equally striking are the findings in experiments in which animals have been reared in a structured visual environment. Kittens exposed throughout the first few weeks of life to only vertical stripes subsequently have great difficulty when they encounter horizontally oriented objects. Physiological recordings made from their brains show that the majority of the cells in the visual cortex respond only to vertically oriented stimuli. Experiments of this kind have already had a profound influence on the way in which human clinical problems such as strabismus are treated and have emphasized the importance of a rich environmental experience during early childhood.

Of special interest for developmental psychologists is the enormous capacity of the brain to adapt to early injury, which to a large extent must depend on its morphological plasticity. Nowhere is this better demonstrated than in the brains of young children who have suffered substantial damage to one cerebral hemisphere. Even after complete removal of the language areas of the cerebral cortex in the normally dominant left cerebral hemisphere, children can learn to speak perfectly normally, language function having been taken over by the corresponding areas in the opposite hemisphere.

The Ontogeny of Behavior Has Been a Neglected Area of Neuroscience

Until recently, developmental neurobiology has been concerned largely with the initial assembly of the nervous system and to a distressing extent has neglected the equally important issue of the emergence of behavior. Inasmuch as the goal of neuroscience is to provide a sound scientific basis for an understanding of all aspects of behavior, including such higher brain functions as thought, memory, perception, and feeling, the study of the ontogeny of behavior is likely to become increasingly important in the near future.

One serious obstacle to progress so far has been the difficulty of identifying good animal models for experimental study. As a result, the problems that have been analyzed—such as the emergence of motor behavior in salamanders and chicks or the development of visual behavior in kittens and young monkeys— have yielded results that are suggestive but hardly definitive. In contrast to most other fields of neuroscience, in this area human studies have led and continue to lead the way, even though by their nature they are usually constrained to observation and description rather than experimental manipulation.

NEURAL PLASTICITY AND ELEMENTARY FORMS OF LEARNING

Learning Is a Major Vehicle for Behavioral Adaptation and a Major Force for Social Progress

Learning is the process by which we acquire new knowledge about the world, and memory is the process by which we retain that knowledge. Most of what we know about the world we have learned, and, in a larger sense, learning goes beyond the individual to the transmission of culture from generation to generation. Learning is thus the major vehicle for behavioral adaptation and also the major force for social change. The ability to acquire and retain new information (that is, to learn and to remember) characterizes all moderately complex animals, and their capacities for learning and memory correlate well with the complexity of their nervous systems. In human beings, these capacities have resulted in a completely new kind of evolution—the evolution of culture.

How does learning occur? What changes occur in the brain when behavior is modified as a result of experience? In a prescient lecture in 1894, Ramón y Cajal proposed that learning might produce prolonged changes in the effectiveness of the synaptic connections between nerve cells and that these changes could serve as the mechanism for memory. More recent research suggested that sensory stimuli could produce two types of changes in nerve cells and their connections: (1) invariant and transient excitability change and (2) a facultative but enduring plastic change.

To determine if synapses undergo plastic changes and to relate these changes to learning, it has been necessary to develop suitable cellular techniques and appropriate experimental systems. Throughout the past decade, significant progress has been made along both lines. During the next decade, it should be possible to extend the analysis of mechanisms more fully to the molecular level and to apply similar strategies to more complex forms of learning.

The Most Instructive Forms of Learning Have Been the Simple Forms
Commonly Referred to as Habituation, Sensitization, and
Classical Conditioning

Habituation, the simplest behavioral modification, occurs in all animals. It is a process in which an animal learns, through repeated exposure, that the consequences of a weak stimulus are neither noxious nor rewarding, and so can be safely ignored. By not responding to stimuli that have lost their novelty or meaning, animals free their attention for stimuli that are rewarding or significant for survival. When Pavlov first described behavioral habituation, he assumed (on the basis of indirect evidence) that it would involve active inhibition. However, in those systems where it has been possible to analyze the synaptic mechanisms of habituation directly, neither presynaptic nor postsynaptic inhibition seems to be involved. Rather, habituation is produced by a simple depression of chemical transmitter released from excitatory synaptic connections. In certain cases, this depression has been related to a reduction in calcium ion influx, but second-messenger systems (Chapter 4) may also be implicated.

Behaviorally, sensitization is the opposite of habituation: It results in a heightened responsiveness after exposure to novel or noxious stimuli. Often a single strong noxious stimulus applied to an animal will enhance its defensive reflexes for periods of up to several hours, and repeated stimulation may lead to reflex enhancement that persists for days or even weeks. Thus, whereas habituation causes an animal to ignore innocuous or trivial stimuli, training that leads to sensitization causes it to attend to the relevant stimuli because they may prove to be either painful or dangerous.

Short-term sensitization has been analyzed in detail in two invertebrate reflex systems, in which presynaptic facilitation is initiated by modulatory transmitters acting through two second-messenger systems: cyclic AMP–dependent protein phosphorylation and calcium ion concentration. The presynaptic facilitation involves a change in the potassium channel that broadens the action potential and increases calcium influx. In addition, an altered intracellular handling of calcium leads to more effective mobilization of transmitter vesicles.

Habituation and sensitization are nonassociative forms of learning—the animal learns about the property of a single event or stimulus. By contrast, classical and instrumental conditioning are associative forms of learning in which the animal learns about the relationship between two events. In classical conditioning, an initially weak or ineffective conditioned stimulus becomes highly effective in producing a behavioral response after it has been paired with a strong unconditioned stimulus. When an animal has been conditioned, it has learned that the CS predicts the US and, by inference, it learns about cause-and-effect relationships in the environment.

In each invertebrate system in which the cellular basis of conditioning has been carefully analyzed, biogenic amines or peptides have been found to serve as the chemical signal for the unconditioned stimulus. Through second messengers, potassium channel function is modulated and the effect of the conditioned stimulus is enhanced. In these cases the effects of the second message are pleiotropic: Many proteins are covalently modified in parallel, and the effects last from minutes to hours.

Similar conclusions have emerged from experiments on learning using a completely different, mutational, approach in *Drosophila*. In *Drosophila*, single-gene learning mutants have now been isolated that are defective in biogenic amine synthesis or in one or another step in the cyclic AMP signaling systems. Thus, the dopadecarboxylase mutant blocks synthesis of the monoamine transmitters dopamine and serotonin, the *turnip* mutation affects calcium stimulation of the adenylcyclase activity, and *dunce* alters or eliminates a cyclic AMP phosphodiesterase isoenzyme. These *Drosophila* mutants, which cannot be classically conditioned, also fail to exhibit sensitization.

Long-Term Memory Involves Genes and Proteins Not Utilized for Short-Term Memory

A variety of studies in animals and humans indicate that newly formed memories are easily disrupted for a short period of time after being formed. Consistent with this view is the well-established clinical observation that head trauma or epileptic seizures produce amnesia for events just preceding the trauma or seizure; memory for earlier events is not affected. Recently, animal experiments with inhibitors of protein synthesis have established that although short-term memory does not require the synthesis of new proteins, memories lasting days or weeks are disrupted by the inhibition of protein synthesis. Particularly important is the finding that long-term memory is most sensitive to disruption during and immediately after training: No deficit in long-term memory is observed if exposure to the protein synthesis inhibitor is delayed by as little as 1 hour after training.

Studies of long-term sensitization in invertebrates have demonstrated that a narrow time window during which new proteins must be synthesized reflects a fundamental property in the storage of information by specific neurons and their connections. With the tools of modern molecular biology, it should now be possible to identify the genes and proteins necessary for long-term memory in simple animals and to use these genes and proteins as probes to explore long-term memory in higher forms.

We Are Beginning to Understand the Mechanisms Underlying Various Simple Forms of Learning and the Short-Term Memory to Which They Give Rise

One of the major findings to emerge from this research is that short-term memory involves second-messenger systems similar to those used for other cellular processes. What is distinctive to nerve cells is the range of receptors that initiate the second-messenger signaling and, even more, the substrate proteins that act as effectors. It should be possible, therefore, to build a bridge between the study of simplified learning and behavior and more general studies of cell biology. The finding that long-term memory is likely to involve alterations in gene expression opens the parallel opportunity to bridge learning and molecular biology. This should make it possible to identify the genes and gene products important for the acquisition and retention of long-term memory in any system and to use those identified in simple systems for probing more complex systems.

NEUROBIOLOGY OF PERCEPTION: VISION

Vision Research Is in a Period of Rapid Progress

In vision research, exciting developments have occurred at many levels of organization ranging from the molecular to the behavioral and a wide range of fundamental questions are being attacked, with an extensive arsenal of techniques. We will therefore use vision as an example to illustrate the principles emerging in the analysis of the mechanisms of perception of all types. This section tries to capture some of the highlights of recent progress in studies of the organization and function of the retina and visual cortex.

The Photoreceptors of the Retina Transduce Light into Electrical Signals

At the photoreceptor level, the most striking advances have been the cloning and sequencing of the opsin genes in several species (including humans) and the elucidation of the transduction process. The successful cloning of the opsins reflects the high degree of conservation in these genes and the relative ease of progressing from one set of genes to another on the basis of sequence homology. The human opsin story is particularly intriguing because it has definitely established the molecular basis of the various forms of color blindness. Our understanding of transduction has been greatly facilitated by applications of patch clamping and related biophysical techniques to the photoreceptors. These methods have provided strong evidence that the elusive second messenger in the transduction sequence is the cyclic nucleotide called cyclic guanosine monophosphate (cyclic GMP). Many critical steps in the overall cascade remain to be worked out in detail and analyzed quantitatively, but the next few years will see rapid progress toward a thorough understanding of the exquisite sensitivity, dynamic range, and other remarkable properties of both vertebrate and invertebrate photoreceptors.

Between the photoreceptors and the output stage of the retina are several classes of interneurons whose functional properties and chemical transmitters are well known because of intracellular labeling methods and immunocytochemistry. The actual output of the retina is carried by distinct classes of retinal ganglion cells that transmit qualitatively different types of information (for example, "on" and "off" channels, color channels, and black-and-white channels) to the brain. Substantial progress has been made in identifying these channels in different species and in delineating the underlying circuitry responsible for their characteristic features. Neural modeling techniques, combined with precise anatomical and physiological data, are beginning to contribute to the analysis of how specific receptive field properties are generated and how the first step in visual perception is organized.

The axons of the visual ganglion cells form the optic nerve and conduct visual information from the eye to a number of subcortical visual centers in the brain, some of which are concerned with various visual reflexes, the control of eye movements, and the regulation of longer-term light-induced hormonal changes, including those responsible for circadian and other behavioral rhythms.

Of greatest interest for visual perception is the relay of information to the primary visual cortex through the lateral geniculate nucleus, in which the different types of ganglion cell axons are spatially segregated.

Visual Areas Occupy More Than Half of the Cerebral Cortex in Some Species and Consist of Two Major Subdivisions, the Striate and the Extrastriate Cortex

The primary visual (or striate) cortex is the largest and best-understood of all cortical areas. It is surrounded by a mosaic of about 20 other visual areas, collectively termed extrastriate cortex, whose organization and function have only recently begun to be understood.

Striate Cortex. The richness and diversity of neural response properties in striate cortex is remarkable. A single neuron can signal detailed information about more than 10 different stimulus features, including orientation, color, depth, motion, and texture. Current evidence suggests that several "functional streams" emanate from the striate cortex, each of which conveys information related specifically to form, color, or motion. The recent success in using optical techniques to monitor activity patterns across the surface of the cortex will provide a powerful tool in the further dissection of these streams.

Despite much new information about cortical circuitry obtained with physiological, anatomical, and pharmacological approaches, we do not know the specific wiring diagram responsible for even a single property of cortical neurons (for example, selectivity for orientation or binocular disparity). The difficultly of resolving these issues may largely reflect the fundamental complexity of cortical circuitry. Several lines of evidence suggest that cerebral cortex in general and visual cortex in particular operate much more as a richly interconnected, highly distributed neural network than has generally been appreciated. If so, neural modeling and other computational approaches will ultimately play an important role in the elucidation of cortical function. And we can look forward to even stronger interactions between experimentalists and theoreticians interested in "computational vision" in the near future.

Extrastriate Cortex. In monkeys and cats, which have been intensively studied, evidence exists for as many as 20 extrastriate visual areas. Obtaining a precise count of the total number of areas along with reliable criteria for their identification remains a formidable challenge. Monoclonal antibodies and other brain-specific markers may provide powerful tools for achieving this goal.

The availability of a variety of sensitive anatomical tracing techniques has led to the identification of a plethora of distinct visual pathways. Nearly 100 cortico-cortical connections alone have been identified in monkeys; with the complexity of the other cortical afferent and efferent connections and the intrinsic connections within each area, it is evident that computer data bases will soon be needed to handle these vast amounts of information. Equally important is the need to elucidate organizational principles that reflect the strategies used for distributing information among various cortical areas. Current evidence supports the notion that visual areas are arranged in such a way that each principally subserves one (or at most a few) specific functions such as motion or color perception. Thus, in the monkey, area MT is concerned mainly with motion-related information, while area V4 is concerned with information about color and form. It is proving difficult to determine exactly how these extrastriate areas process and transform visual messages, as opposed to simply relaying information already extracted in the striate cortex. However, several recent studies have provided intriguing hints about higher-order analysis of motion, form, texture, and color information. More rapid progress on this front can be anticipated as a result of stronger interactions between neurophysiology, psychophysics, and computational science.

Other technical advances will also have major impacts. Positron emission tomography can be used to map functionally distinct regions of the human visual cortex, which will play an important role in linking human psychophysical studies to animal models. Neural recordings with multiple electrode arrays offer the prospect of understanding how neural ensembles process information whose representation at the level of single cells is not explicit. However, fundamentally new concepts in data analysis will be necessary if we are to learn how to interpret such experiments.

The Development and Plasticity of the Visual System Raise Many Intriguing Questions

Until now the major focus has been on the striate cortex, where much has been learned about the interplay of innate instructions and environmental influences in generating the intricate architecture of the normal adult cortex. New principles may emerge from studies of development and plasticity in the extrastriate visual areas, which so far have received comparatively little attention in terms of their development. Again, new techniques, ranging from optical monitoring of activity to area- or cell-class-specific monoclonal antibodies, will greatly enhance the scope of questions accessible to experimental analysis.

Behavioral Studies of Vision in Animals

The effort involved in training animals to make visual discrimination is great. The most impressive contributions of behavioral analyses of visual performance have come in areas in which the neural substrate can be more or less directly linked to the behavior, such as the identification of neurons in the striate and immediate prestriate areas of the cortex that respond only when the images on the two retinas are displaced to an appropriate extent.

The Studies of the Control of Eye Movements and Visual Attention Have Borrowed the Attentional Paradigms of Experimental Psychology and Applied Them to Experiments with Animals

These studies have revealed the existence of complex mechanisms involved in the spatial distribution of attention; companion neurophysiological studies have contributed to our understanding of the neural mechanisms involved.

As indicated earlier, the development of the awake, behaving monkey for central nervous system recording has been, and will continue to be, of the utmost value in studies of the oculomotor system and of the higher functions of the visual nervous system more generally. As we become increasingly able to pose sensible neurophysiological questions of higher cortical areas, these behavioral techniques will become increasingly important.

Animal Models Are Also Used in Studying Visual Development

Rearing animals under conditions of abnormal visual stimulation or visual deprivation leads to abnormal development of the functions of the central visual pathways (amblyopia). Parallel studies that relate the behavioral anomalies in these animals to underlying functional and morphological changes in the visual pathways can be useful in revealing relations between visual functions and the underlying neural substrate, as well as being of general developmental interest in themselves.

One of the Oldest Questions in Neurobiology Concerns the Relation of a Particular Behavioral Performance to the Neural Machinery that Subserves that Performance

The study of vision has often been concerned with this kind of question, particularly with respect to retinal mechanisms. There is now more interest in tackling questions of this kind in the central nervous system. By use of judiciously chosen parallel psychophysical and behavioral experiments, it has been possible to draw reasonably strong conclusions concerning the sites of brain mechanisms that subserve a variety of visual performance.

A reasonably clear example concerns the locus of the spatial frequency channels whose existence is supposed on purely psychophysical grounds. These channels are selective for both stimulus orientation and size; they are sensitive to the direction of stimulus motion; they are binocularly activated and may be sensitive to binocular disparity.

Computational Approaches to Visual Function

Although computer vision has developed without reference to biological visual systems, it has become increasingly apparent that the types of algorithms used in machine vision are likely to throw light on how the brain processes visual information. Although most of the effort in computer vision has so far been devoted to the applied problem of programming computers to deal sensibly with image data, a growing number of workers are turning their attention to biological visual systems and are trying to model different aspects of visual behavior (and the underlying neural mechanisms). Still, relatively few process models take into account the structure and function of the nervous system; much more common are competence models, which attempt to mimic the biological function, but do so in a way that may be completely unrelated to the approach used in the actual biological system. No doubt this situation will eventually be reversed, given the richness of biological vision and the rather barren repertoire of most vision machines.

The Use of the Awake, Behaving Primate and Computational Models Will Show Dramatic Results in the Next Decade

The use of the awake, behaving primate, combined with more sophisticated behavioral and physiological approaches, will be important in increasing our understanding of visual function, especially the broad range of functions subserved by the extrastriate visual cortex. In addition, the application of computational models to the myriad problems of visual processing will increase the sophistication of our approach to studies of the function of the visual system at all levels, but especially at behavioral and psychophysical levels.

NEUROBIOLOGY OF MOTOR CONTROL

Our Understanding of How the Brain Controls the Movements of the Body Is Undergoing Dramatic Changes

For more than a century, the primary goal of those interested in the control of movement was to map the areas of the brain concerned with movement. Now, at last, we are beginning to address such questions as, How does the brain decide when to move? How does the brain select its targets? What are the speed, accuracy, and force of particular movements? The gradual emergence of new techniques that allow effective study of the mammalian brain make it possible to develop increasingly sophisticated and realistic models of motor function. These new techniques include the study of the activity of single nerve cells and populations of cells during real behavior; new anatomical methods for tracing the connections between specific nerve cells of the brain; imaging that allows one to visualize neural activity as it occurs in the living brain; and quantitative methods derived from biomechanics, control theory, and computer science, applicable equally to human subjects and experimental animals. Within the foreseeable future, it should be possible to determine which brain cells play what roles in planning, initiating, and carrying out movements.

Until the recent development of these new methods, the study of motor control was restricted to an examination of primitive reflexes and postural control mechanisms. Experiments were performed on animals that either had large regions of their central nervous systems destroyed or that had been immobilized by anesthesia. Although these studies continue to be necessary to establish some of the fundamental ways the nervous system operates, they have provided little insight into how the normal nervous system operates. Research in motor control now focuses on intact, awake animals and, where possible, on human beings.

Current approaches to the study of motor control derive from techniques for recording from single nerve cells in awake animals performing skilled motor tasks. This method allows us to search for control signals used by different motor regions of the brain. As a result, we now have a general idea of the roles of various regions in producing voluntary movement. Even simple movements require the cooperation of many motor centers in the brain. In controlling movement, the brain acts as a complex parallel computer, which decomposes motor plans into several components, which are then reassembled into a final, effective movement.

As a result of these studies on single neurons, neuroscientists are developing a new model of how the brain controls behavior. This model emphasizes the brain's ability to adapt to new situations and to form new strategies to solve them effectively. It marks a significant advance from earlier views that held the brain to be merely the center for coordinating automatic reflexes.

A second important insight derived from these studies has been the realization that the activity of neurons within a region varies with the task to be performed. The coding of information by single neurons often depends on the behavioral context. For example, one neuron may respond to a specific signal, such as a light, but only if this light tells the animal to expect a food reward. Such selective responding has been observed in nerve cells not only in the brain, but also in the spinal cord. These findings tell us that the awake brain is continuously modulating its activity in meaningful ways to respond to different situations.

While it has become standard to record from single neurons in behaving animals, it remains difficult to simultaneously observe the behavior of different regions of the several neurons in different regions of the brain as they cooperate to shape behavior. The first important techniques used to allow this type of analysis were anatomical ones that can map specific connections between one brain region and another. Using suitable radioactive labels or other markers, it is possible to trace the connections made by individual nerve cells from one part of the brain to another. The selectivity of these tracers makes it possible to examine only those specific classes of brain cells of interest.

The introduction of imaging methods has been particularly important because they can be used in alert human subjects. When the activity of a region of the brain increases, the blood flow to that region also increases and the region uses more glucose, the cellular fuel of nerve cells. By injecting minute amounts of radioactive compounds and then measuring the radioactivity over the scalp, it is possible to measure the blood flow and glucose consumption in different areas of the brain and thereby to reveal, with remarkable clarity, changes in activity of different brain regions when subjects perform different movements. It is even possible to "see" the brain planning a movement. Especially striking has been the finding that the premotor areas of the cerebral cortex become active when subjects mentally rehearse or plan complex motor actions; other areas become active only when the movement is performed. These results are now guiding studies of the activity of single neurons in experimental animals performing similar tasks.

New Techniques Make Possible Collaborative Ventures That Increase the Yield of Scientific Studies

Some of the new imaging techniques have been used to map the location of specific neurotransmitters. Not only can normal brain connections be mapped in this way, but these techniques allow us to identify the abnormalities of neurotransmitters in diseases such as Parkinsonism, Huntington's chorea, and dystonia. These studies reveal the chemical changes of these diseases in the living patient with a minimum of pain or risk and in a manner that goes far beyond the reach of the finest autopsy as carried out even a decade ago.

At the same time as some neuroscientists have been shedding new light on how the brain controls movement, others have been studying the primary effector systems of movement—the muscles and joints. Because muscles have certain intrinsic properties, the brain must adapt its neural language to a dialect the muscles can understand and obey. One major problem for the nervous system, including the brain and spinal cord, is that the muscles distort the commands of the brain. The nervous system has evolved several mechanisms for overcoming these distortions. For example, to produce brief, accurate movements in one direction, the nervous system has developed a strategy in which opposing muscles are precisely activated in rapid succession. This requires coordination of nerve cell signals in hundredths of seconds. Because movements can be impeded or interrupted, the nervous system has also developed means for rapidly correcting errors. Certain automatic reflexes, such as the familiar knee jerk that is tested during routine physical examinations, prevent muscles from distorting neural commands and allow for smooth mechanical action which otherwise would be jerky and intermittent.

Because the tasks performed by the nervous system to control movement are complex, it has been useful to apply techniques developed in engineering to understand what the nervous system is doing. One such technique is control theory, derived from the study of feedback in electrical and mechanical systems. After some initial disappointments, approaches based on control theory have now provided crucial insights into how eye movements are controlled. This theory has predicted the existence of nerve cells with specific properties that were subsequently discovered by experiments in animals. Control theory has also provided a better understanding of disorders of eye movement in human patients. In the future, control theory should provide similar insights into how movements of the arms and legs are achieved.

Artificial Intelligence, Particularly Robotics, Has Also Advanced Our Understanding of Movement Control

Ideas derived from artificial intelligence are currently stimulating research on the role of the cerebellum in the control of movement. This part of the brain is crucial for normal motor coordination, and its regular, modular structure has fascinated brain scientists for more than a century. Recent work in experimental animals has confirmed that the cerebellum helps correct errors in ongoing movement. Artificial intelligence has also influenced studies of the role of the vestibular system, that part of the motor system which orients the head, eyes, and the body and helps to maintain balance. While current attempts to apply models derived from control theory, robotics, or artificial intelligence are unlikely to be the final word in motor control, they are now making an important contribution by sensitizing neuroscientists to the importance of generating testable models of motor control that make explicit performance predictions. These are a necessary step in reaching a more complete understanding of the whole motor system.

An important collaboration has also developed between physiologists and engineers in robotics. Physiologists have learned how to better define the processes necessary to control the human body, which is a complex system with multiple interdependent components. In turn, attempts are being made to apply what has been learned about muscular control to the design of robotic limbs for manufacturing purposes and to the construction of artificial walking machines that can maneuver through rough terrain. Prosthetic limbs for impaired patients may eventually allow some people who currently use wheelchairs to walk again. The development of complex armlike devices has stimulated work aimed at how the brain moves the arm through a specific path in three-dimensional space. This work shows that merely to fix the direction of a movement, many millions of brain cells in several different areas must work together.

NEW TECHNIQUES AND PARKINSON'S DISEASE

One example may clarify the way in which new techniques have combined in recent years to help both our basic understanding of how the brain controls movements and our ability to help patients with motor disorders. Parkinson's disease, a relatively common motor disorder of the elderly, was first described in the nineteenth century, but only in the 1960s was it discovered that a specific chemical neurotransmitter, dopamine, was deficient in these patients. This discovery quickly led to a relatively successful therapy of replacing the neurotransmitter, thereby greatly improving the quality of life and even the survival of these patients. The most exciting recent continuation of this work is the possibility of brain tissue transplants that may restore natural dopamine function. Such transplants have been successful in animals depleted of dopamine and have already been tried abroad in a small group of human patients. In the meantime, new neuroanatomical techniques have greatly increased our knowledge about how the parts of the brain disordered in Parkinson's disease are interconnected. Changes in brain activity have been examined radiologically and the deficiencies of dopamine shown quantitatively in living patients. Studies of the action of single nerve cells in animals have begun to elucidate how disordered nerve cell movement produces the "frozen" limbs of these patients. Increasingly sophisticated studies, based on insights from engineering, have produced a far more accurate characterization of the nature of the motor deficit. Several years ago, the chemical known colloquially as MPTP, which is a by-product of the synthesis of certain illicit drugs, was found to cause a toxic form of Parkinson's disease in a number of drug users. This unfortunate circumstance has been quickly exploited to gain further insight into the mechanism of Parkinson's disease and its possible cause. Already, the study of these patients has made clearer the brain regions damaged in Parkinson's disease and suggested that the disease itself, whose cause remains unknown, might be produced by a subtle toxin and thus might be preventable. It has now been convincingly shown that some animals given the toxin develop a condition virtually identical to Parkinson's disease. Neuroscientists are now applying the new biochemical, anatomical, and physiological techniques to study these animals, and there is reasonable hope that they may solve some of the remaining mysteries of Parkinson's disease in the near future.

NEUROBIOLOGY OF COGNITION

The relation between cognition and the brain has been a topic of philosophical speculation for millenia and a focus of scientific study for more than a century. Recent developments in our ability to monitor brain function and the development of computational models of cognition have abruptly altered the pace of scientific progress in this area.

Methodological Developments Provide Unprecedented Opportunities for Exploring the Neurobiology of the Human Brain

Methods for imaging metabolic correlates of brain function provide a new basis for localization of function. New techniques for analyzing electrical and magnetic signals from noninvasive probes allow analysis of the dynamics of neural activity. These two types of methods have made it possible to relate changes in cognitive function to discrete structural areas in the living experimental subject. They have also made it possible to analyze disease. Perhaps even more important, they are allowing exploration of regional changes in brain activity during perception, thought, and action. There is every reason to suppose that some of those methods will become more precise in the coming years and that additional new methods (for example, magnetic and electron spin spectroscopy) can be applied.

As Our Ability to Explore the Human Brain Has Changed, So Has Our Definition of What Cognition Is and Our Understanding of How It Can Be Studied Experimentally

Computational models of cognition arising from efforts to develop artificial intelligence can provide an analytical basis for neural studies of mental processes. These models picture the nervous system in terms of the sequence of operations necessary to carry out cognitive functions and allow us to view cognition in terms of the elementary mental operations of which it is constituted. Experimental studies of human beings carrying out these operations reveal the dynamics of simple computations with time resolution in the millisecond range. For example, studies of the time needed to determine whether a target item is a member of a stored set indicate a serial search process with a comparison time of 30 milliseconds per item in memory.

We can now relate an individual's complex overt behavior to two different models of how the nervous system functions during thought. One model stresses the time dynamics of serial and parallel computations that occur when human beings execute elementary tasks. The other stresses the anatomical systems in the human brain that become active during thought.

A fundamental goal is to understand how computational models of tasks such as reading, listening, imaging, and problem-solving relate to the anatomical structures and wiring diagrams of regions of the brain known to be involved in these processes. Such an understanding of the complex relations between mental computations and their underlying neural bases seems critical for illuminating the physical basis of changes in mental and emotional life that occur with either normal development or neurological and psychological disorders.

Our current ability to relate images of brain activity to the mental computations found in cognitive models contains large gaps. For example, we do not yet understand how the brain's electrical activity relates to changes in cerebral blood flow. This prevents us from taking full advantage of the opportunity to combine findings based on the more spatially specific neural imaging techniques with those derived from the more accurate, temporally precise recordings of event-related activity. Nor do we know how the activity of individual cells or of such cellular configurations as cortical columns relates to the computations described by computational models. These basic questions of method are common to all areas in which one hopes to understand the relations between brain activity and function.

Several areas of investigation in the study of cognition have already begun to use the new methods to produce important findings relating cognition to neurobiology. These areas include learning and memory, attention, language, and the psychobiology of development.

At Least Two Types of Memory Differ on the Basis of What Is Learned and Where the Memory Is Stored

Cognitive psychology has emphasized that memory for different tasks varies according to the type of knowledge acquired by the subject and on how the subject encodes and recalls the information learned. The two categories are often distinguished as procedural (or reflexive) and declarative. Recent studies suggest the intriguing hypothesis that each of these two forms of memory is processed by a different neural circuit.

Procedural memory is acquired in an automatic or reflexive way without awareness or cognitive processes such as comparison and evaluation. It includes perceptual and motor skills and the learning of procedures and rules. This form of memory is thought to use elementary forms of plasticity and to be stored within the sensory and motor systems employed for the expression of that particular task. For example, the classically conditioned eyeblink response in rabbits can be abolished by specific lesions of certain parts of the cerebellum. When these areas are destroyed, the effective conditioned auditory stimulus no longer produces an eyeblink, although the unconditioned eyeblink response that follows the US (air puff) remains intact. Cells in regions of the cerebellum also show learning-dependent increases in neuronal activity that closely parallel the development of the conditioned behavioral response. The results of these experiments indicate that the cerebellum plays an important role in mediating the conditioned eyeblink and perhaps other simple forms of classical conditioning.

Declarative memory depends on conscious reflection and on such cognitive processes as evaluation, comparison, and inference for its acquisition and recall. Declarative memory encodes information about specific autobiographical events as well as the temporal and personal associations for those events. It is often established in a single trial or experience, and it can be concisely expressed in verbal declarative statements, such as "I read a fascinating book last week." Declarative memory involves the processing of bits and pieces of information that the brain can then use to reconstruct past events or episodes. By repetition, declarative memory may at times be tranformed into the reflexive type. Learning to drive a car at first involves conscious cognitive processes, but eventually driving becomes more automatic and reflexive. Thus, even certain verbal learning tasks, if repeated often enough, are thought to assume the characteristics of reflexive learning because they can be performed without the participation of other cognitive strategies.

Studies in humans suggest that the temporal lobe and closely associated structures of the limbic system including, in particular, the hippocampus, may be critically involved in the acquisition of declarative memory. The structures are not themselves thought to be sites for memory storage, but are somehow involved in the process by which memories are placed into storage or are retrieved and read out from storage.

Some of the first evidence for a role for the temporal lobe in memory came from the study of a few epileptic patients who underwent bilateral removal of the hippocampus and associated structures in the temporal lobes to relieve their epileptic symptoms. These patients are amnesic in the sense that they have lost the ability to store new memories although, for the most part, their early-formed memories are intact. Such patients cannot master tasks requiring declarative memory, but they perform well on procedural tasks. A given learning task often involves aspects of both types of learning, and in these instances patients remember some aspects of the problem, but not others. Thus, if the patient is given a highly complex mechanical puzzle to solve, the patient may learn it as quickly as a normal person but on questioning will not remember seeing the puzzle or having worked on it previously. In other words, amnesic patients can learn a complex skill and yet cannot recall the specific events that allowed them to learn the rules and procedures that make up the skill. This idea helps explain why amnesic patients, when they perform a particular task, are often not aware that they had actually learned it just a few days earlier.

Attention

Visual Spatial Attention Will Probably Be the First Cognitive System to Be Understood in Terms of the Circuitry That Supports It

Highly parallel computations of visual and auditory information have now been described. We are also beginning to understand the neural and cellular bases for selection of sensory information. Clear electrical signs of selection observable even with scalp electrodes can separate the messages that are being attended to from those that are being ignored, within the first 100 msec after presentation. Recording from single cells in alert monkeys has provided a great deal of information about the anatomy of the system that selects information from visual space. We know that an area of the midbrain (the superior colliculus) is important for selection when the animal attends by making eye movements, whereas thalamic (pulvinar) and cortical (parietal lobe) areas are involved when the animal attends covertly to an area of the visual field not currently being fixated. Lesions of these areas produced by strokes or tumors in humans produce deficits similar to those described in the monkey. We are beginning to relate the detailed computations performed when attention is moved from one visual location to another to these anatomical areas.

These signs of selective attention depend on the integrity of the prefrontal cortex. In patients with lesions in this area, electrical activity related to early selection is reduced and performance is impaired. Similar deficits in event-related potentials have been found in schizophrenic subjects who are often described as lacking higher levels of attentional control. Furthermore, blood-flow studies of schizophrenic subjects who are performing tasks that require shifting of attention among different stimulus dimensions of color, form, and number show a deficit in flow in the prefrontal cortex. The deficits in adult patients with frontal lesions are often characterized by difficulty in maintaining coherent programs designed to reach a goal. These same patients are frequently distracted from their goals by sensory events, as if they were less able than normal subjects to control sensory activation. Animals with lesions in frontal areas have difficulty in responding correctly when a delay is imposed between the stimulus presentation and their response. These animals have trouble whenever they are required to select a novel or less typical response. They seem to have difficulty in resisting the momentarily strongest response in order to pursue a goal.

Psychobiology of Development

Studies on the Psychobiology of Development Have Transformed Our Understanding of the Capabilities of the Newborn

The genetic endowment of human newborns provides a considerable capability for perception, learning, and even such higher-level concepts as number. Applications of simple conditioning and habituation methods to infants is providing a basis for exploring differences in cognitive ability. There seems to be some stability from measures in early infancy to later achievement as measured by standardized tests. In addition, temperamental differences among infants in reactivity to external events, emotionality, and inhibitory control are providing a new impetus toward understanding the biological development of personality.

Important shifts in temperament seem to occur during definable time periods, in which the maturation of neural systems change the capability of the developing infants to regulate their own behavior. These critical periods provide important clues to the changes in behavior resulting from maturation of brain regions during development. Studies of the development of monkeys have already enlarged our understanding of the slow maturation of some areas of the brain. The frontal lobes, for example, continue to develop for some years after birth. These developmental processes can now be studied in human infants by observing changes in metabolic activity within regional brain areas. The results so far reported with these techniques fit the time course of shifts found by behavioral studies.

Behavioral changes seem to occur in parallel for monozygotic twins, who become increasingly concordant with age. Thus, at least some of the shifts observed in development seem to have a genetic basis. These new findings set the occasion for reexamining the fundamental issue of how genetic and environmental influences work in concert to shape the social and cognitive development of infants and children. The genetic analysis of the development of behavior promises to provide insight into some disorders. For example, genetic analysis of developmental dyslexia has suggested not only the inheritance of one form of the disorder, but also through linkage analysis provides suggestive evidence for an autosomal dominant locus on chromosome 15.

Fundamental Understanding of the Neurobiology of Cognition Will Have Important Practical Applications

The analysis of reading in terms of elementary cognitive operations has already begun to guide efforts to produce specific remediation techniques in developmental or acquired dyslexia. Of more importance in terms of public health are the efforts to apply these concepts to closed head injuries. Although it is still unclear how successful this kind of cognitive remediation is, the potential benefits are great.

Neural imaging techniques have revolutionized the practice of clinical neurology. In the near future, our understanding of the neural mechanisms underlying selective attention and language should assist the neurosurgeon in the delicate task of avoiding the most critical areas when performing needed surgery. Improved assays of cognitive function should also allow better tuning of drug therapies and replacement or transplant methods.

The combination of great intellectual interest and obvious practical importance makes this area a central one for the future of neurobiology.

BEHAVIORAL ECOLOGY

Behavioral Ecology and Sociobiology Encompass the Study of the Evolutionary Adaptiveness of Behavior

Evolutionary adaptation refers to differences in structure, physiological processes, behavioral patterns, or complexes of traits that increase the inclusive fitness of one organism over that of another organism of the same species. But because the effect of specific traits on the inclusive fitness of organisms is difficult to measure directly, evolutionary adaptations are usually inferred from the organisms' "goodness of fit" to their environments. One of the tasks of behavioral ecology is to investigate this particular correlation: to understand how selection pressure, exerted by the ecological and social environment, favors one behavior over another.

As a consequence of this evolutionary approach, the closely interlocked fields of behavioral ecology and sociobiology have been revitalized during the past 10 to 15 years. The fields have been energized by the merging of ethology, population genetics, and modern evolutionary theory in a manner that has proven effective in generating new hypotheses about the adaptiveness and evolution of behavior.

In the past decade or so, numerous mathematical models have structured the theoretical framework of behavioral ecology. Among these models, which are derived from theoretical population genetics, the most prominent are inclusive fitness theory (now better known as kin selection theory), optimization models (derived from microeconomics), and evolutionary stable-strategy models (derived from game theory).

These theoretical concepts must now be tested by much more extensive and rigorous experimentation. To that end, we need the techniques and methodological approaches developed in experimental ethology, psychobiology, and neurobiology. We need to understand the neurobiological mechanisms underlying behavioral expressions to appreciate the framework that defines the animal's response to environmental conditions. A knowledge of the morphological features and physiological mechanisms underlying behavioral patterns is crucial to our understanding of the evolutionary constraints on behavioral-ecological adaptations. Without an appreciation of the behavioral mechanisms involved in such key phenomena as competition, parent-offspring relationships, communication, and interspecific interaction within ecological communities, an adequate and precise description of ecological organization is not possible. Ecologists need to appreciate more fully the function of behavior as one of the major keys for understanding ecological systems. Conversely, neurobiologists (despite their recent advances at the molecular level) must not forget that these mechanisms are the products of evolution and, in particular, of natural selection acting in specific environmental and social settings.

One of the Most Rewarding Trends in the Study of Behavior Is the Convergence of Field and Laboratory Approaches

As ecologists increasingly realize the importance of behavior, they have begun to turn to laboratory techniques developed by experimental behavioral biologists. At the same time, psychobiologists have increasingly applied their experimental methods to investigations of naturally occurring behavior. They now more fully appreciate the biological and specifically ecological boundaries that affect learning patterns and the development of behavior in general.

The "Umwelt" concept, first presented in 1921, has gained new meaning. We now ask with new methods and vision, What is the perceptual environment of an animal? How is it affected by the animal's developmental stage, social status, motivation, and other behavioral contexts? How does an animal filter out extraneous stimuli or select particular cues from an indescribably rich palette of environmental stimuli? And which cues trigger an animal's predisposition to learn and to store the things learned in its memory? These are important questions not only for understanding adaptive learning mechanisms and sensory-neurobiological processes, but for understanding the significance of the hierarchical organization of cues in such behaviors as orientation, habitat choice, mate selection, kin recognition, and the identification of competition and enemies.

A particularly interesting illustration of the principle is kin recognition. Much of sociobiological theory predicts that animals will behave differently toward close genetic relatives and nonrelatives. As predicted, most instances of apparently altruistic cooperation that have been analyzed reveal nepotism at work. Clearly, animals must be able to recognize their close kin. Animals from many different taxa have this capacity. How they accomplish such often fine-tuned discrimination is now an active area of investigation, and most of the major hypotheses suggest roles for learning, memory, and specific sensory cues. Because adequate tests of these hypotheses will require careful experiments, an array of laboratory techniques in experimental behavioral physiology and psychobiology are being developed and applied.

Studies of Kin Assemblages and Kin Recognition Lead to a New Understanding of Population Structures and Mating Strategies

Kin recognition not only makes nepotistic behavior possible, but bears the responsibility for the avoidance (or optimization) of inbreeding. In highly evolved social systems, such as the eusocial insect societies, kin recognition functions as a social immune system, which accepts individuals that carry the right family label and rejects those labeled with foreign markers or lacking the familiar markers. The strategy of recognition of "self" and rejection of "alien" in such societies, which have been called superorganisms, resembles the strategy metazoan organisms use to protect bodily integrity. Interesting evolutionary parallels can be drawn between kin recognition systems and the immune system.

In most organisms studied to date, kin recognition labels seem to be chemical—probably complex blends of specific chemical compounds that are ultimately genetically determined. These labels apparently have to be learned by kinmates, but the learning process also seems to contain specific temporal patterns. Also, learning seems to be programmed and constrained by templatelike neural mechanisms. Substantial evidence has been adduced of similar mechanisms in invertebrates, lower and high vertebrates, including primates, and even in human infants and children.

A second advance in our understanding of social groups centers on conflicts of interest. A counterpoint to the documentation of cooperation between relatives is the discovery of many instances of subtle and not-so-subtle disharmony in apparently cooperative groups. It follows from the neo-Darwinian "selfish gene" view of evolution that cooperation between individuals should reflect a delicate balance between costs and benefits that could easily tip toward conflict. Here, too, recognition systems play a crucial role.

Conflict and competition appear to constitute a major force in structuring ecologial communities, but little is known about the behavioral mechanisms underlying competition. Central questions that should be addressed concern the role of learning in competitor recognition and the comparative assessment of "self" versus "opponent" in competitive interactions. As in the study of cooperative and competitive interactions within a single species, the role of learning and memory is a central topic for understanding these naturally occurring behaviors.

Direct Links Are Being Made Between Behavioral Ecology and Development Psychobiology

The only significant result of evolution by natural selection is the determination of what genes are preserved, or what new genetic variants persist, and which disappear after numerous generations. But the phenotype on which selection acts is not merely an adult, but a life cycle, and therefore behavioral ecologists are not concerned solely with the genes that affect adults but also with those that affect the whole of development. Such behaviors as foraging, mating, nursing, helping, and fighting are based on short-term decisions. The evolutionary significance of these behaviors will be fully understood only if they can be related to long-term life-history patterns. Life-history theory deals with questions such as how an individual should allocate resources to growth versus reproduction to achieve the greatest fitness. Attempts to integrate typical behavioral-ecological analyses of short-term decisions with long-term approaches of life history theory should increase.

All Social Interactions Involve Communication

The study of communication will continue to be a major topic in behavioral biology and will entail, on the one hand, the investigation of the signal-receptor systems and the neural mechanisms of information processing and, on the other, the comparative study of the evolution and ecological adaptation of communication strategies.

The two main themes of evolutionary biology are adaptation and phylogeny. Both are best examined by comparative methods. Methods developed by taxonomists are now being applied to the comparative study of behavior. Adaptive strategies can often be deduced from analogous mechanisms found in phylogenetically diverse groups of species. As behavioral fossils do not exist, the reconstruction of the most likely history or phylogeny of animal communication is based on comparative studies of organizational levels of communication mechanisms in closely related species.

A current topic of debate is the question of whether signals, or communication displays, were selected during evolution for their efficacy in transferring information or for their effectiveness in persuading or manipulating others. Only detailed behavioral-ecological analyses of communication strategies in animals can provide answers to these questions.

In recent years it has become increasingly clear that communication is rarely characterized by a direct all-or-none response. Communication is not always a deterministic releasing process, but sometimes plays a different and more subtle role, modifying the behavioral properties of the receiver and alerting and focusing the receiver's attention on the situation context. This kind of system has been called modulatory communication. In it, signals do not release specific behavior patterns, but rather modulate the probability of reactions to other stimuli by influencing the motivational state of the receiver. We should expect such modulatory communication to be most frequent in complex animal societies, where many members perform many different tasks at the same time, and where an economically efficent organization of behavior requires that the work force distribute its energy investment among different tasks through an optimum division of labor.

It has recently been argued that the social system itself, by communicatory processes, can develop the properties of problem solving; it can develop what amounts to a cognitive system that encompasses but also exceeds the cognitive capabilities of the individual components. It has even been suggested that we compare the coordinating mechanism active in such superorganisms with the interactive neuronal processes that endow central nervous systems with their acknowledged cognitive capacities.

It is remarkable that in the brain, as in highly social systems, we find mechanisms that set the overall level of arousal. Recent examples in social insect communication illustrate this point impressively. For example, tonic sensory input from a variety of sense organs and spontaneous activity of neural arousal systems perform in the nervous system the functions that unspecific modulating signals serve in social organizations. In both forms of organization we further find more specific regionalized or addressed mechanisms of focusing the attention to a specific subset of stimuli in a given context. We find that, within the larger systems, mechanisms exist that modulate in graded fashion the activity probability of small dedicated subpopulations of neurons or individuals that are thus recruited to perform specific tasks. It is probably more than chance that the neurophysiologist arrives at describing these mechanisms as local modulating interactions between neurons just as the student of social communication in animals independently finds it appropriate to qualify basic processes of social organization in this way.

ABNORMALITIES OF BEHAVIOR

Central to the Study of the Nervous System Is the Desire to Understand the Abnormalities of Behavior Produced by Various Neurological and Psychological Disorders

The goal of the modern study of the nervous system is to understand human behavior: how we sense objects in the world around us, execute skilled movements, feel, think, learn, and communicate with one another.

Study of the nervous system has traditionally provided the scientific and therapeutic underpinnings for neurology and psychiatry. We illustrate this point with two examples: (1) the application of modern molecular genetic approaches to diagnose neurological diseases and (2) the application of modern biochemical and imaging techniques to diagnose and treat psychiatric disorders.

Molecular Genetics and the Diagnosis of Neurological Disorders

A surprising number of serious neurological diseases have a genetic origin. These include neurofibromatosis, Huntington's disease, a subform of Alzheimer's disease, retinoblastoma, and various congenital diseases of muscle. The devastation these diseases produce is great. For example, John Merrick, the "Elephant Man" who lived in the late 1880s, was relegated to a life as a side-show circus freak. He suffered from severe neurofibromatosis. The more than 100,000 sufferers of neurofibromatosis today receive no better treatment than Merrick, other than frequent surgeries (as many as 100 a year in some cases) to remove the offending neurofibromas that can disfigure the entire body, occlude the auditory canal, and extend into the brain and spinal cord to pose an immediate life threat. Neurofibromatosis is an autosomal dominant disorder, which means that each child of a parent with neurofibromatosis has a 50 percent chance of inheriting the disease.

Today, more than half of all nursing homes beds in America are occupied by patients with Alzheimer's disease, an illness considered the fourth leading cause of death in this country. At least 10 percent (and by some estimates a much higher percentage) of cases are autosomal dominant—a 50 percent risk to offspring. There is no treatment for Alzheimer's disease, only a growing number of elderly people in the United States who are at risk.

Huntington's disease is the genetically programmed loss of nerve cells important for mental and motor function, which usually has its onset in midlife. It, too, is an autosomal dominant disorder for which no effective therapy exists. Folk singer Woody Guthrie, who wrote "This Land is Your Land," died of Huntington's disease. He also wrote from Brooklyn State Hospital a poem ending "there is no hope known." But with the advent of recombinant DNA technology, there is, for the first time, real hope.

Within the next decade, it should be possible to know the chromosomal assignment and exact DNA sequence of the genes that cause these and some other pernicious diseases that affect the nervous system. It should be possible to trace from a genetic lesion to the biochemical or regulatory disturbances it produces, from DNA through anatomical and physiological tracts to the expression of an aberrant gene in disordered thought or action. We should be able to chart the pathway from gene to brain to behavior, learning how the tiniest defects can cause the wildest movements, severe memory loss, suicidal depression, or the capacity to hear or see what does not exist. We should also be able to identify the genes responsible for normal brain functioning, which will provide powerful new insights into mental functioning.

Molecular Pharmacology, Modern Imaging, and the Diagnosis and Treatment of Psychiatric Disorders

Diagnosis in psychiatry is less precise than in neurology or in the rest of medicine because most psychiatric diseases cannot as yet be localized to specific regions of the brain, much less to particular proteins in specific nerve cells. Thus, the diagnosis of psychiatric disorders must rely primarily upon clinical symptoms. One major way of grouping psychiatric disturbances is into those which are psychotic and those which are not psychotic. The term psychotic can be loosely defined as reflecting a major loss of contact with reality. Nonpsychotic disturbances include anxiety, neurosis, and character disorder. The major psychotic disturbances are schizophrenic and affective illness, comprising mania and depression. In terms of human suffering and public expense, psychotic disorders present a more serious problem for society than nonpsychotic illness because of the much greater disability caused. Since the major psychoses often commence in early adulthood and persist throughout life, their total cost to society greatly exceeds that of cancer and heart disease. For example, at least 1 percent of the population is schizophrenic, an incidence comparable to that of diabetes. It is likely that the incidence is substantially higher, since many individuals who seem to be schizophrenic are not subjected to rigorous diagnosis.

Recent pharmacological studies have provided important insights into schizophrenia. The effects of drugs have permitted the development of hypotheses about specific neurochemical abnormalities. Many of the drugs that influence schizophrenic symptoms affect the neurotransmitter dopamine. The neuroleptic antipsychotic drugs act by blocking dopamine receptors. Reserpine, which has antischizophrenic effects, depletes the brain of dopamine. Amphetamines, which often exacerbate schizophrenic symptoms, release dopamine. These findings have suggested that an excess of dopamine might be relevant to schizophrenic pathophysiology. Postmortem studies consistently show increased numbers of dopamine receptors in the schizophrenic brain. This result has been reinforced by recent PET studies in patients, which show that schizophrenics have almost twice as many dopamine receptors as control subjects.

It has often been suggested that schizophrenia is a family of diseases that may have different etiologies. One recent classification focuses on the presence of positive or negative symptoms. Positive symptoms refer to florid delusions and hallucinations, while negative symptoms reflect autism and general withdrawal, a "wall-flower" type of behavior. Most neurological drugs are more effective in relieving the positive than the negative symptoms of schizophrenics. Chronic "burnt-out" schizophrenics often display primarily negative symptoms. That patients with chronic schizophrenia have enlarged cerebral ventricles relative to those with the acute forms of the disorder, along with differential drug responses, has prompted the definition of two subtypes of the illness, type I and type II schizophrenia. Type I manifests positive symptoms, a good response to neuroleptic drugs, and no enlargement of the cerebral ventricles, whereas type II schizophrenia is characterized by negative symptoms, a poor drug response, and enlarged ventricles. It is likely that the next five years will witness more characterization of these symptomatic subtypes of the disease and linkages to laboratory abnormalities.

The advent of CAT scanning has led to a greater clarification of the enlarged cerebral ventricles in schizophrenics first noted by pneumoencephalography. The recent development of MRI should permit far more extensive delineation of the ventricular enlargement.

Novel therapeutic approaches to schizophrenia are likely to focus differentially on the positive and negative symptoms. Of the neuroleptics in common use, only the diphenylbutylpiperidines show selective efficacy in relieving negative symptoms. The diphenylbutylpiperidines are just as potent calcium antagonists as dopamine antagonists. Conceivably, centrally active, selective calcium antagonists may have utility in the specific therapy of negative symptoms.

So far, the most direct insight into the genetic contribution to schizophrenia has come from twin and adoption studies. These studies have ruled out the possibility that environmental factors artifactually account for hereditary patterns. For instance, among schizophrenics who had been adopted at birth, the biological parents display a high incidence of schizophrenia, whereas the incidence in the adoptive parents matches that of the general population. Twin studies, however, show that environmental factors must play some role in the expression of the genetic tendency. The concordance rate for schizophrenia in identical twins is about 50 percent and not 100 percent, as would be expected if genetic factors alone accounted for the disease. Several studies examining identical twins discordant for schizophrenia reveal environmental trauma more in the schizophrenic than the nonschizophrenic co-twin. For instance, the schizophrenic co-twins generally have a lower birth weight and a greater likelihood of neonatal infection. More detailed studies of this type in the next decade should tease out specific factors that might be crucial in the transformation of a genetic predisposition into frank schizophrenia. Perhaps the most exciting possibility is the potential identification of the molecular genetic abnormality associated with schizophrenia, as we have discussed in relation to neurological disease. In families with an extremely high incidence of schizophrenia, one might be able to search for specific genetic markers differentiating schizophrenics from the general population by using strategies that have been successful in conditions such as Huntington's disease.

Being depressed is such a common experience that establishing diagnostic criteria for depressive "disease" is difficult. It is thought that as many as 5 percent of the population suffer from major affective disturbance. Both depression and mania are episodic and have been differentiated by the nature of the episodes. Bipolar disorder is characterized by episodes of both depression and mania, and unipolar illness by recurrent depression but no episodes of mania. As with schizophrenia, affective disorders have a strong genetic component. Genetic studies support a fundamental distinction between bipolar and unipolar illness, although there is much overlap. Adoptive and twin studies reveal that affective disturbances and schizophrenia share similar genetic predispositions.

Also, as with schizophrenia, drugs that influence neurotransmitters have provided strong hints as to a possible pathophysiology. Hypertensive patients treated with reserpine evince roughly a 15 percent incidence of severe depression, clinically indistinguishable from endogenous major depression. Reserpine depletes the brain of its biogenic amines—dopamine, norepinephrine, and serotonin. Alpha-methyldopa, also used to treat high blood pressure, selectively depletes the brain of norepinephrine and causes depression in many patients. Beta-adrenoceptor blockers used to treat hypertension also elicit depression, presumably by antagonizing endogenous catecholamines.

The major antidepressant drugs all seem to act through biogenic amines. The monoamine oxidase inhibitors increase brain concentrations of norepinephrine, dopamine, and serotonin. Tricyclic antidepressants inhibit the inactivation by reuptake of these three amines. Since some of the most effective antidepressants do not inhibit dopamine uptake, it is less likely that dopamine is involved in their action, but norepinephrine and serotonin are important candidates. Recently, several antidepressants have been introduced that selectively inhibit serotonin uptake with no influence on norepinephrine or dopamine. Some psychiatrists feel that drugs more selective for norepinephrine relieve depression by enhancing "drive," whereas serotonin-selective drugs act by increasing a sense of well-being. They hypothesize that there may exist two distinct subtypes of depression, one associated with deficits in norepinephrine functioning and the other, with deficits in serotonin functioning.

Lithium relieves both mania and depression and is prophylactic against recurrence of all affective episodes. Conceivably, knowledge of its action may shed light on fundamental aberrations that occur in both poles of affective illness. The interference of lithium in the phosphoinositide cycle may be a valuable clue. If lithium acts specifically by inhibiting a phosphatase in the phosphoinositide cycle, organic chemicals can be developed to mimic this effect. Such agents would not compete for intracellular sodium to cause the typical toxic effects of lithium.

Direct studies of the seratoninergic biochemical system in depression have been particularly promising. Postmortem brains of suicides manifest abnormally low concentrations of serotonin. The spinal fluid of depressed patients consistently shows a bimodal distribution of the serotonin metabolite 5-hydroxyindoleacidic acid. One group of patients has concentrations similar to those of normal subjects, whereas another group of patients of approximately equal numbers has a markedly lower concentration. Several researchers have shown that depressives with the lower hydroxyindoleacidic acid levels are more impulsive and prone to attempt suicide.

It is now possible to image serotonin receptors by PET scanning, which may permit an overall evaluation of serotoninergic neuronal function. Similar techniques will likely be feasible for noradrenergic and dopaminergic neurons. It is hoped that molecular genetic techniques can be applied to the diagnosis and treatment of affective illness, with the ultimate view of identifying specific molecular aberrations that reflect the cause of the illness.

7

The Immune System and Infectious Diseases

THE IMMUNE SYSTEM

Vertebrates Developed the Immune System to Deal with Pathogenic Microorganisms, Malignant Cells, and Macromolecules

The immune system orchestrates a potent defense that consists of the production of specific antibody molecules and lymphocytes capable of reacting with and inactivating foreign agents, either directly or indirectly through the involvement of molecular and cellular inflammatory processes. The importance of the immune system to our survival in the face of the wide variety of disease-causing agents is tragically demonstrated by the devastating consequences of the immunological impairment of acquired immune deficiency syndrome (AIDS) and of congenital immunological deficiencies such as severe combined immunodeficiency (SCID) in infants. On the other hand, the enormous power of the immune system as a protection against pathogenic agents carries with it the price that the action of the system may lead to or exacerbate a series of immunological diseases, including systemic lupus erythematosus, rheumatoid arthritis, and type I diabetes (juvenile onset).

The immune system is composed mainly of lymphocytes and macrophages. The cells of the immune system are found in the blood and lymph, where they are in a recirculating pool, and in the spleen, lymph nodes, and lymphoid tissues associated with the gastrointestinal tract, the brochopulmonary tree, and mucosal surfaces. Lymphocytes and macrophages develop in the thymus and the bone marrow.

Although each cell type in the immune system has a major role to play, it is the lymphocytes that form the unique elements of the system since they display the antigenic specificity that is the hallmark of immunity. Lymphocytes are of two major types, B cells and T cells. B cells are the antecedents of antibody-producing cells, whereas T cells have their major effects in regulating the level of activity of the immune system and in mediating cellular effector mechanisms of immunity, such as destroying virus-infected cells and tumor cells.

The study of the immune system has proven to be of great importance in efforts to understand and to regulate immune responses, both to enhance the system's protective actions against microbes and tumor cells and to control its actions in the development of autoimmune diseases and other disorders with immunological components. In addition, the cells of the immune system are valuable in the study of many aspects of normal biology of mammalian cells since they are easily accessible, have been extensively characterized, and can be easily grown.

Specificity of the Immune Response

Perhaps the Most Remarkable Aspect of the Immune System Is Its Specificity

Each of us has the capacity to make a specific immune response to a vast array of foreign macromolecules: The body develops antibody molecules specific for structures (antigenic determinants called epitopes) on the foreign agent and produces B and T lymphocytes that bear membrane receptors specific for these epitopes. We now recognize that the capacity of the immune system to respond to millions of different foreign antigens depends on the existence of an internal universe of distinct lymphocytes, each capable of mounting an immune response against only one set of structurally related antigens.

The capacity of individual lymphocytes to respond to only a single set of molecules is coupled with a rapid and selective increase in the number of these lymphocytes as a result of the introduction of the specific antigen (Figure 7-1). This increase results in the formation of clones of specific lymphocytes, a property that led to the designation of the clonal selection theory to describe the concept that the enormous specificity of the immune response is achieved by the antigen-mediated selection of individual cells (or clones of cells), each with but a single specificity. Selected clones not only increase in number, but also differentiate into antibody-producing cells (B cells) or into active regulatory cells or killer cells (T cells). Such selection of B cells and the parallel process of T-cell selection forms the basis of protective immune responses.

The capacity of the immune system to make B-cell and T-cell responses to virtually any foreign substance is based on unique mechanisms that allow the assemblage of genes for the extremely large number of alternative forms of antibodies and receptors. Recent progress in the application of molecular biological techniques to the study of lymphocytes has led to a solution of this central biological problem. In order to understand the molecular mechanisms respon-

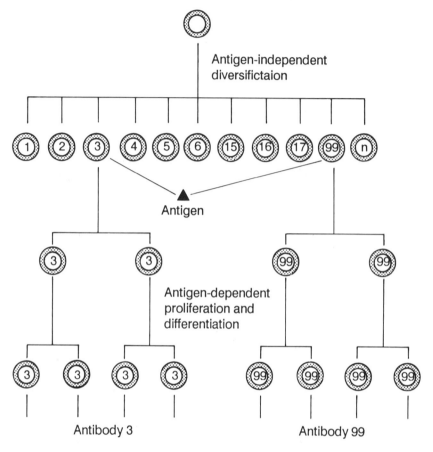

FIGURE 7-1 Lymphocytes undergo antigen-independent diversification and antigen-dependent growth and differentiation. Lymphocyte precursors develop into mature lymphocytes through an antigen-independent process that involves rearrangements of genes for the variable regions of immunoglobulins (for B cells) or for the variable regions of T-cell receptors (for T cells). As a result of this antigen-independent diversification, each lymphocyte expresses a distinctive receptor on its surface. In the example shown, a series of different B cells (expressing receptors designated 1, 2, ..., n) are exposed to an antigen. Two of these (3 and 99) have receptors that can bind to determinants on the antigen, and, as a result, they are stimulated to divide and to differentiate into cells that secrete antibody. That antibody has the same specificity as the receptor; hence, the antibodies are designated antibodies 3 and 99. A similar situation exists for T-cell responses, except that the T cell recognizes a complex of antigen and a class I or class II major histocompatibility molecule and that the T cell does not secrete antibody. T cells mediate their functions as a result of receptor-mediated activation. [Adapted from I. L. Weissman et al., Essential Concepts in Immunology (Benjamin/Cummings, Menlo Park, Calif., 1978), figure 3-1]

sible for the generation of the large number of distinctive antibodies and receptors, a review of the salient feature of the structure of these molecules is necessary. Furthermore, the consideration of the structure of these antibodies, known
as immunoglobulin (Ig), and of the related B- and T-cell receptors is important in
its own right since they represent a majority of biological recognition systems.

Antibody Molecules and B-Cell Receptors Are Proteins Called Immunoglobulins

Although Ig exists in several distinct forms, termed classes, these molecules
have a common structural organization consisting of a unit that is a dimer of a
heavy (H) chain and a light (L) chain pair (Figure 7-2). The H chain contains four
or five segments, or domains, each of approximately 110 amino acids, whereas
the L chain consists of two such domains. Each H-chain–L-chain pair forms an
antigen-binding site, which has elements of both the H- and L-chain amino-
terminal domains. Indeed, the amino-terminal domains of the H and L chains
from different antibodies differ structurally from one another, an observation that
is consistent with the fact that the binding site of each antibody is different. Thus,
these domains of the H and L chains are designated the variable (V) regions (V_H
and V_L). The variability of the V regions is concentrated into three segments of
each chain, designated hypervariable regions, which contain the amino acids that
actually make contact with the antigenic determinant in the binding of antibody to
antigen.

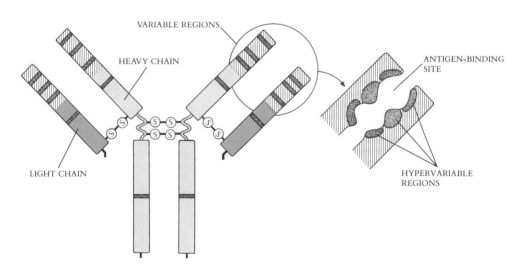

FIGURE 7-2 A schematic drawing of an antibody molecule. Each molecule is composed of two
identical light (L) chains and two identical heavy (H) chains. The antigen-binding sites are formed by
a complex of both H and L chains. [Reproduced, with permission, from P. H. Raven and G. B.
Johnson, Biology (Times Mirror/Mosby, St. Louis, 1989), figure 52-18B]

The remaining (carboxy-terminal) portions of the H and L chains are the constant (C) regions (C_H and C_L). The C_H portion of the Ig molecule determines its biological function. Indeed, Ig exists in a series of distinct classes (IgM, IgD, IgG, IgA, IgE), each of which has a distinct C_H region and distinctive functional properties.

The structural basis of antibody-epitope binding has recently been clarified by determining the crystal structure of antigen-antibody complexes. This structure graphically illustrates the areas of contact of antibody with antigen (Plates 4 and 5). More extensive structural analysis of antibody-antigen interactions may allow the development of a technology to improve the specificity and affinity of monoclonal antibodies.

Receptors of B cells are Ig molecules that have combining sites identical to the antibodies that these cells will secrete upon antigenic stimulation. The receptor and the antibodies differ in that the receptor is a membrane protein, with a specialized carboxy-terminal region of its H chain that anchors it in the cell membrane, whereas the antibody (Figure 7-2) is a secreted soluble protein. These alternative forms of the same molecule are produced through alternative RNA splicing, which produces distinct H-chain messenger RNAs (mRNAs) for secretory and membrane Ig.

T-Cell Receptor Molecules Are Heterodimeric Ig-like Molecules

Most T cells express receptors composed of α and β chains; others express receptors composed of γ and δ chains. T-cell receptor polypeptide chains have amino acid sequence homologies to Ig H and L chains. The α and β chains, which have been studied most extensively, have amino-terminal variable regions and carboxy-terminal constant regions.

The Molecular Genetic Basis of Immunoglobulin and T-Cell Receptor Diversification Relies on DNA Translocation Events

Understanding the structure of Ig molecules and B- and T-cell receptors gives us a basis for considering the genetic mechanisms that develop the information necessary for encoding the vast array of distinct antibodies and receptors. The V_H domain of Ig molecules is coded for by three independent genes, V_H, D, and J_H, which are separated from one another in germ-line DNA, but which are brought together in B cells by translocation events to form a $V_H D J_H$ single gene.

The germ line contains a large number (200 to 1,500) of distinct V_H genes, 12 D genes, and 4 active J_H genes. Since these genes may be chosen randomly for assembly into a $V_H D J_H$ gene and since products of each may contribute to the combining specificity of the resultant V_H domain, the number of functionally different H chains that can be generated by simple combinatorial association may be in excess of 10^4. A similar assembly of distinct genetic elements leads to formation of the gene for the V_L domain. The process by which the V_H and V_L genes are assembled and coupled with random pairing of H and L chains, could lead to the development of more than 10^8 distinct H and L pairs. In addition to these mechanisms for creation of diversity, Ig genes undergo somatic hypermutation in the course of B-cell responses to antigen, leading to a further enlargement in the number of distinct Ig genes that the immune system can generate. Thus, through the use of a substantial (but not enormous) amount of genetic information, the immune system creates an almost limitless array of distinct Ig molecules.

The processes involved in diversifying T-cell receptors resemble those described for Ig molecules, except that little or no somatic mutation seems to occur. Although this limits the number of distinct T-cell receptor genes that may form when compared with the number of Ig genes, the number of possible T-cell-receptor genes is still very large.

B Lymphocytes

B Cells Undergo an Ordered Set of Developmental Processes as They Develop from Hematopoietic Progenitor Cells

The B lymphocytes are derived from precursors in hematopoietic tissue. These progenitor cells develop into pre-B cells, which lack membrane receptors and thus are insensitive to antigenic stimulation. It is within the pre-B cells that the translocation events occur through which $V_H D J_H C_H$ and the $V_L J_L C_L$ genes are assembled and in which their products, the H and L chains, pair with one another to form intact Ig molecules. Once this has occurred, Ig is expressed on the membrane; the cell now has a receptor capable of recognizing a foreign (or self) antigen. The processes through which the individual Ig genes are activated and through which the rearrangements are controlled are currently under intense investigation. The Ig genetic system provides an outstanding opportunity to obtain basic information about the general process of genetic control. Indeed, one of the best-characterized systems of tissue-specific genetic enhancers is that of the Ig H-chain genes.

Binding of antigen to membrane Ig receptors on immature B cells may eliminate or desensitize these cells. This would lead to the functional deletion of those B cells that bear receptors specific for endogenous antigens present on the tissues or in the extracellular fluids of the individual (self- or auto-antigens) and thus should prevent the maturation of B cells specific for self-antigens. The elimination of autoreactive B cells would thus lead to a state of immunological tolerance in the B-cell population. However, the elimination of such autoreactive B cells seems to be incomplete. Indeed, the production of antibodies that can bind to autoantigens is more common than formerly believed. The study of the biology of tolerance induction and of autoantibody production is important because many human diseases seem to be caused or exacerbated by autoantibodies and related autoimmune processes. For example, antibodies to autologous antigens have a serious effect in systemic lupus erythematosus, in which antibodies to DNA cause severe kidney disease through deposition of antigen-antibody complexes in the glomerulus.

The immature state of B cells, in which they appear to be uniquely suscep- tible to induced tolerance, is marked by the expression on their surface of recep- tors of the IgM class only. These B cells mature further and acquire surface IgD in addition to IgM. IgM and IgD have the same V_H and V_L regions and thus have identical antigenic specificity, although their constant regions differ and presuma- bly mediate distinct functions. However, no convincing explanation of the functional difference in IgM and IgD has yet been brought forth. Nonetheless, coincident with the acquisition of IgD by the maturing B cell are the development of a resistance to the induction of experimental tolerance and the development of a heightened ability to become activated and to become antibody-secreting cells, which suggest that IgD may play an important role in these processes.

Receptor Cross-Linkage Mediates B-Cell Activation

B-cell activation in response to antigenic stimulation seems to follow one of two activation pathways; these alternatives may reflect the existence of two distinct types of antigenic substances. Many biologically important antigenic molecules (termed type II antigens) have multiple copies of the same epitope; principal among them are the capsular polysaccharides of many pathogenic bacte- ria. The multiple copies of the same epitope cross-link the receptors on the surface of specific B cells. Such receptor cross-linkage causes rapid biochemical changes within the cell, which set in motion the process through which the cell is stimulated to enter the cell cycle and divide. The earliest event that has been detected is the activation of the inositol phospholipid metabolic pathway. Initial progress in delineating the biochemical events that control B-cell activation raises the possibility that drugs may be developed that can selectively control the process of B-cell (and T-cell) activation and which for this reason may be of therapeutic value.

B-cell activation through receptor cross-linkage also leads to the activation of the c-*fos* and c-*myc* cellular proto-oncogenes. These genes and their products, when properly regulated, play a key role in the cellular activation process. Their deregulated action seems to influence the deregulated growth of malignant cells, with c-*myc* being implicated in certain B-cell tumors.

Among the important tasks that lie ahead are (1) the identification of the membrane molecules through which Ig signals the activation of inositol phospho-

HYBRIDOMAS AND MONOCLONAL ANTIBODIES

One of the most important developments to flow from efforts to understand the cellular and molecular basis of antibody production is the technology for the formation of stable somatic cell hybrids between normal B cells and lines of antibody-producing cells. The resultant hybridomas are capable of producing essentially unlimited quantities of the antibody derived from the normal B-cell partner. Because such antibodies represent the product of a clone derived from a single hybridoma cell, they are referred to as monoclonal antibodies. Monoclonal antibody technology provides reagents of unprecedented specificity for use in virtually every aspect of modern biological sciences. Furthermore, monoclonal antibodies are exceptionally valuable diagnostic reagents and promising as therapeutic agents.

An attractive strategy has been to develop monoclonal antibodies specific for antigenic determinants preferentially expressed on tumor cells. When such antibodies are coupled to a toxic structure, such as the ricin A chain, an immunotoxin is created. Immunotoxins can specifically destroy tumor cells in vivo and may thus prove to be highly specific antitumor agents.

One monoclonal antibody, specific for part of the T cell's antigen receptor complex, has already been licensed for human use in the immunosuppression of transplant recipients. It is anticipated that monoclonal antibody methodology will grow more valuable clinically as additional reagents are developed and brought into general use.

lipid metabolic pathways to initiate the intracellular signaling process and (2) the detailed description of the molecular events that occur as a result of the activation of this signaling pathway. Particularly important will be the identification of the proteins that are phosphorylated in the course of B-cell activation and the delineation of the functions these proteins mediate in the response of the cell.

B-Cell Activation Can Also Occur Through Cognate T-Cell–B-Cell Interaction

Many of the most important antigenic substances against which the immune system must respond have no more than one copy of any individual epitope. These molecules include proteins having many different epitopes but no repetitive elements. Since all the receptors on an individual lymphocyte have the same binding specificity, such antigens cannot cause receptor cross-linkage. Thus, they fail to initiate the set of metabolic activation steps described above. Such antigens can activate B cells only through an indirect mechanism that requires the intimate participation of T cells in cognate T-cell–B-cell interactions. For this reason, such antigens are designated T-dependent antigens. In cognate B-cell activation, the B cell binds antigen through its membrane receptors, but this binding, in and of itself, results in no intracellular signal. A portion of these bound antigenic molecules are taken into the cell by endocytosis, where they are fragmented by proteolytic digestion. Some of the resultant peptides are returned to the surface of the cell as a complex with a cellular membrane protein, generally a class II major histocompatibility complex (MHC) molecule. The association of antigen-derived peptides and class II molecules is based on the specific binding of the antigen by the class II molecule. Receptors of helper T cells are specific for a complex of peptide and class II molecule, rather than for the peptide alone. Thus, helper T cells that have receptors specific for the particular complex formed on the surface of a B cell will bind to that B cell, an event that activates the T cell. In turn, such T cells locally activate B cells either by signal transduction through B-cell surface molecules involved in the T-cell–B-cell interaction or through the secretion of soluble lymphokines such as interleukin-4 (IL-4), interleukin-2 (IL-2), and interferon γ.

The Activation and Proliferation of B Lymphocytes Is a Prelude to Their Differentiation into Antibody-Secreting Cells

The T-cell-derived lymphokines, including IL-5 and the B-cell differentiation factor IL-6, direct the development of B cells into antibody-secreting cells. The recent purification and molecular cloning of both IL-5 and IL-6 should make possible a more precise understanding of control of antibody secretion in the near future. Within the B cell, the differentiation events associated with the development of a B cell into an antibody-secreting cell include a change in the processing of mRNA for Ig H chains. The resting B cell processes the bulk of its H-chain mRNA to produce a membrane form of Ig, in which the H chain has a hydrophobic region near the carboxy-terminal portion of the molecule. This hydrophobic region, which spans the cell membrane, and the additional "cytoplasmic tail" of the H chain are encoded in distinct exons in the gene for the H chain. In the antibody-producing cell, an alternative mRNA-processing mechanism leads to the production of an H chain that is identical to that found in the resting B cell except that it lacks the transmembrane and cytoplasmic regions. This H chain interacts with L chains to form a secretory protein rather than a membrane protein.

An Important Differentiation Event in the Physiology of Antibody Production Is the "Switch" in Expression of Ig Class by B Cells

As has been described, each Ig class seems to play a distinct role in the immune response. Cells producing IgG (there are four subclasses of IgG), IgA, and IgE derive from precursors that initially express IgM or both IgM and IgD on their membranes. The molecular basis of this switching event seems to involve the translocation of an assembled $V_H D J_H$ gene from its position proximal to the gene for the constant region of the μH chain (C_μ) to a comparable position proximal to a C_γ, C_ϵ, or C_α gene. IL-4 and other T-cell-derived lymphokines direct switching to distinct Ig classes. The clarification of the physiological mechanisms controlling class switching both in vitro and in vivo is of great importance because of the different functions of antibodies of different classes.

T Cells

T Cells Recognize a Noncovalent Complex Consisting of an Antigen-Derived Peptide and a Class I or Class II MHC Molecule

In contrast to B cells, whose receptors recognize epitopes on soluble molecules, T cells recognize a noncovalent complex consisting of an antigen-derived peptide and an MHC molecule. The noncovalent complexes that T cells recognize form in and are displayed on the surface of cells with which the T cells interact. Thus, T-cell recognition of an antigen occurs only in intimate association with other cell types, which act as antigen-presenting cells. The subset of T cells that are principally involved in the stimulation of antibody synthesis and of cellular immunity (helper T cells) generally recognize antigen-derived peptides associated with class II MHC molecules, and thus can interact only with those antigen-presenting cells that express class II molecules. The expression of class II molecules is largely limited to B cells, macrophages, dendritic cells, and specialized endothelial cells in the thymus. Under conditions of immunologically induced inflammation, however, class II molecules may be found on other cell types. These cells may thus acquire the capacity to interact with and present antigen to helper T cells. In general, those T cells which co-recognize antigen with class II molecules have a surface molecule designated CD4, which may play an important auxillary role in the T-cell recognition process. Furthermore, it has recently been established that CD4 is the cell-surface receptor for the human immunodeficiency virus (HIV), the virus that causes AIDS. The entry of HIV into cells depends on its binding to CD4, and those cells that express CD4 are susceptible to infection with HIV. This role of CD4 in cellular infection by HIV suggests that strategies to prevent infection by interrupting the binding of the virus to its receptor should be explored seriously. Structural studies of the interaction of CD4 with both HIV and with its natural ligand, the class II molecule, may prove to be of great importance in limiting the spread of HIV.

A second subset of T cells is able to lyse antigen-bearing target cells, such as virus-infected or tumor cells. These T cells are generally referred to as killer cells. Killer T cells mainly recognize antigen in association with a different group of MHC molecules, class I molecules. In contrast to the expression of class II molecules, most tissues express class I molecules on the surface of their constituent cells. Thus, virtually any cell type would be susceptible to the cytotoxic action of a killer T cell if the cell expressed the complex consisting of antigen and the class I MHC molecule recognized by the receptors of the killer cell. T cells that co-recognize antigen with class I molecules express CD8 rather than CD4 on their membrane. CD8 molecules seem to function in a manner generally similar to CD4 molecules, but they are not receptors for HIV. The tendency for CD4 to be expressed on helper T cells and for CD8 to be expressed on killer cells and on T cell that suppress immune responses (suppressor T cells) makes the enumeration of the relative number of CD4- and CD8-positive cells a useful first step in the assessment of the relative state of the immune system in individual patients.

The MHC Molecules Play a Key Role in Several Aspects of Immunity and in the Rejection of Organ and Tissue Grafts

The MHC is a complex of genes located on the short arm of the sixth chromosome in humans. This gene complex contains genes for both class I and class II MHC molecules. Class I molecules are highly polymorphic membrane-spanning glycoproteins with a molecular weight of approximately 45,000. They exist on the membrane in association with a 12,000-dalton polypeptide, designated β_2-microglobulin. The genes for both the class I polypeptide and β_2-microglobulin are homologous with Ig and are members of the immunoglobulin supergene family. The class I molecules have three extracellular domains, each encoded by an individual exon. The two most external domains display substantial structural polymorphism, so that many different allelic forms of each class I gene exist and individual allelic forms have multiple differences from one another. These allelic differences are recognized by the immune system, and the responses to the allelic polymorphisms of class I and class II molecules are principally, but not solely, responsible for the rejection of grafts.

Class II molecules are heterodimers of two chains, designated α and β, both of which are encoded in the MHC. In the human, there are four sets of these molecules.

Helper T cells usually recognize complexes of antigen-derived peptides and class II molecules, whereas killer T cells usually co-recognize antigens and class I MHC molecules. These complexes are formed by noncovalent interaction between the MHC molecule and the antigen-derived peptides. Moreover, distinct polymorphic forms of class I and class II molecules are associated with the capacity to respond to specific antigens; this polymorphism correlates, at least in part, with the capacity of the MHC molecules to bind peptides derived from those antigens. The association of distinct human leukocyte antigen types with susceptibility to diseases such as non-insulin-dependent diabetes mellitus may reflect the capacity of complexes of antigen and specific MHC molecules to give rise to immune responses that destroy the tissues upon which such complexes appear.

The recent determination of the crystal structure of a class I molecule is a major achievement that should be followed up and extended to class II molecules. When this is done, the structure of these molecules in noncovalent association with antigenic peptides and in the ternary complex consisting of a T-cell receptor with antigen and class I or class II molecules should be better understood. This ambitious undertaking should provide critical information for the understanding of the structural basis of immunogenicity and of the mechanisms through which MHC molecules mediate their potential to determine the immunogenicity of virtually all antigenic substances.

An equally important goal of such structural determinations is to develop methods to predict which peptides in a protein will prove capable of forming immunogenic complexes with distinct polymorphic forms of class I or class II molecules. The capacity to make such predictions would be of great value in the design of vaccines, particularly for agents that have displayed low levels of immunogenicity.

T-Cell Development Can Be Studied by Tracking the Expression of Developmentally Regulated Surface Markers

Like B cells, the stem cells that give rise to T cells are found within hemato-poietic tissues. However, T-cell development takes place in a central lymphoid organ, the thymus. It has been possible to follow the development of T cells from their immature thymic precursors into fully mature cells since the expression of a series of surface markers (CD4, CD8, CD2, and CD3) and of antigen-binding receptors on these cells is developmentally regulated. This expression of surface markers allows the analysis of the molecular basis of T-cell development, of the mechanisms by which T-cell specificity is acquired, and of the germ-line T-cell repertoire from which the T-cell repertoire of mature animals is generated. Maturation of T cells in the thymus represents one of the most powerful model systems available for the study of the developmental biology of a highly differentiated set of mammalian cells.

Equally important is the opportunity that the study of T-cell development allows for the understanding of the establishment of immunological tolerance in the T-cell pool. It has been suggested that tolerance induction in T cells is more important than tolerance induction in B cells. Tolerance induction in T cells would prevent cellular (T-cell mediated) immune responses against self-antigens and would also markedly diminish antibody responses against self-antigens since most antibody responses require some form of T-cell help. Furthermore, somatic hypermutation does not seem to operate on T-cell receptor genes in peripheral T cells. Thus, the T-cell receptor repertoire would be fixed after cells leave the thymus, which suggests that T-cell tolerance would not be lost through the development of mutants capable of binding to self-antigens. By contrast, some

ORGAN TRANSPLANTATION

Transplantation of kidneys for the treatment of chronic renal disease has become an almost routine therapy. Transplantation is being increasingly used for failure of function of bone marrow, heart, liver, and lung. Progress in the experimental transplantation of other organs, notably the pancreatic islets, indicates that the field of clinical transplantation will continue to expand. Surgical techniques have advanced remarkably, allowing the successful performance of often heroic operations. The major barrier to long-term function of organ and tissue grafts, however, is immunological rejection of the foreign organ by the host, and, in the case of bone-marrow grafts, of the reaction of the grafted tissue against the host (graft-versus-host disease). Studies in humans and experimental animals indicate that immunological responses to the polymorphisms in class I and class II MHC molecules are a key component of the graft rejection process. Because of the enormous diversity of HLA types (human MHC types) in human populations, the problem of finding matched donors is difficult. The development of monoclonal and polyclonal antibodies specific for the unique antigenic determinants of each allelic form of the HLA molecules has allowed the typing of tissues to determine the relatedness of HLA molecules of a potential donor-recipient pair. Recently, molecular genetic techniques have made possible tissue typing by the examination of structural polymorphisms of HLA genes. Nonetheless, the relative success of organ grafts from nonrelated individuals (usually cadavers) is not strikingly related to the degree of tissue matching; when donor and recipient are related, however, tissue matching predicts a better outcome. This suggests that unrecognized HLA or other differences found in unrelated pairs play an important role in graft rejection. Thus, further improvements in typing techniques may be prerequisites to greater success in transplantation of cadaver-derived organs, but the continued use and improvement of immunosuppressive techniques also seem essential.

The development of effective immunosuppressive drugs, such as cyclosporin A, has been of great value in improving transplant acceptance and success. Nonetheless, nonspecific and often chronic immunosuppression has the major drawback of decreasing the overall state of immune function in the recipient.

Efforts are now under way to attempt to develop means to induce immunological tolerance in the recipient to the transplantation antigens of the donor, in the hope that this specific unresponsiveness will allow long-term graft survival while in no way diminishing the overall immune reactivity of the recipient. If a successful, clinically applicable strategy for the induction of specific tolerance becomes available, the possibility of using organs from other species (xenografts) will need to be considered. The successful use of xenografts would largely ameliorate the current severe shortage of suitable organs for transplantation.

evidence suggests that one mechanism through which autoantibodies emerge is somatic mutation in immunoglobulin genes in mature B cells. This process may be a major factor in the development of pathogenic autoantibodies.

T Cells Are Activated by Signals from a Receptor Complex and the Expression of IL-2 Receptors

The membrane receptors through which T cells recognize antigen associated with class I or class II MHC molecules are composed of two distinct chains, which are Ig-like in that they bear sequence homology to Ig and have a generally Ig-like structural organization. Most T cells express a receptor composed of α and β chains, both of which are glycosylated peptides with molecular weights of about 45,000.

The antigen-recognizing polypeptide chains of the T-cell receptor are part of a complex of polypeptides, designated T_i-T3. The generation of intracellular signals as a result of the binding of the antigen and MHC molecular complex to the T-cell receptor is believed to be mediated through the associated polypeptide chains. Intracellular signaling events in T cells are similar to those in B cells. Two activation events of particular importance have been identified: the expression of membrane receptors for IL-2, which is a potent T-cell growth factor, and, in a subset of T cells, the production of IL-2. These events imply that T cells are capable of autologous stimulation of growth since they both make and respond to IL-2.

Activated T Cells Produce Many Distinct Lymphokines Having Important Functions in the Immune and Hematopoietic Systems

Lymphokines are generally made in small amounts and in many cases seem to act locally by binding to target cells having receptors for the lymphokine. Indeed, those lymphokines that assist in the interactions of T cells with antigen-presenting cells, such as macrophages, B cells, and related cell types, are likely to be directionally secreted and seem to act almost exclusively on the cell with which the T cell is engaged. This exclusivity provides a mechanism through which antigen-specific responses may be mediated by the action of factors that are not specific for antigen. In interactions with specific T cells, these lymphokines would be concentrated only in the immediate vicinity of the cells that participate with the T cells in specific immune responses. Other factors important in the immune response may act more like endocrine hormones. Interleukin-1 (IL-1), a factor made by activated macrophages as well as by many other cell types, has a broad range of functions in inflammatory responses, including the activation of T cells and the elevation of body temperature by its action upon hypothalmic cells. These functions suggest that some of the actions of IL-1 are mediated by action at a distance from the cell that secreted it.

The Action of Lymphokines on Their Target Cells Is Mediated by Their Binding to Cell-Surface Receptor Molecules

The interaction between lymphokines and receptor molecules results in the creation of intracellular biochemical signals through which the action of the lymphokine on the target cell is achieved. The most intensively studied lymphokine receptor is the IL-2 receptor, which exists in both high- and low-affinity forms; the bulk of current evidence suggests that the high-affinity form is largely responsible for the biological actions of IL-2.

Since adult T-cell leukemia cells have large numbers of IL-2 receptors, antibodies to the receptor may provide an important tool for the therapy of this highly aggressive lymphoid malignancy. Immunotoxins based on antibodies to the IL-2 receptor are now being tested as therapeutic agents. If successful, such a therapy would suggest that other malignancies displaying heightened numbers of lymphokine receptors would be candidates for a similar antibody-receptor-immunotoxin therapeutic approach.

Killer T Cells and Natural Killer Cells Destroy Cells That Cause Disease

Lymphokine production is one mechanism through which T-cell responses are mediated. Although lymphokines can function as effectors, most of their actions are regulatory. Killer T cells, by contrast, have as their principal role the destruction of antigen-bearing target cells, such as cells infected with viruses and those bearing on their membranes viral proteins. These antigens form complexes with class I (in some cases class II) MHC molecules, which are recognized by the receptors of the killer T cell. The recognition of the antigenic complex on the surface of the target cell activates the killer cell, which in turn sets in motion the process that destroys the target cells. Although controversy still exists on this subject, T-cell-mediated cytotoxicity seems to involve killer-cell production of substances called porins, which polymerize, integrate into the target-cell membrane, and cause cell lysis. Activated killer T cells may also produce esterases, which may also play a critical role in the destruction of the target cell by specific killer T cells or may help the killer cell detach from targets.

Cellular cytotoxicity is also mediated by a second cell type, designated natural killer (NK) cells. These cells are large granular cells resembling lymphocytes. The molecular mechanisms of killing by NK cells and killer T cells seem to be the same, but the means through which the cells identify their targets differ. NK cells lack T-cell receptor molecules and do not display the type of specificity generally seen in B or T lymphocytes. Nonetheless, they seem to distinguish and to preferentially lyse certain types of tumor cells. The identification of the molecular nature and specificity of the NK cell receptor is a goal of considerable importance. NK cells seem to mediate an important element of nonspecific

POTENTIAL CLINICAL APPLICATIONS OF LYMPHOKINES

The potent actions of lymphokines on a wide variety of cellular targets and their roles in the control of many aspects of the immune response and of inflammation have suggested that these agents may be of great importance in therapy of diseases in which the immune response or the hematopoietic system normally plays a major role. This possibility has led many biotechnology and pharmaceutical companies to undertake substantial research and development programs aimed at developing lymphokines and related biological response modifiers for clinical purposes. Efforts to treat various malignancies with interferon have attracted considerable attention. It is now clear that interferon is a useful tool in the treatment of a limited number of malignancies, most notably hairy-cell leukemia. IL-2 has recently been used to treat a wide range of metastatic malignancies. Lymphocytes from the blood of a cancer patient are grown in tissue culture to generate lymphokine-activated killer (LAK) cells and are then returned to the patient, who is next treated with large amounts of IL-2. This therapy causes regression in some metastatic malignancies, but it has serious side effects that limit its use.

Other potential uses of lymphokines are not limited to the treatment of malignancies. For example, treatment of infectious diseases and the re-population of the immune and hematopoietic systems in immunosuppressed patients may be aided by the use of T-cell-derived hematopoietic and lymphoid growth and differentiation factors produced by recombinant DNA techniques. The use of these agents to hasten cellular repopulation in individuals who have received cancer chemotherapeutic agents and in re-cipients of bone-marrow grafts may be particularly important. If it proves possible to find methods to treat AIDS patients by preventing cell-to-cell transfer of virus, T-cell-derived hematopoietic growth factors may be of great importance for the reconstitution of the immune system of the recovering patient.

immunity, mainly against arising tumor cells, but possibly also against infectious agents.

Regulation of the Immune Response

Immune Responses Are Subject to Both Positive and Negative Regulatory Control

A key example of regulatory control is the action of T lymphocytes as helper cells in antibody responses. The production of antibody by B cells requires either intimate contact between T and B cells (cognate interactions) or the presence of T-cell-derived soluble products. It is now recognized that helper T cells from mice can be subdivided into two sets, T_{H1} and T_{H2} cells, on the basis of the lymphokines they produce. The lymphokines produced by T_{H1} cells promote cellular immune responses, such as macrophage activation and delayed hypersensitivity, and those produced by T_{H2} cells promote antibody production. The two forms of helper T cells are also specialized in terms of the relative degree of cellular immunity that emerges in immune responses in which one or the other predominates. One of the most important issues to be resolved is the means through which the relative numbers of T_{H1} and T_{H2} cells participating in a given immune response are determined. It is most likely that the T_{H1} and T_{H2} cells have distinctive activation requirements and that immunogens may differ in their ability to activate cells of the two types. For example, antibody responses to many parasitic agents are marked by the production of very large amounts of IgE. It seems most likely that responses to these parasites are dominated by the action of T_{H2} cells, with some feature of the parasite leading to the selective activation of the T_{H2} cell.

A second major aspect of immunological regulation is the inhibition or suppression of responses. This mechanism is necessary to keep immune responses from overwhelming the system; it may also play a role in the maintenance of immunological tolerance. Suppressor cells seem to be a distinctive subpopulation of T cells. Many of them express the CD8 surface marker. Their mode of action has been the subject of intense research interest. In addition, in many autoimmune diseases, a failure of suppressor cell function may account for some of the response to self-antigens. Suppressor mechanisms may also limit immune responses to tumor cells and thus diminish the effectiveness of the immunological control of tumor cell growth.

A particularly intriguing mechanism through which immune responses may be regulated involves the "immunological network," which is based on the capacity of antibodies and receptors to recognize one another. The variable regions of antibodies and receptors are themselves antigenic determinants, which may thus elicit the production of specific antibodies. The unique antigenic determinants of antibodies and receptors are referred to as idiotopes, and the collection of idiotopes on an individual antibody is its idiotype. Anti-idiotope antibodies, through their recognition of structures on the variable regions of antibodies and receptors, may be thought of as surrogates for the antigens to which the antibodies and receptors "normally" bind. Thus, such anti-idiotope antibodies may be regarded as "internal images" of conventional antigens, and idiotope–anti-idiotope interactions may determine the level of action of the immune system against exogenous antigens. Idiotope-based network interactions may be either stimulatory or inhibitory, depending on the conditions of the interaction. What remains to be established is the relative importance of network-based regulation in determining the level of activity of the immune system.

Complement and Other Effector Molecules

The Complement System Can Destroy Antigen-Bearing Microorganisms or Cells

The interaction of antibody molecules with antigens on the surface of a microorganism or a malignant cell may prevent the organism from entering a host cell or may block the normal metabolism of the cell. More often, the antibody by itself has only a limited effect on the infectivity of the microorganism or the behavior of the malignant cell. However, the interaction of antibody with antigens locally activates a series of enzymes, the complement system, that may directly or indirectly destroy the antigen-bearing microorganism or cell. In addition to direct destruction of target microorganisms or cells through attack on the cell membrane, the activation of the complement system leads to the formation of several highly active fragments of complement components. Among these are the anaphylatoxins that lead to the release of vasoactive amines from mast cells and basophils, resulting in vasodilatation and striking local inflammation.

Finally, macrophages and polymorphonuclear leucocytes bear receptors for the components of complement (complement receptor molecules); the engagement of this receptor triggers phagocytosis. In this way, the deposition of complement components on the surface of a microorganism markedly enhances phagocytosis and the destruction of the infectious agent.

The complexity of the complement system and its high degree of internal regulation have made it a difficult system to study in detail. Nonetheless, great progress has been made as a result of purification of each major component of the system and cloning many of the complement genes. Efforts to understand the molecular and cellular basis of complement action, the regulation of the level of activity of the complement system, and the means infectious agents use to evade the system are producing exciting results.

Hypersensitivity, Inflammation, and Phagocytosis

The Study of the Cellular and Chemical Mechanisms of Inflammation and Phagocytosis Is Crucial to the Development of Anti-inflammatory Drugs

A common feature of the cellular response to foreign agents is a local inflammatory response. Inflammation that occurs as a result of immune responses is often referred to as hypersensitivity. The immediate type of hypersensitivity is mediated by the action of antibodies of the IgE class. IgE interacts with

AIDS:
DESTRUCTION OF IMMUNE FUNCTION
BY T-CELL INFECTION WITH HIV

The hallmark of AIDS is the loss of immune responsiveness and the consequent susceptibility of patients to overwhelming infections, often with microorganisms that are not pathogens for healthy young individuals. The immune systems of AIDS patients display a variety of defects that contribute to their failure to make protective immune responses. Central to many of these defects is a reduction in the number and function of T cells that express CD4 molecules on their surface (CD4$^+$ T cells). The CD4$^+$ cell population contains both the T cells that help B cells make antibodies and the T cells that are important in cellular immune responses. The role of CD4$^+$ T cells includes the activation of CD8$^+$ killer T cells and of macrophages, which allow them to destroy bacteria and other microorganisms that they take up through phagocytosis. HIV infection destroys CD4$^+$ T cells, resulting in an immune system deficient both in regulatory control and in responsiveness to new antigenic stimuli. Moreover, even those CD4$^+$ cells that remain in infected individuals seem to be abnormal. These cells fail to make in vitro immune responses to conventional antigens such as tetanus toxoid, to which virtually all normal individuals respond. Abnormalities in the transcription of the IL-2 gene in HIV-infected T cells may explain the abnormal function of these cells. Immunological abnormalities of AIDS patients also include profoundly depressed activity of cytotoxic T cells, although patients do have precursors of cytotoxic T cells that can express normal amounts of cytotoxic activity in the presence of a source of IL-2.

Major efforts are now under way to determine how a protective immune response to HIV can be induced through the use of HIV components in the hope that this understanding will point the way to the development of an AIDS vaccine. A major impediment in this effort is the lack of an experimental model of HIV infections leading to an AIDS-like condition. The availability of such a model either in a naturally occurring animal or in an animal made susceptible through cellular or genetic engineering might markedly speed the pace of vaccine development. Efforts to ameliorate the destructive effects of HIV infection through the use of antiviral agents, such as azidothymidine (AZT), now known as zidovudine, must also be emphasized to provide treatment for infected individuals.

receptors on the surface of mast cells and basophils (Fc_ε receptors) and, when cross-linked by antigen, causes these cells to release histamine and other highly active molecules. These substances cause local vasodilatation and may cause bronchoconstriction as part of systemic reactions that occur in severe cases, termed anaphylactic reactions.

The inflammatory response that occurs both in hypersensitivity and in direct irritant responses includes the entry into the tissues of neutrophils, macrophages, and eosinophils. These cells are produced in the bone marrow and are distributed throughout the body by the blood. They arrive at sites of inflammation by migrating between postcapillary endothelial cells. Neutrophils are the first cells to enter inflammatory sites. They are phagocytic cells that also release products that activate complement and recruit monocytes and additional neutrophils to the inflammed site.

Monocytes are the predominant cells at wound sites within 24 hours. Over a period of days, they develop into macrophages capable of phagocytosis, microbicidal activity, and the release of mediators, such as IL-1 and tumor necrosis factor. Macrophages play a particularly important role in the destruction of intracellular parasites; such organisms can cause tuberculosis, leishmaniasis, and toxoplasmosis. In response to infection with these microbes, macrophages often fuse to form multinucleated giant cells and granulomas. This process is believed to help contain and destroy pathogens.

Eosinophils are normally present in small numbers, but they increase strikingly in frequency in parasitic infestations and in allergic diseases. These cells are believed to destroy certain parasitic agents, possibly through the recognition by Fc_ε receptors of IgE antibodies coating the parasite, leading to the release of cytoplasmic granules that contain enzymes that are highly toxic for the parasite.

These inflammatory cells can also phagocytose particles. Phagocytosis is enhanced when the cell binds the particle through its cellular receptors for the Fc portion of IgG (Fc_γ receptors) and through its CR1 surface receptors. Thus microorganisms or cells coated with antibodies or complement are much more efficiently taken up by phagocytic cells than are uncoated particles. The function of the Fc_γ receptors is mediated through their action, when aggregated in the plane of the membrane, as voltage nonselective channels for monovalent cations. Through this mechanism, Fc_γ receptor aggregation elevates the concentration of calcium ions, an absolute requirement for the membrane internalization that is a key feature of phagocytosis.

Phagocytosis brings microorganisms (or cells) into lysosomes, which are rich in degradative enzymes. If sensitive to these enzymes, the phagocytosed agent is destroyed. Phagocytic cells have other means of destroying ingested microorganisms, including a complex enzyme system for generating toxic oxygen products, including hydrogen peroxide and hydroxyl radicals. In addition, these cells produce hypochlorous acid, hypochlorite, and chlorine, which oxidize and halogenate microorganisms and tumor cells. These metabolites can also damage normal cells and injure tissue as is seen in chronic inflammatory conditions. Since the study of the cellular and chemical mechanisms of inflammation and phagocytosis forms one of the principal arenas for the development of drugs that limit tissue damage in chronic inflammatory diseases such as rheumatoid arthritis, it will continue to require strong emphasis, by both academic and industrial laboratories.

INFECTIOUS DISEASES

In Recent Years Our Understanding of Mechanisms of Microbial Pathogenicity Has Significantly Advanced

Microorganisms constitute an extraordinarily diverse group of organisms, which are found in virtually every environmental niche, infect all living creatures, and can cause disease in their hosts. When infection occurs, it may lead to a nonpathogenic outcome, in which the infecting agent does not penetrate the host barriers or is effectively controlled by the host immune system. Many bacteria and viruses coexist with the host, causing disease only rarely and only when the host is compromised by impaired immune defense mechanisms. By contrast, the most virulent pathogens produce disease in almost every infected host. After being infected with a pathogenic microbe, the host organism may recover completely and eliminate the pathogen or may develop a persistent or latent infection, resulting in subsequent illness or relapse after a prolonged interval.

Pathogenic microorganisms are a highly heterogeneous group of agents; they include some of the smallest and simplest of all biological life forms, from viroids, which consist solely of nucleic acids, to helminths and protozoa, which are highly complex, sometimes multicellular eukaryotic organisms.

For most pathogenic microorganisms, our understanding of the precise mechanisms of pathogenesis (the capacity of the microbe to cause disease) are poorly understood. The degree of pathogenicity of a microorganism is termed virulence. Virulence is a multifactorial property determined by the product of more than one gene. For bacteria, virulence factors include such properties as growth, motility, chemotaxis, adherence to host tissues, and resistance to lethal host defense mechanisms (for example, phagocytosis and bacterial antibodies), elaboration of toxins, and penetration of host cells. For viruses, virulence factors include the capacity to spread, replication in cells, survival in tissue fluids, interaction with proteases and host-cell receptors, and resistance to host defense (macrophages, T cells, and interferon). The interaction between the invading pathogen and the host's immune system, together with the concerted effects of many distinct microbial gene products, is ultimately responsible for inducing disease in the host.

Entry into the Host

The Mammalian Host Presents a Number of Barriers to Early Events of Pathogenesis

The portals through which pathogens gain entry into a host include the conjunctiva, the mouth and gastrointestinal tract, the nose and respiratory tract, and the genito-urinary tract. In addition, the body is covered with skin, a natural barrier that may be broken by a number of means (trauma, insect or animal bites, needles). For each of these sites of entry, different microbial strategies have evolved that allow microbes to infect the host.

After entering the host, the microbe undergoes primary multiplication. For viruses and intracellular parasites, multiplication occurs inside cells. For extracellular parasites, adherence of the microbe to cell surfaces is an important event that allows the colonization of specific tissues.

Many Colonizing Microorganisms Seem to Prefer Certain Tissue Sites over Others

The specificity of the association of particular bacteria with various host tissues was suggested by ecological studies of the indigenous (normal flora) and pathogenic bacteria colonizing oral mucosal surfaces and the various niches of dental tissue. For example, the actinomycete bacteria *Streptococcus mutans* and *S. mitis*, both of which promote tooth decay, were found in large numbers in dental plaque but sparsely on the surface of tongue epithelial cells. The reverse was true for *S. salivarius*, an organism normally found in abundance on the tongue but not at all on the teeth.

The contrast between two common pathogenic bacteria reinforces the concept of specificity. *Escherichia coli*, the most frequent cause of urinary tract infections, is abundant in periurethral tissues, but is seldom found in the upper respiratory tract. In contrast, group A streptococci, which are virtually limited to colonization of the upper respiratory tract and skin, are seldom associated with urinary tract infections. The specificity of bacterial adherence and colonization is further supported by the observation that certain bacterial infections are limited to one animal species.

The Chemical Nature of Some Bacterial Adhesins and Their Corresponding Receptors on Host Tissues Is Known

A large number of studies performed during the past decade have established that many bacteria have surface structures that bind to specific macromolecules on host cells in a lock-and-key (or induced-fit) fashion analogous to the combination of an enzyme with its substrate or an antibody with its antigen. The terms adhesin and receptor describe the corresponding molecules on the surfaces of the bacteria and the animal cells, respectively. In general, the bacterial adhesins are composed of proteins in the form of fimbriae or pili (hairlike fibers), whereas the recognized part of the receptor is composed of carbohydrates. Because of their specific interactions with carbohydrate residues, most of the bacterial adhesins can be considered lectins, and like plant lectins, many of them agglutinate red blood cells.

Bacterial agglutinins (or lectins) were first reported in 1908, following the observation that cells of *E. coli* agglutinate erythrocytes. Only recently, however, was it established that carbohydrate-lectin interactions constitute the molecular means by which most bacteria adhere to animal cells. These interactions occur in three ways.

1. Bacterial surface lectins can bind to the carbohydrates on the surface of animal cells. Since the 1950s it has been known that mannose and methyl-α-D-mannoside specifically inhibit the adherence of many gram-negative bacteria to eukaryotic cells. In 1977, bacteria were suggested to have surface lectins, which serve as adhesins that bind the organisms to mannose residues on human cells. Since then, both gram-positive and gram-negative bacteria have been found to express specific lectins on their surfaces. Several different sugars that may serve as sites for attachment for these lectins are characteristic constituents of cell-surface glycolipids or glycoproteins.

The lectins are organized on bacterial surfaces in either fimbrial or nonfimbrial structures. Fimbriae, long tubelike projections, are assembled as hollow fibers by the polymerization of the monomeric subunits composed entirely of protein. On the basis of their sugar specificity, several types of fimbrial lectins have been distinguished.

Some bacterial strains are genotypically able to turn on and off the expression of fimbriae. The phenotypic expression of these organelles seems to be under the control of an on-off switch in the DNA itself and is minimally affected by environmental influences. This switching results in a relatively constant shift from fimbriate to nonfimbriate (or vice versa) at a frequency of one per thousand bacteria per generation, which is about 10,000 times as frequent as a true mutational event. The ability to undergo such rapid phenotypic variation in vivo is probably a key determinant of the survival of bacteria on mucosal surfaces and of their pathogenicity once they invade deeper tissues. Evidence for such rapid phase shifts has been reported for *Proteus mirabilis* (responsible for induced urinary tract infections).

2. Extracellular lectin molecules can form bridges between carbohydrates present on the surface of the bacteria and their host. The bridging type of carbohydrate-lectin interaction was first noticed in the interaction of the plant bacterium *Rhizobium trifolii* with the root hairs of clover. Clover roots secrete a carbohydrate-binding protein, trifolin A, which recognizes carbohydrate structures on the surfaces that are shared by the bacteria and the root hairs and ligates them by forming a bridge between their common sugar residues.

Recent data suggest that a mechanism of adherence similar to that of *R. trifolii* may occur between bacterial and animal cells. The bridging lectin may have endogenous (produced by the host) or exogenous (acquired mostly from food) origins. For example, the intestinal secretions of guinea pigs is characterized by lectin activity specific for glucose and fucose. This lectin agglutinates *Shigella flexneri* and may mediate its attachment to mucosal surfaces of the gut.

Since lectins are abundant in food, the possibility was examined that these exogenously acquired lectins may become associated with mucosal cells as a cell-coat component and thus function as receptors for bacterial adherence. Indeed, buccal epithelial cells scraped from persons shortly after they eat raw wheat germ contain high amounts of wheat germ agglutinin on their surfaces. Because wheat germ agglutinin is an *N*-acetylglucosamine-specific lectin, the coated cells bind an increased number of *Streptococcus sanguis* organisms, which contain *N*-acetylglucosamine residues on their surface. Certain dietary lectins may also reach the alimentary tract in a functional form and thus may potentially mediate bacterial adherence.

3. Bacterial surface carbohydrates can bind to a lectin contained within the animal cell membrane. Evidence accumulated during the past 10 years indicates that animal cells exhibit lectins on their surfaces. Such lectins have been found on many cell types. Liver-cell lectins have been found to serve as receptors for serum asialoglycoproteins, asialoerythrocytes, and various glycoproteins, and are thus responsible for the clearance of these tissue elements from the blood. Similarly, when bacteria were injected into the blood streams of animals, they become lodged in the liver in less than an hour. Recent studies demonstrated that bacteria were trapped in the liver by hepatic receptors (lectins), which recognize corresponding sugar residues on the bacterial surfaces.

Carbohydrate-Lectin Interactions Play a Central Role in the Mutual Recognition Between Many Bacteria and Host Cells

One must be cautious in drawing general conclusions about the significance of lectin-carbohydrate interactions in vivo on the basis of in vitro studies alone. It is not unlikely that the physiological environment influences the orientation and relative accessibility of the sugar moieties in the oligosaccharides involved in the interactions in vitro; a change in orientation theoretically would markedly affect not only the affinity but also the specificity of the bacteria-host-cell interaction.

Nevertheless, with this caution in mind, further characterization of an increasing number of unique bacterial and host cell lectins and oligosaccharides should lead to a better understanding of their functional significance and the mechanisms by which their synthesis and expression are controlled.

Sugars Are Not the Only Determinants of Recognition Between Bacteria and Epithelial Cells

Studies have been aimed at clarifying the molecular basis of the host-cell attachment of streptococci. These organisms inhabit the skin, the mouth, and the throat and may cause a variety of infections and postinfectious complications, such as acute rheumatic fever or acute glomerulonephritis. Transmission electron microscopy of streptococci attached to buccal epithelial cells has revealed the presence of fibrillar structures extending from the bacteria to the surface of the epithelial cells. These fibrillar structures are composed of proteins complexed with lipoteichoic acid, an ampiphathic polymer produced by the bacteria: It consists of repeating units of glycerol phosphate capped at one end with a glycolipid moiety. The latter moiety is responsible for the binding of the strepto-cocci to specific receptors on oral epithelial cells. Indeed, isolated lipoteichoic acid binds to receptors on a wide variety of mammalian cells. Moreover, the purified polymer blocked the adherence of epithelial cells of streptococci, but not of other bacteria, such as *E. coli*.

Since the fatty acid of lipoteichoic acid mediates the epithelial cell attach-ment of streptococci, a protein or glycoprotein with fatty-acid binding sites may serve as the receptor. The receptor now seems to be fibronectin, a ubiquitous glycoprotein present on the surface of many host cells—including oral epithelial cells. Binding of lipoteichoic acid to fibronectin occurs at a site on the glycopro-tein that specifically binds fatty acid residues with a binding affinity two orders of magnitude higher than the binding of lipotichoic acid to fatty-acid binding sites of serum albumin. These results indicate that in addition to lectin-carbohydrate interactions, protein-lipid interactions are centrally involved in the specific at-tachment of certain bacteria to host cells.

Flagellates, which are unicellular eukaryotes, invade host cells by the specific interaction of parasite and host molecules. Malaria parasites invade erythrocytes by interacting with specific ligands on the red cell membrane. The ligand involved in invasion by *Plasmodium vivax*, a human malaria, and *P. knowlesi*, a simian malaria, is the Duffy blood group antigen; this identification explains the resistance to *P. vivax* of West Africans who lack the Duffy antigen on their red blood cells. The comparable molecule for *P. falciparum*, the major human malarial parasite, is sialic acid on glycophorin. The parasite receptors that bind these ligands are now being identified. The value of these receptor molecules as vaccine targets will depend on their variability in the parasite population and the presence of alternative pathways for invasion.

Spread in the Host

When microbes enter into tissues beyond the epithelial surfaces, they are transported into the lymphatic system, where they are presented to the immune system. This interaction may limit the infection or may disseminate it to regional lymph nodes and eventually to the blood. Some microorganisms invade blood vessel walls and enter the blood directly. Spread through the blood can be rapid and can result in generalized infection.

A Number of Pathogenic Bacterial Species Can Penetrate and Grow Intracellularly Within Host Cells

Intracellular bacteria resist host immune mechanisms essentially by hiding within host cells. Vascular circulation of host cells infected in this manner can then disseminate the bacteria further into the lymph and blood, which in turn seed other organs and tissues of the host.

Some bacteria and protozoa gain entry into the host cell cytoplasm passively by being phagocytosed by macrophages. Once inside the phagocytic cells, these organisms evade the intracellular killing activity of the macrophage by mechanisms that include resistance to lysosomal microbicidal activity, inhibition of phagolysosomal fusion, or simple escape from the confinement of the lysosomal vacuoles and multiplication free in the cytoplasm of the host cell.

Other bacteria induce their own uptake by phagocytic cells such as epithelial and endothelial cells. The microbial gene products that mediate the binding to host cells and that trigger the endocytic uptake of bacteria have been partially characterized by genetic methods (gene cloning and mutagenesis).

Microbes Can Spread Through Local Extension into Neighboring Tissues and Through Peripheral Nerves

For microbes that do not enter the bloodstream or peripheral nerves, spread may occur by extension to adjacent cells or release into fluids on body surfaces. Nonspecific host defenses and local anatomical features may play central roles in limiting infection. The spread of bacteria from local sites is enhanced by spreading factors, particularly enzymes. But whether these enzymes play additional roles in the pathogenesis of bacterial disease remains obscure. Viral spread by successive infection of adjacent cells is best illustrated with viral infections of the skin (warts and vaccinia).

An alternative route of spread is through peripheral nerves. Certain viruses (rabies and herpes simplex) and toxins (tetanus and diphtheria) spread from the periphery of the body to the central nervous system through nerves. The exact neural pathway and the microbial determinants remain undefined. For neurotropic reoviruses, the viral attachment protein is responsible for allowing the neurotropic strain to enter peripheral nerves, whereupon subsequent movement takes place in the fast axonal transport system (see Chapter 4).

The specific microbial and host factors that determine the capacity of the microbe to spread and choose one or another pathway in the host are poorly defined. The initial battle between the host and the parasite takes place in lymph nodes and results in either a virulent or nonvirulent outcome. Virulent bacteria entering a lymph node often kill macrophages or resist being taken up by them, produce products that dilate blood vessels and affect inflammatory cells, and produce substances that allow the bacteria to continue to divide and spread. Similarly, some viruses resist macrophage destruction and thus have the capacity to replicate, rather than be destroyed in macrophages or lymphocytes. These viruses can then spread within the host as part of the cells' normal migration. Whereas viruses—including those that are relatively avirulent—spread with relative ease through the host via the bloodstream, most bacteria, fungi, and protozoa are blocked from spread by lymphatics. Understanding the factors that determine whether a microorganism can evade the filtering action of the lymph nodes is a goal of major importance.

After Entering the Blood, Microbes Are Efficiently Transported Through the Body

Although most bacteria or fungi do not regularly invade the blood, many viruses, rickettsiae, and protozoa do. These microbes may be carried in several compartments: free in plasma, in leukocytic-associated compartments (mononuclear phagocytes), and in erythrocytic-associated compartments. Removal of microbes from the blood may be efficient. Since the reticuloendothelial system with its population of fixed macrophages is heavily concentrated in the liver and spleen, most clearance occurs at these sites. Infected leukocytes may also be trapped at these organs, which would help to arrest infection. Transient bacteremia or viremia are thus usually of little consequence. But, when host resistance is impaired or when especially virulent microbes circulate, a sustained viremia or septicemia may occur.

The liver is the main organ responsible for clearing foreign particles, including bacteria, from the blood. Studies of virulent strains of bacteria may shed light on the mechanisms by which pathogenic organisms escape being recognized by hepatic lectins and hence being cleared from the blood during bacteremia or septicemia.

The blood may also deliver microbes to distant organs. The blood-tissue junction plays a central, albeit poorly defined, role in determining spread into an organ. For a microbe to enter an organ, it must first bind to the surface of the endothelium of the blood vessel, usually in capillaries. It must then reach the tissue by moving through breaks or leaks in the vessel wall, growing through the wall, or being carried passively in association with some structure (or cell) passing through the wall. Our ability to study this phase of the infectious cycle has improved dramatically in recent years with the development of methods to culture blood vessels and to study the interaction of microbes with endothelial cells.

Interactions of Viruses with the Host Cell

Viruses Exist Either as an Extracellular Virion Particle or as Intracellular Forms Undergoing Replication

For inactivation of a virus, the infectious virus particle must be neutralized by antibodies or its replication must be inhibited within the host cell. To be therapeutically useful, any agent that inhibits replication must specifically affect a viral replication process and not the host cell.

Once a virus has bound to the cell receptor molecule, the virion must cross the plasma membrane to enter the host cell. Nucleocapsids of some lipid-enveloped viruses cross the plasma membrane by direct fusion of the lipid envelope with the cell plasma membrane. Nucleocapsids of other types of viruses cross the membrane into the cytoplasm after being taken into endocytotic vesicles. Once inside the host cell, the viral genome is transported to its cellular site of replication—often the nucleus for DNA viruses and the cytoplasm for RNA viruses. One of the next events is the synthesis of viral mRNA if the viral genome cannot be used directly for translation. With some DNA viruses, host-cell enzymes produce the viral mRNA. For other viral genomes, such as double- or single-stranded RNA genomes, no cell enzymes exist that can transcribe the genomes into mRNA. Therefore, these viruses encode a transcriptase enzyme, which is often encapsidated in the virion. After viral proteins are translated, the input viral genome is amplified or replicated and progeny virions are assembled. Any of these viral-specific stages of replication are potential targets for antiviral agents.

Cellular Molecules or Structures May Be Appropriated by Viruses and Modified to a New Form for the Replication of the Virus

If the virus utilizes a host-cell component for its own replication or inactivates a host-cell function to promote its replication process, injury to the cell can result. For example, poliovirus inactivates one of the host-cell translational initiation factors during its replication cycle. This inactivation is believed to favor the translation of viral mRNAs, but it is also one of the factors that eventually leads to cell death and lysis. Similarly, DNA viruses such as herpes simplex utilize parts of the cell's nuclear transcription and DNA replication apparatus for their replication, and they inhibit these cell processes. Herpes simplex virus may even utilize preexisting nuclear sites of cell DNA synthesis as sites for assembly of viral DNA replication complexes. Such perturbations of the host cell may lead to the gross pathological changes in the cell known as cytopathic effect.

If the virus stimulates the expression or activity of certain cell gene products and the cell survives infection, the cell may emerge with an oncogenic, or cancerous, phenotype. In this case, part or all of the viral genome becomes integrated into the cell genome. For example, infection of mouse cells with simian virus 40 (SV40) leads to a low frequency of transformed cells that retain part of the SV40 genome and continue to express the large T antigen that interacts with the host cell to alter its growth controls.

Two other possible outcomes of viral infection are the establishment of a latent or a persistent infection. In a latent infection, no infectious virus is found in the cell, but upon reactivation, the viral genome within the cell is replicated and infectious virus is produced. In a persistent infection, the infected cell survives, but the cell continuously produces infectious virus at some detectable level.

Cell and Tissue Tropism

A Hallmark of Microbial Infection Is the Localization of Infection in Specific Tissues or Cells

Localization gives rise to the specific symptom complex of infectious diseases and is perhaps the most characteristic feature of microbe-induced diseases. The factors determining the localization of microbes in certain tissues—their tropism—have been under intensive investigation. Although we still do not know the details of all the factors involved, certain aspects of major significance in host-parasite interaction are emerging. These factors include the presence of cell-surface receptors recognized by surface structures of microbes, tissue-specific proteases, and tissue-specific enhancer DNA or other regulatory sequences in the viral DNA. Although tissue localization has been studied in all classes of microbes, most of the recent insight into tissue specificity has emerged from studies with bacteria and viruses; the findings concerning bacterial adhesion have already been described.

That the specificity of certain viruses (for example, poliovirus) for human cells results from the interaction between the viral capsid and host-cell receptors exposed at the cell exterior has been well known since the late 1950s. The presence or absence of receptors for viruses on the cells' plasma membrane is of central importance in initiating infection. For a number of viruses, the attachment protein on the virus particle recognizes sugar molecules on host cells as bacterial lectins do, suggesting that many strategies used by bacteria to colonize and adhere to host tissues resemble those used by viruses.

The Identification of Receptors for Viruses and Other Microbes Is a Focus of Biological Research

In addition to receptors containing sialic acid, considerable progress has been made in identifying cell-surface proteins that are both physiological receptors and receptors for viruses. These cell-surface proteins include the acetylcholine receptor for rabies virus, the β-adrenergic receptor for reovirus 3, the T4 lymphocyte antigen for HIV, and the CR_2 receptor for Epstein-Barr virus. Given the central role of receptors in pathogenesis and tropism, one should stress the importance of determining receptor identities and characterizing the molecular interactions involved in the binding reactions between virus and host receptor. Although considerable work is being done on the receptor binding of some viruses (such as influenza, polio, and HIV), the role of receptors, especially the question of receptor identity, has been neglected in most other virus systems.

In addition to illuminating microbial pathogenesis, the isolation of receptors should provide insights into the structure and function of many normal cell proteins. The possible importance of cell-surface receptors for protozoa is illustrated by studies with malaria. Malaria sporozoites attach to liver cells, possibly in response to specific receptors. Whereas the merozoite responsible for initiating and perpetuating the erythrocyte stage has a complex surface, the sporozoite has a dense surface coat protein that contains two regions conserved in all malarias. These regions may be conserved for binding to liver cells or to mosquito salivary glands. One of these regions is highly homologous to the soluble human protein thrombospondin, which cross-links host cells.

Other Determinants of Tissue Tropisms Include Tissue-Specific Host Enzymes and Regulatory Factors

The cleavage of viral glycoproteins from an inactive precursor form to an active product is an important determinant of pathogenicity for myxovirus and paramyxovirus. The cleavage of viral proteins by tissue-specific proteases is necessary to the production of infectious virus. The nature of this cleavage determines whether there will be high titers of virus. Thus, host proteases probably play a specific role in the organ-specific activation of viral proteins.

Increased research on specific steps in viral replication and the factors regulating those steps will undoubtedly enhance understanding of the molecular biology of viral virulence. In the context of tissue tropism, certain regulatory factors may be cell- or tissue-specific. Thus, for example, enhancer sequences active only in certain cells could lead to enhanced replication and high titers of virus in specific cells. Such factors may play an important role in the localization of viral growth and tissue injury.

How Is Injury Mediated?

Many Infectious Agents Can Multiply in the Host Without Causing Significant Damage

Infectious diseases are the most important outcomes of multiplication of microorganisms in the host. Many infectious agents, however, multiply in the host without causing significant damage. For viruses, this is probably the rule rather than the exception, even for viruses that are capable of causing disease. For some viruses, damage at the morphological level may be undetectable, even when effects on aspects of cell function or secretion may be profound. This type of effect on the cell, whereby the cell is capable of replicating without any detectable histological abnormality, but loses a differentiated function, has been termed a loss of "luxury" functions. Cells that secrete important products for host physiology (such as hormones) are especially important in this regard. Thus, disease production by viruses may not be associated with more than minimal overt pathology.

In contrast, when bacteria, fungi, protozoa, or rickettsiae invade tissues, overt tissue injury, slight or profound, is the rule. Microorganisms may damage cells directly by replicating in a particular cell, or, in the case of bacteria and fungi, by exposing cells to toxins that are part of the bacterial cell wall (endotoxin) or that are released from the replicating microbe (exotoxins).

In addition to the direct injury by microorganisms or their products, the damage may occur indirectly. The host response itself may lead to pathological outcomes, which may be relatively nonspecific (as when proteases are released from macrophages in an inflammatory response that subsequently results in injury to normal cells) or may result in specific injury from cellular immune responses or the action of antibodies.

Direct Damage of Cells by Viruses Can Occur in Different Ways

In general, viruses shut down host macromolecular synthesis while modifying the cell to synthesize viral proteins. Cytopathic viruses generally cause cell death only after they replicate, suggesting that the primary effects of viral infection involve the cellular synthetic machinery and that later effects on the cell are secondary. The later effects include, depending on the virus, profound effects on the cytoskeleton, injury to membrane and leakage of ions, fusion of membranes (resulting in multinucleated giant cells), or damage to lysosomes with release of lysosomal enzymes.

One of the best-studied viral systems is poliovirus. Poliovirus causes host protein synthesis to decline and ribosomes to dissociate from host mRNA. Polio infection inhibits the host by inhibiting the cellular cap-binding complex required for attachment of "capped" mRNAs to ribosomes. Since polio RNA is uncapped, unlike cellular mRNA, it is unaffected by the loss of the cap-binding protein.

Other viruses inhibit the host in different ways. For example, inhibition of host-protein synthesis by mengovirus results from competition between viral and host mRNAs.

These studies illustrate the feasibility of gaining insights into details of the molecular basis of viral injury. Undoubtedly, this will be a major focus for the next decade and will aid in identifying new approaches for developing antiviral drugs.

Infection Can Result in Indirect Damage from the Response of the Host

Inflammatory Response. The host response to microbial infection generates inflammation. This response, clasically resulting in redness, swelling, pain, and loss of function, may damage the host more than the replicating microbe. The early inflammatory response has both cellular components (macrophages and natural killer cells) and humoral components (complement, interferon, and tumor necrosis factor). The release of proteases from macrophages with injury to normal tissue (the so-called innocent bystanders) illustrates one way by which such injury may be mediated.

Hypersensitivity. In addition to early inflammatory reactions, the classic reactions of hypersensitivity may be involved in microbial immunopathology. Anaphylactic reactions depend on reactions of antigens with IgE antibodies attached to mast cells that lead to the release of histamine and heparin from mast cell granules, and they may be severe enough to lead to urticaria, bronchospasm, or shock. Helminth parasites induce severe hypereosinophilia, as in tropical (filarial) pulmonary eosinophilia, in which the toxic contents of the eosinophile granules are responsible for tissue destruction and resultant pathology.

In the type of hypersensitivity reaction termed antibody-dependent cellular cytotoxicity, antibody combines with antigen on the surface of the cell and results in cell lysis through the action of killer lymphocytes, polymorphonuclear leukocytes, or macrophages. This type of reaction may be important in certain viral, parasitic, and bacterial infections. Although cytotoxic reactions can be shown in vitro, they need further study to define their role in vivo.

Tissue Damage Mediated by Immune Complex. Circulating immune complexes are probably common in most viral infections, but they also occur with

SLOW VIRUSES, POORLY UNDERSTOOD,
MAY CAUSE SEVERAL DISEASES

The causative agent for the transmissible spongiform encephalopathies (scrapie, kuru, Creutzfeld-Jakob disease, and Gerstmann-Straussler syndrome) remains a mystery. Possibly associated are mysterious agents called slow viruses. The causes of these diseases challenge our concepts of conventional microorganisms since, to date, they have not been shown to contain RNA or DNA. Infectivity derived from brain or lymphoid tissue homogenates has several unusual biological properties, such as exceptional resistance to inactivation by formaldehyde or by ultraviolet or x-irradiation. Attempts to purify the agent indicate that infectivity remains associated with cellular membrane fractions and thus lacks homogeneous biophysical properties. Futhermore, evaluation of purification protocols has been hampered by the inaccuracy of assays used to quantify infectivity. The recent use of the incubation period assay to quantify infectivity is quick and requires fewer animals but, if anything, is less precise in discriminating quantitative differences. Thus, both because of the inaccuracy of the assay and the nonhomogenous physical nature of the agent, it remains extremely difficult to design an effective purification protocol for the infectious agent.

Recently several research groups have partially purified infectivity on sucrose gradients. In infectious fractions, macromolecular fibril-like structures have been visualized with the electron microscope, which raises the possibility that these might be the agent or a component of the agent. Protein components of these fractions were further purified under denaturing conditions. Since infectivity was destroyed in these procedures, it was impossible to relate the purified proteins directly to the infectious agent. Partial amino acid sequence analysis of the major polypeptide obtained, known as prion protein, has led to the cloning and sequencing of the complementary DNA encoding the prion protein. Hybridization experiments with these clones showed that prion protein mRNA was not found exclusively in scrapie-infected animals, but occurred to the same extent in the tissues of uninfected animals. Thus, the relationship of the prion protein gene and its expressed protein to the infectious agents of scrapie and other spongiform encephalopathies is as yet uncertain.

parasites and bacteria. Immune complexes may lead to pathological outcomes, either in the vascular spaces or in tissues.

Autoimmune responses have recently been identified for viral and bacterial antigens. In such cases, the host responds not only to the microbial antigen but also to normal host components. A variety of mechanisms for autoimmunity have been proposed. With our increased capacity to study components of the immune system as well as to define the individual microbial proteins or components involved in recognition of the immune system by microbes, we should in the next several years gain striking insights into the etiology of autoimmune reactions to microbes.

Bacterial Toxins Are Studied More as Probes for Specific Cell Functions Than They Are for Their Roles in Pathogenesis

Toxic proteins and peptides, which are involved in a wide variety of biological phenomena (such as bacterial pathogenesis, attack by venomous insects and animals, and the toxicity of certain plants), have classically been of interest from a medical perspective. Microbiologists in the 1880s and 1890s were intrigued by toxins because they provided a potential explanation of bacterial disease. Moreover, they soon found that repeated injection of sublethal doses of toxins into animals induced specific resistance to those toxins, a finding that represented the discovery of humoral immunity. They developed methods of immunization, which culminated in the mass inoculations and successful immunizations against diphtheria and tetanus toxoids.

From a biological standpoint, toxins represented little more than curiosities until recently. Within the past two decades, interest in these substances and the mechanisms whereby they affect cells of the animal or plant host has revived. Toxins are excellent probes of important biochemical and cell-biological processes; they are interesting from the point of view of structure and functions of proteins and with respect to the theoretical analysis of microbial host-parasite interactions and their genetic regulation. Moreover, the progress made in understanding toxin structure and mode of action has stimulated new potential applications in medicine. The coupling of cytotoxic proteins to monoclonal or polyclonal antibodies provides a new approach to chemotherapy for cancer and other diseases that call for the elimination of a specific class of cells.

Finally, the cumulated knowledge on classical toxins may provide an important base of information for understanding yet another aspect of the immune system, namely cellular immunity. Evidence is increasing that the killing of cells recognized by the cellular immunity system involves the action of cytotoxic proteins. Similar molecules may also mediate tissue regression and other ontological phenomena.

Persistent Infections

Microbial Persistence Is a Common Sequel to a Large Number of Infections

Persistent infections may be important in the maintenance of microorganisms in host populations. In certain circumstances they may be activated in immunosuppressed patients, and they may be associated with neoplasms. Persistent infections include those in which the microbe is latent (as in a noninfectious state in some location in the host) or persistent (as when it multiplies more or less continuously for long periods of time).

Parasites Are Able to Remain in the Blood (Exposed to the Immune System) for Months to Years

Plasmodium malariae, one of the four human malarial parasites, can persist in the bloodstream for the life of the host, reinvading new red cells every three days. The adult schistosome, a trematode, lives in venules for many years, copulating and laying eggs that spread the infection and trigger the disease. Within days after the schistosome enters the host, its tegument becomes resistant to complement lysis and to cell-mediated killer mechanisms. The African trypanosome *Trypanosoma gambiense*, a flagellate that infects animals and humans, undergoes antigenic variation to escape destruction by antibody-dependent complement lysis. Each parasite clone contains multiple genes capable of expressing antigenically unique surface proteins. The control of expression is not fully understood, but gene duplication and translocation to an expression site is one mechanism. The trypanosome continually expresses new surface proteins from its large repertoire of genes, which are further expanded by mutation. The American trypanosome *Trypanosoma cruzi*, uses a different evasive maneuver: It elaborates a molecule similar to a decay-activating factor that speeds the breakdown of a critical complement component and thus may prevent complement-dependent lysis by the alternative pathway. Despite the many strategies of parasites, the host does eventually become immune to most parasites. The challenge to immunologists is to develop vaccine strategies that attack parasites in susceptible stages or by mechanisms that they cannot resist.

CONCLUSION

Studies on the Immune System and Infectious Diseases Have Made Enormous Progress During the Past Decade Because of the Introduction of Powerful New Technologies

The applications of molecular biology, monoclonal antibodies, flow cytometric analysis, modern methods of cell culture, and powerful new techniques of protein chemistry have led to many of the insights described in this chapter. Improvements during the forthcoming decade promise to bring us the solutions to many of the central unresolved problems of immunology and pathology as well as to provide a new level of insight to all aspects of cellular biology. Equally, the central role of the immune system in resistance to infectious diseases, in the elimination of tumor cells, and in the pathogenesis of many chronic disease makes it an obvious target for clinical efforts. During the next decade, substantial emphasis needs to be placed on developing a quantitative understanding of the contribution of the various components of the immune system to the function of that system in humans and other animals, both in normal situations and in various disease states. Such information together with our growing understanding of the basic cellular and molecular mechanisms of immunity should make possible the development of new vaccines for the prevention of many of the still-uncontrolled infectious diseases such as AIDS and malaria and should allow the introduction of rational therapy for a host of immunologically based disorders.

8

Evolution and Diversity

Evolution and diversity result from the interactions between organisms and their environments and the consequences of these interactions over long periods of time. Organisms continually adapt to their environments, and the diversity of environments that exists promotes a diversity of organisms adapted to them. In recent years, new techniques and approaches have opened exciting new avenues of investigation of the processes that generate evolution and diversity. As a result, greater opportunities exist now for advancing knowledge than during any period since the 1930s and 1940s, when evolutionary biology and genetics became united in what came to be called the modern synthesis of evolutionary biology.

The Processes and Results of Evolution Are Exemplified in the Evolution of Insecticide Resistance in Insects and Antibiotic Resistance in Bacteria

The first synthetic organic insecticide to be adopted for practical use was DDT, which was introduced in 1941. DDT appeared to have many advantages because, in proper dose, it was toxic to insects but not to humans. As a consequence, DDT was quickly employed worldwide to control houseflies, mosquitoes, and a variety of other insect pests. After the initial success of DDT, many other exotic chemical compounds were introduced as insecticides. The introduction and widespread use of each of these was quickly followed by the evolution of resistance in large numbers of insect species. In fact, more than 200 species of insects had become resistant to DDT by 1976; some species have evolved multiple resistance to four or more groups of chemical insecticides.

In many cases, the insecticide resistance results from the action of a single gene, although multiple other genetic changes that can modify the response to insecticides also occur. In the common housefly, resistance results from the presence of an enzyme called DDTase, the natural function of which is unknown. Mutant forms of the enzyme convert DDT into the relatively harmless compound DDE. Resistance in the mosquito *Aedes aegypti* is also associated with a DDTase enzyme, but not the one found in the housefly.

The evolution of resistance to insecticides is so common because the insect populations often contain rare mutant variants that are already resistant. Exposure to the insecticide gives an advantage to these mutants, and over several generations, they gradually increase in frequency at the expense of the normal types until very few of the normal sensitive types remain.

A remarkable principle in population genetics states that insecticide resistance can be expected to evolve in approximately 5 to 50 pest generations, irrespective of the insect species, geographical region, nature of the pesticide, frequency and method of application, and other seemingly important variables. The phenomenon occurs because the time required to evolve significant resistance depends on the logarithm of the total increase in frequency of the resistance gene as a result of the pesticide application, which over a wide range of realistic values is effectively limited to 5 to 50 generations. The rapid, repeated evolution of insecticide resistance in many parts of the world reflects the operation of this simple mathematical principle.

A similar situation accounts for the repeated evolution of antibiotic resistance in bacteria: Rare bacterial types containing genes for resistance are favored in the presence of the antibiotic and eventually displace the normal sensitive types. In this case, the overuse of inexpensive antibiotics, not only in medicine but in animal feed, fish culture, and agriculture, has promoted the evolution of antibiotic resistance in a wide spectrum of microorganisms. In many cases, the resistance genes are contained in mobile genetic elements that can be transmitted from one organism to the next, and their spread has resulted in the wide dissemination of the resistance genes among pathogenic and nonpathogenic forms.

The molecular evolution of antibiotic resistance is similar to the process that bacteria have used for millenia to evolve resistance to naturally occurring antibiotics and to soil contaminated with lethal concentrations of heavy metals. A resistance gene that evolves in one bacterial species can potentially be disseminated to many others by means of molecules known as plasmids, which are transmitted among suitable hosts by cell contact. These plasmids occasionally pick up transposable DNA sequences that contain genes resistant to antibiotics, and they confer resistance upon host cells. When the antibiotics are widely used and present in the environment, cells containing the resistance plasmids are favored, and the plasmid spreads. In many cases resistance plasmids have acquired genes for simultaneous resistance to five or more chemically unrelated antibiotics. For some pathogenic bacteria, such as gonorrhea, antibiotic resistance has become so widespread that clinical treatment is severely compromised.

The evolution of insecticide resistance in insect populations, antibiotic resistance in microbial populations, herbicide resistance in plant populations, and

EVOLUTION OF INSECTICIDE RESISTANCE

Some of the most dramatic examples of evolution in action result from the natural selection for chemical pesticide resistance in natural populations of insects and other agricultural pests. In the 1940s, when chemical pesticides were first applied on a large scale, an estimated 7 percent of the agricultural crops in the United States were lost to insects. Initial successes in chemical pest management were followed by a gradual loss of effectiveness. Today, more than 400 pest species have evolved significant resistance to one or more pesticides, and 13 percent of the crop of yields in the United States is lost to insects.

In many cases, significant pesticide resistance has evolved in 5 to 50 generations in spite of great variation in the insect species, the insecticide, and the method of application. Theoretical population genetics helps us understand this apparent paradox. Many of the insecticide resistances result from single mutant genes. The resistance genes are often partially dominant, so the change in the frequency of the resistance gene is governed approximately by the equation

$$\ln\left(\frac{p_t}{q_t}\right) = \ln\left(\frac{p_0}{q_0}\right) + \left(\frac{s}{2}\right) t,$$

in which p and q are, respectively, the gene frequencies of the resistant and sensitive genes, initially (time 0) and at time t generations after insecticide application, and s measures the degree to which resistant insects are favored over sensitive ones.

Prior to application of the pesticide, the gene frequency p_0 of the resistant mutation is generally close to 0. Application of the pesticide increases the gene frequency, sometimes by many orders of magnitude, but significant resistance is noticed in the pest population even before the gene frequency p_t increases above a few percent. Thus, as rough approximations, we may assume that q_0 and q_t are both close enough to 1 that $\ln(p_0/q_0) = \ln(p_0)$ and $\ln(p_t/q_t) = \ln(p_t)$. Using these approximations, the equation implies that $t = (2/s)\ln(p_t/p_0)$. In many cases, the ratio p_t/p_0 may range from 1×10^2 to perhaps 1×10^7, and s may typically be 0.5 or greater. Over this wide range of parameter values, time t is effectively limited to 5 to 50 generations for the appearance of a significant degree of pesticide resistance. Details in actual cases will depend on such factors as the size of the insect population and the extent of genetic isolation between local populations, and the evolution of polygenic resistance may be expected to take somewhat longer than single-gene resistance. Nevertheless, the example demonstrates the predictive power of mathematical approaches in evolutionary biology.

heavy-metal tolerance in plant and bacterial populations has been demonstrated repeatedly. In every case, genetic variation and natural selection provide an amazingly effective process for promoting the adaptation of organisms to their environments. The study of evolution and diversity of life on earth is concerned with the tempo, mode, and patterns of such adaptations.

Some of the Most Exciting Current Research Opportunities in Evolution and Diversity Result from Technical Innovations in Molecular Biology

The techniques of molecular biology have revealed a rich level of detail in studies of DNA variation and its analysis. They have uncovered an unexpected avenue of genomic evolution through the activities of transposable elements. They have opened up the transfer of individual genes between species as a majoi new tool for the study of the mechanisms and subsequent events in speciation. And they have made possible an integration of the techniques of molecular biology with questions of field natural history, such as in the use of mitochondrial DNA polymorphisms to study population structure and migration in fish and other organisms.

Application of the techniques of molecular biology has made possible, for the first time, the beginnings of a synthesis of microbiology and evolutionary biology. These two fields have developed in almost complete isolation from each other. Microbiology is among the least evolutionarily oriented of biological disciplines, and evolutionary biology is the evolutionary biology of metazoans. Studies at the interface of these disciplines will result in the definition of new evolutionary principles and a deeper understanding of principles already established. Perhaps the most surprising initial result of studies in microbial evolution has been the discovery of what some scientists regard as a distinct kingdom of organisms, the archaebacteria, which combine some features of more familiar kinds of bacteria (eubacteria) with others characteristic of eukaryotic organisms.

Indeed, paleobiologists now believe that the earth's biota was composed solely of bacteria for at least two-thirds of its total history. Many evolutionary innovations have been powered by changes in intracellular biochemistry rather than by changes in the shape, size, or physical organization of organisms. Moreover, the global biota, especially bacteria, with their diverse physiological capabilities, have interacted with and changed the global environment in numerous crucial ways, such as in the creation of our oxygen-rich atmosphere.

The application of molecular techniques has also contributed to the current revolution in systematics. Molecular studies of DNA and proteins are now used routinely to distinguish species and to estimate phylogenetic relations among closely related species. Direct DNA sequencing is providing phylogenetically useful data almost faster than they can be analyzed. The inferred genealogical relations based on macromolecules are usually consistent with those based on morphology, but molecular studies often help to resolve relations that are morphologically ambiguous.

Overall similarity in macromolecules provides a reliable measure of evolutionary time only when the molecules being studied evolve at much the same rate in different lineages and at different times. Whether DNA sequences actually evolve with regular rates like molecular clocks is still much debated, but the data so far suggest at least moderate regularity. The concept of the molecular clock has provided a unique and powerful time dimension in evolutionary studies and has augmented as well as complemented inferences from the fossil record. However, not all of the evidence is consistent with the hypothesis that molecular evolution occurs at a nearly constant rate, and further evidence is needed to establish the validity of the hypothesis and to determine its range of application.

The explosive increase in knowledge of DNA sequences has created an acute need for new kinds of computational technologies and algorithms, as well as new statistical approaches, so that the data can be interpreted to maximum advantage. Appropriately analyzed, the new kinds of data will reveal, with a level of detail never before possible, patterns in the history of evolution; the new data will thus shed light not only on the evolution of macromolecules, but also on the processes of evolution of morphology, life histories, and physiology. Regrettably, the analysis of sequence data, which must bring together experts in statistics, computer science, mathematics, molecular biology, population genetics, developmental biology, and systematics, has lagged behind as ever more data have accumulated. At the same time, more extensive data on the extent of DNA sequence variation within species are badly needed.

Technical Innovations That Have Transformed Studies of Evolution and Diversity Are Not Limited to Molecular Biology

Studies in biomechanics, ecology, and behavior have profited tremendously from improved techniques of photography and telemetry, and in almost every area of study, modern computers facilitate sophisticated simulation modeling and data analysis. Paleobiology has benefited from methods of organic geochemistry that enable the determination of the nature and isotopic characteristics of biologically derived organic compounds preserved in ancient sediments and also from new techniques of radiometric age determination and new data regarding the geologic and plate-tectonic history of the earth. These approaches, when combined with information derived from molecular biology, promise to promote new knowledge about such fundamental evolutionary events as the origin of invertebrates, vertebrates, plants, and human beings. Even earlier Precambrian events, such as the advent of photosynthesis, oxygen-dependent respiration, nucleated cells, eukaryotic sexual reproduction, and the modern type of anaerobic-aerobic global environment, may be within the reach of the new approaches.

Progress in the study of evolution and diversity does not require technical innovations, although it frequently benefits from them. Advances also come from the synthesis of previously disconnected areas, from new ways of looking at problems, or from new concepts. Therefore, in evolution and diversity, too much stress on technical virtuosity and trendiness runs the risk of promoting a kind of brush-fire pattern of scientific advance, with great activity and excitement near the front but little behind in the area where the practical applications of basic discoveries are often developed.

Although many exciting directions in evolution and diversity have been opened by advances in molecular biology, numerous fundamental problems occur at levels of biological organization above that of molecules. The evolution of populations of organisms is affected by the interactions with the environment of physiology, development, and behavior at levels that are not amenable to molecular analysis. Molecular biology is an aid but not a panacea in the discovery and classification of organisms. And the processes of speciation and extinction, while fundamental to evolution and diversity, are population, not molecular, processes.

THE EVOLUTIONARY PROCESS

Population Genetics Continues to Emphasize Genetic Variation — Its Nature, Causes, and Maintenance in Populations

Studies of population genetics or genetic variation have become significantly more sophisticated with the use of molecular techniques and new types of material, including microbial organisms and chloroplast and mitochondrial DNA. Progress in molecular biology has been especially helpful for population genetics and promises to aid in the resolution of several outstanding problems in the field. Genetic variation can be resolved at the ultimate level of the DNA sequence. With this level of resolution, it becomes possible to determine whether genes that are highly variable within populations also evolve rapidly. The distribution of DNA polymorphisms within and among species results from the operation of evolutionary forces that are in many cases too weak or too difficult to measure in the laboratory or field; it may be possible to infer their magnitude from analysis of the sequences themselves. The rich possibilities of inferences that may be made from DNA sequence data warrant significant efforts in this direction.

The Technique of Site-Directed Mutagenesis Also Opens New Possibilities for Population Genetics

Traditionally, inferences about evolutionary constraints on molecular structure have been gathered from comparisons of homologous molecules among species. With site-directed mutagenesis, the inferences can be tested directly by

DNA SEQUENCES IN EVOLUTIONARY STUDIES

Comparisons of DNA and protein sequences have revolutionized the reconstruction of the evolutionary relations among organisms because the sequences themselves contain information about their ancestral history that can be extracted by appropriate statistical methods. Equally powerful inferences can be drawn from comparisons of sequences among individuals within species. This is possible because the sequences also contain information about the evolutionary forces that molded them, which can be studied to make inferences about the magnitude of natural selection, the importance of random processes, the role of recombination, and so on.

Two studies of DNA sequences among natural isolates of the bacterium *Escherichia coli* underscore the power of the molecular approach. One study focused on DNA sequences in the *gnd* gene, which codes for the enzyme 6-phosphogluconate dehydrogenase. The purpose was to estimate the fraction of observed amino acid polymorphisms that are selectively neutral. This has been a central issue in population genetics for more than a decade, but it has defied resolution because most statistical tests of observed gene frequencies and most laboratory experiments lack sufficient power to detect selection coefficients of the relevant magnitude. Although most random amino acid substitutions might be expected to be harmful, only a small proportion of harmful mutations ever become established as polymorphisms in natural populations. A significant proportion of alleles that become polymorphic might therefore be expected to be selectively or nearly neutral.

The idea behind the study of DNA sequences is that nucleotide polymorphisms at silent sites, which do not change amino acid sequences, can be used as internal standards for comparison with amino acid polymorphisms in the same gene. When 768 nucleotides in the *gnd* gene in seven strains of *E. coli* were compared, 12 amino acid polymorphisms and 78 silent polymorphisms were found. All 12 amino acid polymorphisms occurred in singleton configurations (meaning that six strains shared a common amino acid at the site and only one strain was different), whereas only about half of the silent polymorphisms exhibited this configuration. Based on this difference alone, one can conclude the no more than six of the amino acid substitutions could be selectively neutral. Alternatively, if all amino acid polymorphisms are assumed to be mildly harmful, the amount of selection necessary to account for the preponderance of singleton configurations is only about 1.6×10^{-7}, an amount of selection much too small to detect except by means of DNA sequence comparisons.

The second study concerned the occurrence of genetic recombination among natural isolates of *E. coli*. Evidence for recombination was found in a region of 1,871 nucleotide pairs around the *phoA* gene, which codes for alkaline phosphatase, in 10 natural isolates. The region contained 87 polymorphic nucleotide sites, of which 42 were shared by two or more strains. Comparisons of the shared polymorphisms gave clear evidence for intragenic recombination in that polymorphic nucleotides common to two or more strains tended to be spatially clustered within the gene. The putative exchange events involved short stretches of DNA on the order of several hundred nucleotide pairs. Although reproduction in *E. coli* is thought to be largely clonal, clonal reproduction is nevertheless consistent with recombination involving short stretches of DNA because most genes are still transmitted uniparentally. This is yet another example of how a DNA sequence can contain information about its history that cannot easily be inferred from direct experiments.

deliberately altering parts of the molecule of interest, reintroducing the gene for the altered molecule into living organisms, and studying the effects of the changes. Such experiments reveal not only which changes affect the molecule, but also the magnitude of these effects. For the first time, population geneticists are able to study a collection of mutant molecules that are well characterized at the molecular level.

The process of mutation, which until recently seemed to result from an essentially simple process of nucleotide substitution or rearrangement, is now appreciated to include mechanisms for creating evolutionary novelty through the movement and other activities of transposable elements. Indeed, virtually all proteins may have been created in evolution by the rearrangement of exon units, which code for smaller structural domains able to fold autonomously and carry out elementary functions such as ligand binding. If true, this would mean that the evolution of new functions cannot be likened to the proverbial monkey pecking away at a typewriter in hope of creating something meaningful; the analogy should rather be to a monkey that can shuffle complete words and entire sentences and paragraphs.

Recombination, traditionally viewed as important from the standpoint of creating genetic variation through new combinations of genes, has taken on a new dimension in population genetics because of its conservative role in maintaining similarity between members of multigene families. However, little is known about the rate of gene conversion in multigene families or about the role of intragenic recombination in creating new genetic variation.

The Study of Natural Selection Remains One of the Principal Preoccupations of Evolutionary Biologists

The understanding of the mechanisms of selection in natural populations is still inadequate. At the molecular level, it is necessary to understand how changes in protein molecules affect fitness and to critically evaluate the contribution of selectively neutral mutations to molecular evolution. These problems are ideal for the application of site-directed mutagenesis in experimental organisms such as bacteria, yeast, and *Drosophila*. At the phenotypic level, it is necessary to understand how genes affecting quantitative characters respond to natural selection. This is an area in which substantial advances in the theory have been made recently and in which further progress can be expected. Analysis of multifactorial traits is essential to understanding the genetic basis and inheritance of many genetically complex disease traits in humans, including the most common birth defects and adult disorders. It is also important in evolution and diversity in interpreting the evolution of such multifactorial traits as morphology.

Significant methodological problems in natural selection include difficulties in measuring reproductive components, including fertility selection and sexual asymmetry in selection, nuclear-cytoplasmic gene interactions in fitness, and the elaboration of statistical models and experimental designs to estimate fitness components when there is inbreeding (as in some plants). Studies of selection in natural habitats are often hampered by lack of a rigorous, quantitative approach to studying the environment and its variation.

Evaluation of the role of population structure in evolution is also marred by important unresolved problems, such as the need to improve methods of estimating migration rate, to define the role of interactions between genotypes in selection, and to evaluate the significance of selection among demes (a local population of closely related organisms) in the genetic divergence and transformation of populations. Genetic differentiation results in variation among populations, and methods for the statistical analysis of such spatial patterning are now being developed.

Progress Is Being Made in Our Understanding of Speciation and the Evolution and Maintenance of Diversity

Organismal diversity is a direct and inevitable outcome of speciation, the process whereby a single species evolves into two or more distinct ones. The conditions required to initiate, promote, and complete the speciation process are still poorly understood and hence controversial. To resolve this problem, a major effort has been made in recent years to examine the biological and genetic attributes of closely related taxa actively undergoing various degrees of differentiation. Three approaches are taken in these investigations: field, experimental, and theoretical.

Significant advances have come from the analysis of the genetic variation and structure in natural populations. Many of these studies are of insects. For example the Hawaiian *Drosophila*, which have proliferated rapidly on the emerging islands of the archipelago, serve as an outstanding model system for examining the relations among geographic isolation, population size, sexual selection, and genetic divergence. The fact that the islands can be accurately dated in geologic time provides a unique opportunity to ascertain how the species have evolved. Founder events followed by repeated population expansions and contractions accompanied by strong sexual selection appear to have promoted the rapid divergence of isolated populations of these flies.

The causes of speciation are different in *Rhagoletis*, a group of economically important flies whose larvae infest the fruits of a wide range of plants. Within the past 150 years, species of these flies have formed genetically distinct host races on introduced plants, in the absence of any geographic barriers to gene flow. These races appear to be in the early stages of speciation. Detailed behavioral, ecological, biochemical, and molecular research has revealed that because mating in these flies occurs on the host fruit, genes that govern host choice directly affect mate choice.

Another approach to the study of speciation in natural populations focuses on the genetic and biological outcome of hybridization in zones of overlap either between previously geographically isolated, but closely related, populations that have reestablished contact, or in zones of transition across a sharp ecological boundary between populations adapting to different habitats. These investigations are being carried out on a wide range of animals and plants. The objective of such studies is to establish whether different mate recognition systems and reproductive isolation can evolve in zones of contact as a result of a selective process called reinforcement or develop as a by-product of genomic divergence in isolation. The increased genetic resolution recently provided by molecular techniques is contributing significantly to our understanding of how hybridization affects the speciation process.

A third approach to the study of speciation involves direct laboratory selection experiments. Such experiments suggest that considerable progress toward speciation can occur rapidly, even in the face of considerable gene flow. This experimental approach offers promise for testing some hypotheses of speciation mechanisms now being generated from studies on natural populations.

In recent years, theoretical population genetic models, using analytical and computer stimulation approaches, have been developed in an attempt to understand under what conditions species evolve in nature. These models have become increasingly more sophisticated and biologically meaningful and have yielded insights into the speciation process as well as models for exploring, in nature or in the laboratory, the conditions under which speciation can occur.

The study of speciation, one of the most important fields of research in evolutionary biology, has a direct bearing on our understanding of the origin of organismal diversity in the past, the present, and the future. It has left the descriptive, comparative phase that predominated in the past for a more empirical approach to the study of speciation mechanisms. Sufficient evidence has come from recent studies to indicate that we are on the threshold of resolving some of the most intractable problems concerning modes of speciation. The increasing interest in microbial evolution also encourages a new analysis of the species concept and species formation in prokaryotes.

The Study of Evolution of the Organization and Composition of the Genome Is Still in Its Infancy

Even though we know that the overall genetic organization of the chromosome in certain groups of bacteria is strongly conserved, the reasons are obscure. Similarly, in eukaryotes, no principles are known that govern conservation or changes in chromosome structure or organization. Genomic evolution also includes unknown contributions from various localized and dispersed highly repetitive DNA families and numerous types of transposable elements with different characteristics and evolutionary implications. In a wider sense, genomic evolution also includes that of viral genomes and the interactions with the host genomes. Recently it has become clear that certain plant genomes undergo a novel and potentially major mechanism of evolution in response to environmental stress. For example, plants under stress manifest marked phenotypic changes that are associated with heritable changes in copy number of several multigene families including ribosomal DNA sequences. New methods of manipulating and cloning large DNA molecules will be critical to the study of evolution at the level of the chromosome.

Although ambitious, the synthesis of disciplines that characterize modern evolutionary biology should be extended to embrace areas such as developmental biology, neurobiology, and behavior. Little is known about possible developmental sequences available to organisms with particular genotypes, or about new kinds of developmental pathways that are accessible by mutation from genotypes already existing. In addition, virtually nothing is known about the genetic determination of complex animal behaviors and the manner in which these behaviors feed back on the evolution of molecular and morphological traits.

THE RESULT OF EVOLUTION

The Study of Adaptation Is Still a Pervasive Theme in Biology

The most dramatic result of the evolutionary process is seen in the adaptations of organisms alive today. One of Darwin's chief accomplishments in *The Origin of Species* was to show that the exquisite adaptations of organisms that "so justly excite our admiration" could be explained by the purely mechanistic process of natural selection.

Important research opportunities in studies of adaptation derive from both technical and conceptual innovations during the past several decades. Some of the technical advances have been mentioned. As an example of conceptual innovation, it is now generally agreed that traits do not typically evolve for the good of the group or species as a whole, but for the direct or indirect advantages they confer on their possessor. Interdemic selection may provide an exception to this generalization, but the overall importance of interdemic selection to changing the genotypic composition of a species is unknown. The search for theories other than group selection to explain puzzling traits has led to a rich proliferation of concepts regarding, for example, selection acting not on individuals themselves but through increased fitness of their kin, and the trade-offs between fecundity and mortality in life-history strategies. However, some phenomena remain puzzling, such as that parthenogenesis does not rapidly replace sexual reproduction even though its rate of reproduction is theoretically higher.

Other conceptual advances have also enriched the study of adaptation. One is the realization that organisms often buffer themselves against changes in selection pressures. For example animals can choose species-specific microhabitats, and seeds can germinate in response to cues that signal favorable conditions. Another important concept is that of developmental constraint: the manner in which certain adaptations close off other possible paths of adaptation, thereby constraining the further evolutionary potential of the species. For example, the exoskeleton of arthropods provides attachment sites for muscles enabling rapid movement, but it also limits the maximum size of the animals.

The study of adaptation has also benefited from the integration of previously separated fields. For example, ecology and behavior are becoming increasingly integrated into evolutionary biology. By examining the genetic and phylogenetic aspects of physiological, morphological, and biochemical traits, biologists are forming bridges among evolutionary biology and physiology, development, and molecular biology.

Among numerous promising research opportunities in adaptation are studies of evolutionary and functional morphology, which increasingly includes biomechanics. Application of quantitative engineering principles combined with computer modeling has moved this field from descriptive to analytical studies. The approach enables the analysis of the specific mechanical properties of biological materials, the relation between the design of organisms and their environments, and the understanding of repeatable historical patterns in the evolution of design and the constraints placed on design by evolutionary history.

Physiological adaptations of plants and animals to factors including temperature, aridity, and osmotic stress have been abundantly analyzed by physiological ecologists, whose approach is becoming increasingly evolutionary. Indeed, some workers have begun to examine individual variation in physiological traits and to apportion the variation into genetic and nongenetic causes in attempts to determine physiological mechanisms.

Important Advances Have Been Made in Behavioral Ecology and Evolutionary Biology

The understanding of such phenomena as habitat use, food selection, social aggregation, cooperation, cannibalism, and ritualized conflict has greatly increased in the past decade. Sexual selection has become a major topic in both behavior and population genetics, and the reality of sexual selection by female choice in birds has recently been demonstrated. The next step is to test the prediction that male characteristics evolve in concert with female preference for even more exaggerated male characteristics, virtually without limit. The coevolution of male-female mate recognition characteristics may play a key role in animal speciation.

The evolution of life histories provides an active area of contact between the fields of ecology and evolution, as does the study of how interacting species adapt to each other and how such coevolution affects the structure of ecological communities. During the past decade, such studies have expanded beyond the previous emphases on competition and predation to embrace, among others, parasitism and mutualism.

The study of adaptation has been invigorated by the infusion of new concepts and theories, by an increasingly experimental and analytical approach, and by the increasing communication among fields. The incorporation of population genetic theory and phylogenetic analysis into the study of adaptation has only begun and promises to be instructive. Several hurdles must be overcome to ensure success. Tests of theories in natural populations often require considerable time—often years—before they acquire real substance; in some areas, such as physiology, techniques must be developed to automate the measurement of numerous individuals.

Although modern molecular techniques promise to contribute to an understanding of numerous unresolved questions related to the processes and history of evolution, equally important contributions will emerge from new conceptual, statistical, and technical approaches in areas such as population genetics, phylogenetic analysis, and developmental biology. Foremost among the poorly understood areas in evolution are the relations between evolutionary processes at the population level and the longer term evolutionary changes involved in the origin of species and higher taxa. A bridge between the almost separate domains of population genetics on the one hand and systematics on the other is sorely needed. Progress in building such connections may have to await advances in developmental biology and imaginative new approaches in genetics and development, but some advances in these areas hold out the promise that population and historical studies can inform each other.

For example, we may anticipate that, by the use of molecular sequences or large numbers of morphological traits or both, reasonably reliable phylogenies of groups of related species will soon be abundant. In groups that are amenable to genetic or developmental studies, the conjunction of genetic and phylogenetic or paleontological analysis offers the opportunity for studying numerous open questions. These include issues such as (1) whether rapidly evolving characteristics are more variable genetically than slowly evolving features; (2) whether genetic correlations exist between characteristics that evolve in concert across phylogenies, or whether observed phylogenetic correlations result from coadaptation and natural selection only; and (3) whether correlations among species result from common ancestry rather than adaptation or genetic correlation. These are some of the rich fields that are available at the organismic level for the exploration of evolution and diversity.

The relation between population genetics and long-term evolution will also be strengthened as evolutionary biologists turn to developmental biology and developmental genetics. The greatest progress will come when the mechanisms of development are more fully understood. Even now we can hope for some understanding, perhaps by developmental comparisons and experiments not only between distantly related kinds of organisms such as frogs and salamanders, but also between closely related species in which hybridization or experimental transfer of genetic material may prove feasible. Among the neglected questions coming to the fore once again are, What is the mechanistic basis of the sterility of species hybrids? Are few genes responsible for hybrid sterility, or many? Why are mutations of large effect generally deleterious and what are their pleiotropic effects? Conversely, what processes are altered when gradual, polygenic changes yield a viable phenotype that may resemble the nonviable phenotype of a single mutation of large effect? What is the developmental nature of invariant or evolutionarily conservative traits, and what is their relation to the concepts of canalized phenotypes that develop in constant ways in a wide range of environments? What relations exist among the functional, phenotypic, genetic, and developmental correlations among traits?

Research in Functional Morphology and Biomechanics Has as a Major Goal the Analysis of Patterns of Diversity at the Level of Whole Organisms

Functional morphologists study mechanisms of integration of organisms, usually within both phylogenetic and evolutionary frameworks. Complex organisms are highly integrated, and the basic pattern of organization of most major taxa is conservative. This conservatism probably arises from couplings, or interlinkages among the parts of organisms that stabilize morphology. These links may be genetic (pleiotropy, genetic correlations, and so forth), developmental (inductive interactions), functional (physiological, behavioral), or structural (direct part-to-part connections). Functional morphologists examine organisms to describe such linkages. Once understood, such couplings can be used to explain why evolution is likely to proceed in certain directions rather than in others and why certain structures and functions have not evolved in the past and are unlikely to appear in the future. Thus, many functional morphologists are concerned with constraints on evolution and on opportunities that arise when such constraints are removed. Furthermore, certain evolutionary phenomena can lead to uncouplings, which may be followed by the incorporation of novelties and adaptive radiation. For example, certain salamanders lost lungs as an adaptation to live in rapidly flowing streams; the hyobranchial system was thereby uncoupled from its role as a respiratory pump and evolved into a high-speed, long-distance projectile tongue. Modern functional morphology uses a large array of methods, including high-speed video, kinematic, and x-ray cine systems for visualizing movement and behavior, electromyographic and other physiological approaches for characterizing patterns of movement, neurobiological methods such as modern staining methods for tracing neural components of integrated systems, and even quantitative genetic methods of analyzing patterns of interaction for analysis of variation within individuals.

Application of principles from the fields of materials science, engineering, cell and developmental biology, ecology, and evolutionary biology to the study of the structure and function of plants and animals has progressed rapidly and holds promise for the future. The field of biomechanics is relatively young; it differs from functional morphology in having a focus on details of structural organization and in having application from the level of cells to that of whole organisms facing the environment. The kinds of studies undertaken range from investigation of the structure of the cytoskeleton to those of the collagenous fiber wrapping of the dermis in whales and fishes and the meaning of these structures for function. Recent discoveries include the biomechanical significance of spicule arrangement in the bodies of sponges, reasons for the organization of the holdfast in giant kelps, and the means by which sea anenomes survive the battering they receive in tidal zones. Biomechanical approaches have led to new understanding of the organization and function of the notochord, of the significance of osteogenic patterns, and of the organization of muscle. Some workers span the small gap to functional morphology, while others extend their interests into surgical and other medical uses of biomechanical perspectives.

Systematics Is a Key Discipline in Evolutionary Biology

In a Chinese proverb, calling things by their proper names—systematics—is the beginning of wisdom. Modern systematics, which is basic to the study of adaptation, stresses the basic recognition and naming role, but simultaneously reaches out to all other disciplines concerned with biological diversity. Systematics comprises taxonomy—that is, surveying, recognizing, naming, describing, and making identifiable the kinds of organisms—and the development of classifications of organisms, placing them into taxa from the population to the kingdom levels. At another level of analysis it embraces the study of the relations, origins, and histories of these taxa, including the factors that led to their origin and shaped their histories.

Systematics, gradually transformed by principles and techniques from other disciplines, has the chief responsibility for analyzing diversity and putting such knowledge into a more accessible form. Cataloguing of organisms is still so incomplete that we do not even know to the nearest order of magnitude the

FUNCTIONAL MORPHOLOGY AND EVOLUTIONARY ECOLOGY

Three of Africa's Great Lakes—Victoria, Tanganyika, and Malawi—are home to three remarkable species assemblages. Each lake contains 150 to 200 endemic species of the family Cichlidae, small to medium-sized sunfish-like fishes. The total of nearly 600 species represents an astonishing 3 percent—perhaps more—of the world's fish species. Why are there so many of them? Does their morphology have anything to do with their remarkable multiplicity of species?

In general, when species coexist, they are partitioning some resource—such as food or space—in such a way as to reduce competitive overlap. Such specialization is often accompanied by diversity of some physical or behavioral features of the species. Studies of the African cichlids reveal enormous diversity in two attributes: their behavior, including feeding, and the morphology of their jaws. Their feeding habits range from scraping algae from the underside of rocks to eating the scales of other fishes. One species frequently bites the eyes out of other fishes, others scrape algae from the leaves of higher plants, some eat whole fish, some eat invertebrates, and so on. Their jaws show remarkable variety in shape, size, and dentition. In addition, cichlids, like most bony fishes, have pharyngeal jaws—"throat teeth." But in cichlids, the pharyngeal jaws are more specialized and variable than in most other bony fishes.

For many years, the Great Lakes cichlids have been regarded as the showcase example of adaptive morphology associated with adaptive radiation. The unusual morphological adaptability of the jaws has been thought to have permitted the remarkable adaptive radiation observed. Certain morphological characteristics of their pharyngeal jaws have allowed those jaws to become adapted to some of the functions usually performed by the mouth jaws. This has freed the mouth jaws to become diversified to perform unique food-gathering functions, almost like a hand; the mouth jaws also have characteristics that seem to allow greater diversification of function than those of most other bony fishes.

The above interpretation is the more plausible because radiation has occurred three separate times in the three lakes. There is even a natural control: Several other families of fishes that lack the cichlids' jaw adaptations inhabit the same lakes but have not radiated similarly.

But recently, puzzling questions have been raised. For example, the cichlid family is represented in the African lakes by two subfamilies, both of which have the specialized pharyngeal and mouth jaws discussed above, specializations presumed to have allowed the great speciation observed. But the explosive speciation has occurred primarily in one subfamily, the Haplochromines. The tilapine subfamily has relatively few species. Why? Another question concerns the characoid fishes of Amazonia, a group of fishes containing the piranhas. They, like the African cichlids, have enormous numbers of species, but they lack the specializations of the cichlids' jaws.

Finally, it has recently become clear that the behavioral and morphological diversity of the African cichlids is strongly influenced by environmental factors. This means that differences observed in nature might not be entirely—or even mostly—genetically based.

The African cichlids remain, as they have long been considered, of enormous evolutionary interest. But, rather than being a textbook example of any one particular phenomenon, they seem to represent a natural laboratory for studying evolution, ecology, and morphology. And that study is still in its exciting early stages.

number of species on earth. Although approximately 1.4 million species of all kinds of organisms have been formally named since Linnaeus inaugurated the binomial system of species identification in 1753, this figure grossly underestimates the diversity of life. Considering the prodigious variety of insects alone and the underrepresentation in the catalogue of many types of organisms, such as microbes, it is reasonable to guess that the absolute number of species of all groups on earth falls somewhere between 5 and 30 million.

Proper species classification is important because a species is not like a molecule in a cloud of molecules, but is rather a unique population of organisms, the terminus of a lineage that split off from the most closely related species thousands or even millions of years ago. Species have been shaped into their present forms by mutations and natural selection, during which certain genetic combinations survived and reproduced differentially out of an almost inconceivably larger total. No two species, no matter how closely related, are any more interchangeable than are two Mozart sonatas. Each species of organism is incredibly rich in genetic information. The genetic information in the constituent bases that make up the DNA in a single mouse cell, if translated into ordinary letters of printed text, would nearly fill all 15 editions of the *Encyclopedia Britannica* published since 1768.

Since other evolutionary disciplines, including ecology, biogeography, and behavioral biology, depend on systematics, an entire hierarchy of important problems must be addressed. Two stand out in the sense that progress toward their solutions is needed to put the other disciplines on a permanently sound basis. The first problem is to define the magnitude and causes of biological diversity, and the second is to determine the most reliable measures of homology and their implications for phylogenetic relationships.

In defining the magnitude and causes of biological diversity, systematics will undoubtedly fall short of obtaining a complete catalog of life on earth, but a determined effort would pay many dividends. A greater understanding of biological diversity promises to resolve some of the conflicts in current theory and at the same time to open productive new areas of research. In addition, the answers will influence a variety of related disciplines, affect our view of the place of humans in the order of things, and open opportunities for the development of new knowledge of social importance. For example, control of mosquito-borne diseases such as malaria has profited from the ability to define as separate species mosquitoes that are morphologically almost identical but that differ in behavior and in their ability to transmit the diseases.

Systematics is also a discipline with a time limit because much diversity is being lost through extinction caused by the accelerating destruction of natural habitats. This is especially true in the tropical moist forests, where more than half of the world's species are thought to exist. Although extinction rates are difficult to estimate, in part owing to inadequate systematics, current rates of extinction seem to approach or exceed 1,000 times the average rate in past geological time. Tragically, and perhaps ominously for human welfare, most of the tropical forests, and with them many thousands of species of plants and animals, seem destined to disappear during the next 30 years. It is not too much to say that humanity is locked into a race in which systematics must play a crucial role.

From a practical standpoint, plants provide many critical medicines and pharmaceuticals, many species contain genes for disease resistance and other desirable traits that can potentially improve agricultural varieties, and many could potentially be developed into important crops themselves. For example, the taxonomy of plants has stimulated and in turn been invigorated by the discovery of more than 10,000 secondary compounds scattered among a vast array of species. These substances (alkaloids, terpenes, phenolics, cyanogens, and glucosinolates) are equally crucial to the understanding of plant evolution and to the improvement of human welfare. Thus, the study of biological diversity and the desire for its preservation are not based on esthetic principles alone.

Systematics Also Includes Studies of the Interrelationships Among Organisms

Studies on phylogenetic relationships among organisms aid in the development and evaluation of theories about evolutionary processes. Models of the origin of species have been stimulated as well as guided by the development of the species concept in systematics. Phylogenetic information is important in many areas of evolutionary biology. For example, in biogeography, the occurrence of flightless ratite birds (ostrich, rhea, emu, and others) in Africa, South America, Australia, and New Zealand is apparently inconsistent with morphological and molecular data indicating that the birds share a common ancestry. The paradox is resolved by the knowledge that the birds all diverged from a common stock that inhabited the great southern continent of Gondwanaland before it split and became dispersed through continental drift. Phylogenetic information is the basis of the comparative method for the study of adaptation. A positive adaptive value for a particular characteristic is suggested when two or more distantly related organisms have undergone parallel evolution in that characteristic, for example flower shape or color. Phylogenetic analysis is also necessary for understanding the sequence in which characteristics have undergone evolutionary transformation and for estimating rates of evolutionary change, be it morphological or molecular.

One of the chief tasks of systematics is the elucidation of phylogenetic, or genealogical, relationships among organisms. Inference of genealogy is a desirable goal both for fossilized forms and for living organisms whose ancestry is poorly documented in the fossil record. The reconstruction of phylogenetic history is often of great interest in itself, for example in determining the ancestral relationships among humans and other higher primates. But the reconstruction of phylogenetic trees has numerous other uses as well. Indeed, phylogenetic data are the source of almost everything we know about the patterns of evolutionary change over the course of millions of years, including convergent evolution, parallel evolution, adaptive radiation, and mosaic evolution. Phylogenetic studies have been essential to understanding how species have arrived at their present geographical distributions and to interpreting processes and rates of change at the level of the DNA.

Only in the past 20 years have the logical and evidential criteria for establishing phylogenetic relationships been articulated. Through these sometimes controversial developments, systematics has become a highly sophisticated, rigorous science in which mathematics, statistics, and molecular biology play leading roles. Modern systematics differs greatly from what it was even 10 years ago and poses extremely complex questions.

An important step in this revolution was the development of methods of classification that allowed treelike diagrams expressing the similarity among organisms to be derived by objective criteria through the use of appropriate mathematical expressions evaluated by computer algorithms. These kinds of clustering procedures first developed for biological classification have since been used in many other applications, for example, in linguistics.

The treelike diagrams derived from clustering procedures do not necessarily reflect the genealogical relationships in phylogenetic trees unless the similarity of two species is directly proportional to how recently they diverged from their common ancestor. Proportionality does not exist when many characteristics undergo convergent evolution or when different evolutionary lineages evolve at different rates. However, methods have also been developed that aim to infer the correct phylogenetic relationships among species, although these methods are sometimes difficult to apply in practice because of uncertainties and ambiguities in the data. An important area of current research is the development of statistical techniques to evaluate the degree of uncertainty in estimates of phylogenetic relationship. Just as estimates of numerical quantities should be accompanied by confidence intervals giving the precision of the estimates, inferred patterns of phylogenetic relationship must be accompanied by some kind of measure of their reliability.

Traditionally relying on the data most readily available, usually the morphological characteristics of preserved museum specimens, modern systematics also includes other sources of data, such as ecology, behavior, genetics, and biochemistry. The power of systematics has recently been augmented by data from molecular biology. Electrophoretically distinguishable proteins are now routinely used to distinguish species and to estimate phylogenetic relationships among closely related species, and restriction enzyme digests of DNA sequences such as mitochondrial DNA provide numerous systematic characters.

EVOLUTIONARY HISTORY

The Fossil Record Makes Special Contributions to Evolutionary Biology and to Knowledge of Present-Day Diversity

Although questions of both process and result are central in evolution and diversity, the history of evolution has only one source of primary direct evidence, one court of last resort, which is the fossil record. Studies in paleobiology therefore directly affect all aspects of evolution and diversity.

The fossil record provides the vital time dimension for the understanding of biological diversity and the history of life. The current data base of paleobiology consists of records of some 250,000 extinct species of plants, animals, and microorganisms occurring in deposits spanning more than 3.5 billion years of earth history. Although the record comprises only a small fraction of all the fossil taxa that ever lived, systematic collections in museums and universities contain tens of millions of documented specimens, in many cases with good representation of individual species in space and time.

With respect to the history of diversity, the fossil record can be analyzed to determine whether diversity is higher now than in the geological past, whether the evolution of diversity might be expected to reach a steady-state level, and whether community structure has changed over geological time.

Paleobiology is unique in being the only source of data about certain evolutionary processes and events. For example, although the observational and experimental work of most biologists is necessarily limited to processes and phenomena that are relatively rapid or common, paleobiologists capitalizing on the depth of the geological record have access to much rarer events.

The geological record also documents a unique and lengthy natural experiment in adaptation. Many biological innovations originated, flourished, and died out long before the modern biota emerged. Studies in paleobiology can shed light on when these lost adaptations originated and whether they were better solutions to functional problems than are found among living organisms today.

Adaptive radiations—bursts of speciation in which the number of species in a biological group or adaptive zone increases exponentially during a relatively short time, with accompanying expansion in the diversity of structure and function—is

280   OPPORTUNITIES IN BIOLOGY

THE RISE OF THE ANTS

In 1967, Harvard University received the first known specimens of fossil ants of Mesozoic age, two beautifully preserved specimens in the clear orange amber from a redwood tree that grew 80 million years ago in New Jersey. These specimens were something of a breakthrough in the study of insect evolution. Until then the oldest known fossils were about 30 to 40 million years old (from the Oligocene epoch) and quite modern in aspect. The main features of ant evolution had already been fleshed out. The only phylogenetic tree that could be drawn from such evidence was the canopy, with the trunk and roots cut off. The Mesozoic ants provided what appeared to be a piece of the trunk.

Soon afterward, Soviet paleontologists began to describe a long series of other antlike fossils, also about 80 million years old, giving a separate scientific name to almost every specimen. When all these bits and pieces were fitted together and the New Jersey fossils added in 1986, a remarkable picture emerged: The specimens fell into three classes, representing the worker caste, the queen caste, and the male of the most primitive ants. The workers lacked wings and had proportionately small abdomens, the hallmarks of a sterile caste. These fossils made it possible to conclude that social life had been established in the ants by 80 million years ago, a startling conclusion in view of the earlier lack of such ancient fossils.

A close examination of the American and Soviet fossils showed them to be similar to what had been expected for ancestral ants. Their anatomy was a mosaic of traits, some typical of nonsocial wasps and some more modern—but still typical of generalized ants. They provided the first clue concerning the group of wasps from which ants arose.

The Harvard collection recently obtained the first ant fossils of mid-Eocene age, from Arkansas this time. Chinese and Soviet paleontologists were close behind, discovering Eocene specimens from Manchuria and Sakhalin, respectively. These ants are thought to be 50 to 60 million years old, and most of them are very different from the Mesozoic fossils. They are diverse, representing both modern taxonomic groups and (in one case) a stock not too distant from the Mesozoic ants. It thus seems that the ants, like the mammals, crossed a threshold around the end of the Mesozoic era. For some reason not yet understood, they expanded into a richly various, world-dominant group.

Entomologists and paleontologists continue to search avidly for fossils from Mesozoic and early Cenozoic deposits. The questions we hope to answer include when, where, and from which wasplike insects the ants arose; exactly when they radiated into their modern aspect; the directions they took when spreading around the world; and, not least, what traits contributed to their spectacular success.

well documented in the fossil record, but it is not clear whether these grand adaptive radiations are analogous to the smaller-scale bursts of speciation observed, for example, among Hawaiian drosophilids and African cichlid fish.

Extinction Has Been the Fate of Almost All Species That Have Ever Lived

Extinction, as a biological process, is difficult to study in modern environments. Although the background rate of extinction is low—estimated as about one global extinction per million species per year—extinction is not only frequent on a geological time scale but has been responsible for many complete turnovers in the biological composition of the earth. A proper understanding of the evolutionary process is impossible without knowledge of rates of extinction, quite apart from the importance of such knowledge in evaluating the magnitude of the increase in rates of extinction resulting from human activities in modern times.

Understanding the environmental causes and evolutionary implications of the occasional, brief periods of mass extinction in earth history is a key problem in paleobiology. The most severe mass extinction occurred 250 million years ago and eliminated between 75 and 95 percent of the species then alive. In short, the global biota had a close call with total annihilation. Somewhat less severe mass extinctions are scattered throughout the fossil record. Recent work on the likelihood that some mass extinctions were caused by meteorite impact shows promise of establishing strong connections between biological evolution and the cosmic environment. When combined with the more speculative possibility that impact-induced extinctions are regularly periodic, this hypothesis opens the possibility for major shifts in the way the evolution of the global biota is interpreted.

Within the past 2 million years of the fossil record, constituting the Pleistocene epoch, are special opportunities for studies of biological diversity. During this period, the biota was essentially modern but subjected to the effects of well-documented major changes in climate and geography that set the stage for the present distribution of plant and animal species. Modern tropical rain forests, to pick just one example, can be understood only by knowing the historical underpinnings that led to their present distribution and composition. This understanding is critical in developing a strategy for dealing with the effects of human activities, especially in the moist tropics.

Paleobiologists Have Made Major Progress in the Past Two Decades

Research results have been astounding, at the other end of the time scale, in deciphering the oldest records of life on earth. Not only did life begin far earlier than biologists had previously envisaged, but, perhaps even more surprising, the earth's biota was composed solely of bacteria over such an extended period. These fossil discoveries have recast concepts of evolution and diversity and have reemphasized the fact that, when viewed over the long sweep of geological time, a significant part of evolutionary progress has resulted from changes in the intracellular biochemistry of bacteria.

Paleobiology includes several research areas that have special promise of making significant contributions to evolutionary biology and to other fields of the natural sciences. Among these are the origin of life itself, including not only when life began, by what processes and in what types of environments, but also whether life might exist elsewhere in our solar system or in the universe. Such issues are ripe for exploration during the coming decade because recent progress in studies of ancient Precambrian fossils has extended the known record of life on earth to more than 3.5 billion years. Studies of even more ancient deposits, coupled with laboratory studies of chemical reactions that can occur in a lifeless environment and biochemical studies of existing microbial organisms, promise to provide new evidence of the beginnings of life and of the environment in which life originated.

Organisms Alive Today Are Well Adapted to the Vagaries of Their Present-Day Environment

Environmental conditions such as atmospheric composition, day-night light regime, and temperature conditions have changed markedly over the course of geological time. Until about 1.7 billion years ago, well after the origin of living systems, the atmosphere contained too little oxygen to sustain obligately air-breathing forms of life. Day length has progressively lengthened as the distance between the earth and moon has gradually increased, and there is good evidence that the earth's average surface temperature has changed markedly. Each evolving species became adapted to the environment in which it originated, and as the environment changed, life evolved and built on foundations that had become established under earlier regimes. Therefore, recorded in the genetics, biochemistry, cellular structure, and gross anatomy of living organisms may be a coded history of their evolution. For example, analyses of growth bands in fossil corals and mollusks have made it possible to track the changes in day length caused by tidal friction. Even more spectacular has been the recent recognition of Milankovich cycles of climatic change over the past 700,000 years, which almost certainly were responsible for the pulses of continental glaciation during the Pleistocene epoch.

Deep-sea drilling during the past two decades has provided continuous sections in which important population-level analyses of evolutionary changes are feasible. This increased resolution in the fossil record introduces a time scale comparable to that of microevolutionary change in population genetics, and it opens a more complete synthesis of these two disciplines. The oceanic fossil record is excellent for the last 160 million years, and the deep-sea cores provide a rich source of information on the evolution of single-species lineages. Statistical analyses have already documented important patterns of morphological change and the not-uncommon lack of such change known as stasis. But the surface of this field has only been scratched by the investigations carried out thus far, and we have much more to learn about the tempo and mode of evolution.

Much also remains to be learned regarding the timing and nature of major evolutionary events. Some of them, such as the origin of invertebrates, vertebrates, flowering plants, angiosperms, and humans, have been recognized as important research problems since the mid-nineteenth century. Other events, such as the advent of photosynthesis, oxygen-dependent respiration, the anaerobic-aerobic global ecosystem, nucleated cells, and eukaryotic sexual reproduction, have been addressed only recently with the upsurge of interest in the Precambrian fossil record. Future studies will promote a better understanding of the timing and context of major evolutionary events in the history of life on earth.

CURRENT STATUS OF RESEARCH

Contemporary Research in Evolution and Diversity Features Several Exciting Growing Points

A particularly promising field spanning the synthesis of molecular biology and evolutionary biology is expected to reveal new evolutionary principles even as it resolves some longstanding issues. Important as this new synthesis is, it must be emphasized that not everything in evolution and diversity can be reduced to molecular biology.

Many central issues of evolution at the organismic level require different kinds of approaches. These include the study of the evolution of complex multifactorial traits within populations and the evolutionary role of selection among populations. Innovative approaches to uniting physiology, behavior, and development should also be encouraged.

Those who set research priorities in evolution and diversity must also recognize the continuing importance of cataloguing the diversity of life on earth and understanding its origins through speciation and its disappearance through extinction. Apart from the scientific value of such research are the many potential practical applications of the findings in medicine, agriculture, and biotechnology. Groups of organisms that are already relatively well known, such as vertebrates, plants, and butterflies, are important to study because of the light further information about them would shed on overall biogeographic problems. In addition, economically important groups of organisms, such as legumes and mosquitoes, should be emphasized in choosing priorities for study. Areas of vegetation that are already decimated and those that are being destroyed rapidly but that contain large numbers of endemic species should also receive special emphasis. Concerted efforts to survey more or less completely the biota of selected places, especially in the disappearing forests of the tropics, would be much more rewarding than miscellaneous sampling of poorly known groups over wide areas. Greater attention should also be given to groups that are especially tractable for the solution of basic problems in ecology, population biology, and evolution. To accomplish this, additional systematic biologists must be trained and employed, since the current world supply is much too limited to attack the millions of species of unknown or poorly known organisms profitably.

Paleobiology also presents significant new opportunities for breakthroughs in understanding the history of life on earth, including its earliest history in the Precambrian, its diversification and geographical distribution, and its extinction through, in some cases, global processes.

Collections and Special Facilities

Museums Are One Logical Place to Concentrate the Effort to Encompass Diversity

These institutions are already the repositories of vast numbers of priceless specimens, often representing species that are endangered or recently extinct. Yet most of the collections are fallow, and the halls of some of the leading research museums are largely empty of qualified researchers. The same is true of zoos and botanical gardens, which are in effect museums filled with living specimens. One of the premier tropical botanical gardens in the continental United States has purchased no major items of research equipment in 20 years. Although it averages only one postdoctoral fellow per year, it could easily accommodate six.

An additional need exists for regional or international centers for the storage and analysis of fossil pollen and other microfossils, which are vital in the reconstruction of evolutionary histories and past environmental change. For example, we are only now becoming aware of the considerable extinction of species that has been caused by human disturbances, especially on islands, lakes, and other geographically restricted habitats. One of the most promising domains of research is the detailed analysis of this impoverishment during the past several thousand years, with an emphasis on the factors that make certain species more vulnerable than others.

The future of systematics and its contribution to evolutionary studies depend on collaboration among workers in different fields, funding of interdisciplinary studies, and mutual education. Museums, the traditional home of systematics, will find it necessary to expand their facilities and personnel to encompass statistical, molecular, and experimental approaches. The traditionally modest sums granted to systematists will not support molecular investigations, and it will be useful to set up facilities for molecular systematics that can be used by multiple workers. Above all, university biology departments, in their staffing and curricular decisions, must take into account the growing impact of the new systematics on the study of evolution and its implications throughout biology, from molecular biology to ecology.

Museums are also vital to the continued health of research on the fossil record by maintaining and developing systematic collections. These collections are the lifeblood of research progress. Research questions change continually, and it is important that museum collections remain an effective source of empirical data and that the data be actively studied and described by competent specialists.

The paleontological collections of the United States are in reasonably good shape, thanks to many years of financial support from the Biological Research Resources program of the National Science Foundation. Continued support is critical to sustain active research programs of relevance to broader problems of evolutionary biology. Museum collections are becoming especially critical in some areas because of the phasing out of support for collections by many major research universities.

Collection and Conservation of Germplasm Is Crucial for Improved Agricultural Production

Human activities associated with modern civilizations are causing a loss of diversity at all levels of biological organization. Once lost, this store of genetic diversity can never be recovered. A case in point is the loss of genetic diversity associated with the primitive land races (plants that are adapted to a region in which they evolved) and wild relatives of our crop plants. These genetic resources have repeatedly provided genes for disease resistance when agriculture has been challenged by serious disease epidemics, and these resources constitute an important source of novel phenotypes in conventional plant improvement. They must be found and conserved for our common good.

Modern agriculture is characterized by extensive plantings of genetically uniform monocultures. For example, genetically uniform hybrid corn is widely grown in the United States. Genetically uniform populations have the advantages of high yields, uniform size, and uniform dates of maturity, and these features have played a major role in the great increase in productivity of agriculture in the United States during the past 50 years. Uniformity of size and maturity are also required by highly mechanized agricultural practices.

On the downside, genetically uniform crops are vulnerable to large losses from pest or disease outbreaks because monocultures may lack the genetic variability for resistance to the pathogens. In 1970, a fungal pathogen raced through the U.S. corn crop in the corn leaf-blight epidemic. It was quickly discovered that susceptibility to the fungal pathogen was associated with a particular mitochondrial genotype that had been widely incorporated into breeding stocks. Luckily, other mitochondrial genotypes conferred resistance, and the resistant genotypes were introduced into commercial lines of corn. By 1971 corn varieties resistant to the leaf blight had largely replaced the susceptible type in agricultural production, and the corn crop was protected. The resistant types were available because an international effort had been made to conserve plant genetic resources for just such contingencies.

A second major disadvantage associated with the wide, and in some cases nearly global, adoption of monocultures is that these plantings supplant and drive to extinction the wild relatives and primitive cultivated forms of crop plants, which provide a source of genetic variants for future breeding efforts. Genetic conservation is faced with two problems—how to save and maintain useful plant germplasm and how to evaluate plant gene pools in order to preserve as wide a sample of potentially useful genetic variants as possible. The problem of evaluation is particularly difficult because we have no way of predicting which kinds of novel genetic variants the future may require. At present, the best that can be done is to evaluate the plants of interest for a wide range of genetic traits and select a sample for conservation that includes as much diversity as possible. Little is known about the adequacy and scope of contemporary germ-plasm collections. Genetic screening procedures and statistical sampling plans need to be developed for this task.

Finally, what little effort is expended to protect plants is almost entirely devoted to crop plants and their wild relatives—about 150 species out of the more than 260,000 kinds of plants known. Botanists estimate that tens of thousands of kinds of plants could probably be developed into useful crops—not only for food but also as sources of medicines, oils, waxes, and other chemicals of industrial importance—if we would carry out the appropriate investigations, identify them, and develop them according to their cultural requirements. Virtually no effort is being expended in such investigations; yet fully a quarter of all plant species, along with a similar proportion of animals and microorganisms, are in danger of extinction. Even if the techniques of genetic engineering are fully applied to the development of new kinds of crops, there will need to be a source of appropriate genes; the plants that we are passively allowing to become extinct could well provide such genes, and we should find and conserve them while they still exist.

9

Ecology and Ecosystems

INTRODUCTION

Ecological Problems Are Challenging and Complex

As we enter the last decade of the twentieth century, we face greater environmental problems than humans have ever faced. We are confronted with changes in the distributions and exchanges of elements on broad scales, with the alarming loss of biotic and habitat diversity, with the consequences of species invasions, with toxification and contamination of our aquifers and other systems, with the disposal of hazardous wastes, and with the collapse of resource systems. As never before, we need to improve our understanding of basic ecological principles: of the factors governing the interrelations between organisms and their environments, of the mechanisms governing the structure and functioning of ecosystems, and of the patterns of response of ecosystems to stress. Our ability to deal with environmental problems will depend on learning to manage systems, which must ultimately be based on advances in basic science.

Ecology occupies a unique position in biology because it relates directly to issues and concepts that are widely viewed as being in the public domain. Although most other branches of biology also have great relevance to society, the concepts they deal with are less a part of everyday experience. The earliest ecological studies were those by naturalists interested in organisms and their relations to their environments, and this kind of work remains the core of basic research in ecology.

Ecologists must be concerned with all levels of biological organization: cells, organisms, populations, communities, ecosystems, landscapes, and the biosphere. They work with cross-disciplinary approaches and are concerned with phenomena that are inherently complex. Ecologists study the highest level of biological integration and provide the tools by which humanity is able to manage the biosphere.

The diversity of organisms and of the interactions among them has made the study of ecology fascinating but difficult. There is no unequivocally correct way to reduce that diversity to a set of easily understood and applied rules. Instead, one must look across a wide range of diverse examples and seek common unifying principles. In almost every instance, determining the domain of applicability of the results of experiments or comparative surveys is difficult. Nevertheless, several important principles exist. Central among these is evolutionary theory, which describes the ways natural selection, stochastic factors, and historical constraints have interacted to determine the patterns we see in nature. Other concepts—such as succession and community—provide organizing principles.

Many ecological problems require a comparative approach for their solution and are not always amenable to experimentation. Often, for aesthetic and logistic reasons, ecological experimentation can take place only on small temporal and spatial scales. However, opportunities for large and long-term experiments do exist, and they have led to important insights. In the most general sense, ecology uses the comparative approach extensively in its attempts to order complexity. Its scientific basis is the description and elucidation of pattern. Many of the most interesting and relevant problems in the field are so complex that they cannot be solved by a reductionist approach. In this sense, ecology differs from most of the rest of biology.

In ecology, the transition from explanation to prediction is a large jump. Thus, our experience over the past several decades in managing ecosystems and in dealing with environmental hazards has been punctuated with the accumulation of unique experiences that seem to fit no pattern. Despite these surprises, we are developing an increasingly robust predictive theory. The challenge to ecology generally is to develop further rigorous bases for classification of phenomena and to construct a framework that can accommodate our past experiences, summarize our vast but often anecdotal knowledge, and serve as a basis for prediction.

IDEAS AND APPROACHES IN ECOLOGY

The Responses of Organisms to Environmental Variation

Environmental Variation Profoundly Influences the Distribution and Adaptation of Organisms

Ecology is concerned with the interrelations among organisms and their environments, with the organization of organisms into populations, with the organization of those populations into communities, and with ecosystems. Autecology is concerned with how organisms adapt to their environment through specific biochemical, morphological, and physiological mechanisms.

For plants, examples include changes in leaf shape in different environments (for example, some desert plants have thorns instead of leaves); the evolution of an impressive array of complex chemical defenses against herbivory (stinging nettles, hemlock, poison ivy), and adaptation to water stress (succulents). These adaptations extend from the biochemical to the physiological and whole-plant performance levels.

For animals, similar progress has been made in understanding responses to environmental factors, as well as in understanding the constraints imposed by morphological, thermoregulatory, and behavioral features. Examples of such adaptations include the antifreeze proteins found in the blood of Antarctic fishes; the modified cardiovascular system of seals, which allows them to remain submerged for long periods; the impressive array of chemical defensive (such as tetrodotoxin) and offensive (snake venom) weapons; and the giant versions of otherwise small marine species that have been recently found living at deep-sea hydrothermal vents.

Environmental factors also play a primary role in determining how organisms are distributed. Although in some cases a single factor seems to correlate well with the success of the animal or plant, the basis for that success may actually be complex. The classical studies of ecological races in different plant species confirmed that adaptation in altitudinal races is complex—many genes are involved in determining such features as frost tolerance or time of flowering. An increased understanding of the mechanistic basis of tolerance can come from integrating studies of whole organism-integrated responses with studies at cellular and subcellular levels. Recent advances in subcellular physiology and molecular biology are providing new tools, and the prospects of establishing physiological and genetic bases for adaptation are within reach. The practical consequences of applying improved understanding in this field to crop productivity are obvious, but the insights will also lead to increased understanding of evolution and ecology.

Specific genes associated with stress tolerance, such as those coding for antifreeze proteins, can be identified, cloned, and studied in detail. In turn, the ways in which the products of these genes interact at the developmental, morphological, and physiological levels can be determined. The full understanding of such responses will require the integrated efforts of ecologists, molecular biologists, and physiologists. Ultimately an understanding of the molecular biology of adaptive metabolic features will result. Such knowledge can then be applied directly to evaluate potential new crops as well as to develop kinds of plants and related agricultural practices that are efficient at using limited water and nutrient resources.

As discussed in more detail in a later section, anthropogenic environmental changes also produce stresses and provide opportunities for study, as do natural catastrophes. The large-scale application of pesticides has caused insect pests to evolve and has been a most instructive seminatural experiment. The eruption of

Mount Saint Helens has allowed some unique studies of ecological succession. Studies of the effects of the Glen Canyon Dam on the vegetation along the Colorado River have also provided ecological insights.

The relations of organisms to their physical environment is only one aspect of ecology. Their relations to other individuals of their own or other species are also critical in determining their roles in nature. Combining the two aspects leads to the study of how populations of plants and animals are regulated and structured.

Structure and Regulation of Populations

Population Ecology Is the Study of How Populations of Organisms Are Regulated, How They Behave, and How They Evolve

The most basic question in population ecology is how natural populations are regulated. Why do some species suddenly increase in numbers while others suddenly decline? Why do some organisms reproduce only once and then die, whereas others reproduce repeatedly, perhaps dozens of times? Why do some organisms produce only one offspring at a time and others produce millions at a time?

Gypsy moths can have sudden outbreaks after being at low population densities for years. During these outbreaks, millions of acres of forests can be defoliated in a few weeks. After a year or two, the population density suddenly declines and gypsy moths are not noticed in that area again for many years. Although much has been learned about the factors regulating these pests—predators, food supply, and climate—it is still not possible to predict the outbreaks more than a year in advance. But the regular inundation of areas of the eastern United States every 13 or 17 years by immense hordes of cicadas is well understood. These creatures live underground, feeding on tree roots, for 13 or 17 years. Then, all at once, billions of adults appear; they make streets slippery as they fall from trees and they produce an almost deafening noise. Their unusual life-history pattern appears to be a method of escaping predation. When they emerge in such immense numbers, there are too few predators to seriously dent their populations even though the predators lucky enough to be in an area of emergence can eat to satiation. But for the predators it is a one-time feast. They cannot make a living off this vast banquet because by the time the predators have produced their offspring, the cicadas will have vanished for another 13 or 17 years.

Not only insect populations fluctuate in this way. Hares, lemmings, lynx, and many fish species, such as herring, striped bass, bluefish, spot, and tilefish, fluctuate in numbers over time. In the 1920s, spot became so numerous that they clogged the cooling-water intakes of New York City's power plants. In a few cases it is possible to identify environmental changes responsible (for tilefish, a cold snap seems to have devastated the population in the late nineteenth century) but often the list of possible causative or contributing factors is so long that understanding each case can require a large research program. Many cases simply are not understood.

Unraveling the myriad causes of population fluctuations—and population dynamics in general—and applying that knowledge is an important challenge. Causal and contributing factors in population ecology are either biological or nonbiological, and their study involves many disciplines.

Understanding biological factors requires understanding individual life-history patterns and predator-prey, host-parasite, community, or evolutionary relations; sometimes all need to be understood at the same time. The complexities involved have made successes enormously rewarding; even our failures have been interesting.

Concepts of Population Ecology Are Important in Managing Hunting, Fishing, and Agriculture

Generally speaking, a large class of applied ecological problems has to do with production. How much of something can we get, or how much can we limit something that we don't want? In fisheries, the problem consists of knowing how large the populations are and then trying to understand the ways in which the characteristics of population growth affect the size of a suitable catch per unit time. Estimating the size of populations of fish is difficult, but important conceptual advances have been made with the assistance of models.

The lengths of life cycles are likewise important for management practices. Long-lived, slow-growing species (such as ocean perch, king crab, redwood trees) require different management or agriculture than do fast-growing, short-lived species (shrimp and corn). It is all too easy to mistake abundance for high production, as the story of the passenger pigeon illustrates.

Trees, obviously, lie at one extreme, but many of the problems with managing other long-lived organisms such as whales stem from the same features—long prereproductive spans and low recruitment of juveniles. In whales, these features are due to low reproductive rates, whereas among trees they are due to low seed and seedling survival rates in a world dominated by well-rooted adults.

Biological pest control is even more complex, but the application of ecological studies has led to great advances. For instance, California red scale, a serious pest of citrus, has been controlled successfully in many areas as a result of the thoughtful application of a detailed knowledge of predator-prey relations. Other successful examples include the control of rabbits in Australia by the myxomatosis virus and the control of prickly-pear cactus in the same continent by a cactus-feeding moth. Great care is needed in making such introductions, however, lest the organism that is introduced become a pest itself.

MODELING AGRICULTURE AFTER NATURAL SYSTEMS

Naturally occurring ecosystems have many traits that would be desirable in agricultural systems: They tend to use resources—light, water, nutrients, and carbon dioxide—effectively and efficiently; when undisturbed they tend to resist invasions by competing species; and they seldom succumb completely to pest attacks.

Any realistic view of agriculture for the future must take into account three considerations. First, the spectacular gains in agricultural productivity of this century have resulted from increased use of fossil-fuel derivatives, especially nitrogen fertilizers. Since petroleum reserves are finite, continuous gains in yield cannot be obtained by increasing our applications of petroleum-based fertilizers indefinitely.

Second, agricultural lands everywhere—including the United States—are being degraded by improper husbandry. Techniques to maintain site quality, as well as to restore the productive capacity of already degraded lands, must be developed before the degradation becomes irreversible.

Finally, as the world's population surges past the five billion mark, people are being forced onto lands unsuited for agriculture. These incursions usually result in the irreversible destruction of natural communities followed by the short-term agricultural exploitation of the land that supports them. By designing communities patterned after natural ecosystems, it may be possible to devise land-use schemes that are more sustainable and subsidy-free, while still maintaining an acceptable level of productivity. Such agroecosystems should improve human welfare and reduce the pressure on natural communities that harbor the earth's legacy of evolution.

Chemical Ecology

Many Ecological Interactions Are Mediated Through Chemistry

All organisms are chemosensitive, and each is the source of substances to which other organisms respond. In the course of evolution, this potential for interactions has been thoroughly exploited, and organisms of the most diverse kinds have entered into chemical interdependencies, both mutualistic and antagonistic, that are central to the fabric of life itself. Chemical ecology focuses on such interdependencies. It brings the molecular dimension to our understanding of biological relations—those between animal and plant, parasite and host, predator and prey; between the multicellular and the unicellular, the social and the nonsocial, the kin and the nonkin. Chemical ecology deals with the chemical messengers of nature, defining their functions and ecological roles, and elucidating their chemistry. The discipline is thriving on many fronts, with biologists and chemists joined in an exciting venture of exploration and discovery.

Progress in chemical ecology is being accelerated by recent technical innovations in analytical chemistry. Vastly improved procedures have been developed for separating complex mixtures into their individual components, as well as for quantitating and chemically characterizing naturally occurring compounds. New methods of structure determination both for small organic molecules and for biological macromolecules have been developed, and the amount of sample needed for analysis is constantly decreasing. Since most chemical substances of signal value are produced and "broadcast" by organisms at low—sometimes vanishingly low—concentrations, these refinements in analytical sensitivity and efficiency have proved invaluable.

Although biologists have long recognized the ubiquity and fundamental importance of chemical interactions, they have tended to underestimate the subtlety of roles mediated by chemical ecological factors. Virtually every primary activity of an organism, be it related to growth and development, food acquisition and defense, or sex and reproduction, may be subject to regulation by chemical factors produced either by the organism itself or by other living sources. Pheromones are the best known of these factors. Defined as intraspecific chemical messengers, they have been most thoroughly studied in insects, in which they regulate courtship; in social species they also regulate many of the basics of communal life (foraging, kin and nest recognition, and caste determination). Pheromones have proven useful in applied control programs, both for trapping of pests and for monitoring their densities, and such use is likely to expand as our knowledge of these substances increases.

Pheromones also play important roles in higher animals, including mammals. In mice, for example, information on sex, state of male dominance, and degree of genetic relatedness may all be conveyed by pheromonal cues. A male mouse may even, through its sheer chemical presence, prevent implantation in a female of eggs fertilized earlier during a mating with another male. Chemical induction of infanticide by a male who by killing the offspring of another opens increased reproductive possibilities for itself.

Biologists are only beginning to envision the full scope of functional possibilities of pheromones. Courtship in insects, for example, as in animals generally, involves more than the mere recognition of and attraction to the opposite sex. At close range, males and females may subject one another to a process of appraisal, in which specific fitness criteria are quantitatively assessed. In certain butterflies and moths, the males transmit certain alkaloids to the females, which the males initially sequester from plants. Receiving these toxic molecules, the females transmit them to the eggs they lay and thus protect these eggs from their predators. Prior to mating, the male provides the female with a measure of his intended nuptial gift by releasing a pheromone that is biochemically derived, in quantita-

HUMAN PHEROMONES

Animal pheromones surely—but *human* pheromones? The notion that we might communicate chemically like insects, microorganisms, and other "lower" forms of life, had always met with considerable skepticism. We are primarily visual and acoustic in our conventions, the argument went, and unlikely to respond chemically to one another in any major behavioral context. This view is changing now with the unexpected discovery in humans of extraordinary olfactory capabilities and of interactions that are clearly pheromone-mediated. Nursing infants, for example, show olfactory recognition of the mothers. They may wake and show suckling motions when presented with the mother's breast pad, but ignore pads from other mothers. Discrimination tests with worn clothing showed mothers to be able to recognize their children by odor and adults to differentiate between the sexes and to recognize their individual sexual partners. Both sexes can identify males and females on the basis of breath and palm odor.

Most remarkable are the effects elicited by chemical extracts of underarm secretions. It had long been known, for example, that women who work or live together tend to synchronize their menstrual cycles. The effect is chemical: Extracts from the underarms of women applied to the upper lip of others caused the cycles of the recipients to shift toward synchrony with that of the donors. An intersexual effect has also been demonstrated. Regular sexual activity with a man may normalize the menstrual cycle of the woman. Male underarm extract may induce the effect: Application of the extract to the upper lip of women with abnormal cycles and no current sexual relationship normalized the rhythm. Neither the chemistry of these pheromonal factors nor their mode of action—that is, whether they are inhaled or topically absorbed—is known. Interest in the substance involved is considerable since pheromones could obviously be used in fertility studies.

tive proportion, from the alkaloid. Males of high pheromonal titer are selectively favored by the females.

Plants likewise use the chemicals they produce for a variety of ecological purposes. For example, allelopathic phenomena, involving growth inhibition of plants by chemicals released into the soil by conspecific or heterospecific neighbors, have long been of interest to ecologists and chemists alike. Most of the "unusual" molecules that plants produce, however, are used to deter herbivores or disease-causing agents.

Much remains to be learned also about chemical interactions in aquatic organisms. A vast array of natural products has been isolated from marine organisms, but the function of most of these compounds, aside from those that play a defensive role as toxins or feeding deterrents, remains unknown. Here, too, subtle actions are bound to be uncovered. Some rotifers, for example, grow large protruding spines only when certain of their predators, other rotifers, are present in their ponds at high densities. The spines are defensive, and their outgrowth is prompted, as if by cross-specific embryonic induction, by chemicals emanating from the predators themselves.

The prospect of eventual characterization of human pheromones is especially intriguing. Recent studies have shown pheromonal factors to be at play in a variety of human interactive behaviors, but in no case have mediating chemicals been isolated or identified. The characterization of human pheromones is likely to pose special problems since the substances tend to occur as complex mixtures, subject to both individual variability and variability over time.

Natural products, the very substances that are the subject of chemical eco-logical studies, have proven invaluable to humankind. They constitute the treas-ure trove from which most compounds of technical or medicinal use have been derived, yet the treasury has only begun to be explored. Relatively few kinds of organisms—certainly fewer than 1 in 25—have been examined chemically at all, and thousands of kinds of new compounds, some of them completely unexpected, await discovery. To find the full array of chemicals that exist in nature, however, chemical exploration must be greatly accelerated owing to the quickening pace of extinction throughout the world. The practical potentialities of this area likewise provide another reason to emphasize conservation both in nature and in stock centers.

Behavioral Ecology

Behavioral Ecology Is a Growing Field

Behavior is the study of how animals sense, react to, and manipulate their social and ecological environments. The modern science of behavior arose from the marriage of comparative psychology and ethology, with some admixture of sensory and motor physiology and endocrinology. In the past two decades, the focus has shifted to research on the adaptive basis of complex individual and social behavior and has led to the growth of sociobiology, which seeks to under-stand the evolution of behavior in its social and environmental context.

Over the past decade, the major topics of research in behavioral ecology have included (1) studies of communication, with an increasing emphasis on chemical communication; (2) foraging behavior, focusing on habitat selection, movement patterns, and prey choices in a patchy resource environment; (3) sexual behavior, especially the evolution of behavioral traits under sexual selection; (4) the roles of the sexes in an ecological and evolutionary context; (5) the ways in which conflicts of interest between organisms are resolved ontogenetically and phylogenetically; (6) kinship studies, focusing on recognition of and behavioral differences toward relatives and nonrelatives; and (7) learning, studied as an adaptation in an ecological context. Other important topics have included parental care, predator-prey interactions and strategies, the ecological and social context of aggressive behavior, and the behavior of social insects.

Significant Progress Has Been Made in Understanding the Neural Basis of Simple Motor Patterns and Reflexes in a Number of Animals

These include insect walking and flight behavior, attack behavior in the octopus, the swimming behavior of leeches and sea slugs, and optomotor responses in horseshoe crabs. The search for neural mechanisms to account for more complex behavior has been slow; consequently, many neurobiologists have turned to more tractable research questions, such as those in molecular and developmental neurobiology. Of considerable interest is the interface between sensory physiology and the study of animal communication; in this area, new technical means are now open for understanding the structure and action of chemical communication signals, such as pheromones. Collaborations are also developing between physiologists interested in metabolism, energy regulation, and water use and behaviorists studying behavioral energetics such as behavioral thermal and water regulation and the energetic costs of reproduction. There will also be new research to tie the physiology of digestion, nutrition, and detoxification to foraging behavior in relation to dietary requirements and secondary plant chemistry. Very little is known about how animals satisfy their nutritional needs while minimizing intake of the toxic substances that are so abundant in many of their natural food sources.

The Adaptive Basis of Behavior in Habitat Selection by Animals Is a Growing Area of Research

Life history and population growth vary with habitat, so the behavioral basis of habitat selection can have a profound effect on population processes. At one level, for example, behavioral physiologists have known for years that animals prefer particular temperatures and will seek out these temperatures on thermal gradients in the laboratory. In the field, behavioral thermoregulation has been demonstrated many times. For example, the body temperatures of day-active desert ground squirrels actively cycle. The squirrels forage above ground to cool down. Behavioral physiologists have rarely considered the longer term fitness consequences of habitat selection, however. It has not been demonstrated that animals can optimize their thermal environments, given those available, in the sense of choosing those which maximize growth, survival, and reproduction. Indeed, research has been lacking on almost all aspects of behavioral habitat selection, particularly those involving complex biotic factors rather than simple abiotic ones.

Even though habitat selection is presumed to be adaptive, the behavioral decisions involved are often complex and indirect; many challenging questions remain. For example, an animal may not be able to meet all of its requirements, or at least achieve its optimal performance, in one habitat at one time. When demands on time and energy reserves conflict, behavioral priorities must be set to aid in habitat selection.

Theoretical Studies Have Led to Insights into Foraging Behavior

Until the 1970s, the behavior of animals foraging for food was poorly understood; foraging patterns seemed complex and unpredictable. Then a series of theoretical insights brought order to chaos. It turns out that foraging rules are governed by simple principles, one of the most important of which is that of maximizing food harvest per unit time. However, food is not uniformly distributed in the habitat, animals usually do not have prior knowledge of where it will be found, and they have to search for it. Models have been developed that predict how animals should search, given their sensory capabilities and the distribution of food in the environment, to maximize their food intake per unit time. These predictions are testable, and the best models are remarkably accurate.

Diet choice is another part of behavior predicted by foraging theory models, and it is the aspect of such models that has been tested most thoroughly. The application of foraging theory to a diverse set of organisms has made possible the emergence of a general theory of foraging, which applies to organisms as different as bumblebees and moose and to processes as diverse as growth patterns in plants "foraging" for light and the sexual behavior of males "foraging" for mates.

Recent work on parental house wrens foraging for food for their nestlings under risk of predation illustrates the promise of behavioral hierarchy studies in the context of habitat selection and life-history research. In the absence of predators, parental birds forage until a large prey item (insect) above a critical size is found before returning with it to the nest. Preferred large insects are rarer than small insects, take longer to find, and are generally farther from the nest and in different habitats, so foraging trips and time away from the nest are relatively long. When a natural potential predator of nestlings such as a snake is experimentally placed in a visible location near the nest box, however, the birds make much shorter foraging trips, return frequently to the nest, and spend considerable time watching or attacking the snake. Parental birds have successfully driven off snake predators on several occasions. When parents make short foraging trips in the presence of the snake, the average prey size returned to the nest is smaller and the total amount of nestling food collected per unit time is lower. Nestlings have high, constant demands for food, so a reduced feeding rate is a real threat to their growth and survival. If the parents devote all of their time to fighting predators, the young will be deprived of their essential food. But if they ignore the predators and continue foraging for large insects, the predators might eat the nestlings. The parental behavior is a compromise between these two undesirable results.

Genetic Ecology

A Genetic Approach Often Leads to Greater Understanding of
Ecological Questions

A significant advance in the field of evolutionary ecology has been the recent application of quantitative genetics theory and technique to the understanding of evolutionary processes. Historically, quantitative genetics was the domain of breeders concerned with crop and livestock response to artifical selection for improved growth, yield, and food value. In the past decade these techniques have been transferred to evolutionary biology and have provided new insight into the operation of natural selection. Most traits of organisms are the product of large numbers of genes that individually contribute small additive amounts to the expression of the trait. Moreover, many of these genes affect many traits simultaneously. The success of quantitative genetics has been to demonstrate that a relatively simple theory of additive gene effects can often accurately predict how a complex suite of traits will respond to selection. By studying how a set of traits genetically covary in organisms, we can improve our understanding of how selection operates in nature and also come to a more mechanistic understanding of the limits to adaptation and potential response to selection.

An example is a recent quantitative genetics study of a species of fly that is a crop pest. The study provided insight into the evolution of host-plant preferences in insects and cleared up a long-standing ecological controversy. For years it was thought that insects were behaviorally conditioned through experience to choose the host plant they were reared on as a larva, but tests of this hypothesis yielded conflicting results. The quantitative genetics study revealed why: The genetic strength of the conditioning response varied among fly families. The fly in question is a pest of beans and tomatoes in the southeastern United States. The progeny of some flies showed positive genetic conditioning: If the mothers were reared on tomatoes, the offspring were predisposed to choose tomatoes. On the other hand, the progeny of some flies showed negative conditioning: Flies fed on tomatoes had offspring genetically predisposed to choose beans, and vice versa. The progeny of still other flies showed no conditioning effect. The prevalence of these conditioning types corresponded to the local crop rotation practices where the flies were collected. In areas where bean and tomato crops were rotated every year, the flies showed negative conditioning, presumably because offspring with positive or no conditioning were less successful at finding preferred food plants in the next crop season. In contrast, in areas where crops were seldom rotated, the flies exhibited positive or no conditioning. Studies of this sort in behavioral ecology and evolutionary genetics can have an important practical benefit because they reveal how pests have evolved adaptively to feed on crops—knowledge that can be used to help defeat their attack.

Molecular genetics is also likely to play an increasing role in answering questions in evolutionary ecology in the next decade. As one example, techniques have recently been developed for using noncoding "satellite" DNA regions that are highly variable between individuals for paternity analysis in bacteria and human beings. There are promising signs that similar techniques can be developed for general use for kinship analysis in plants and animals. One of the major unknowns for most organisms is the breeding structure of their populations, an important question in evolutionary biology. Knowledge of breeding structure gives information valuable to understanding natural selection (variation in fitness) and the social systems and effective breeding population size of organisms. This knowledge will have a major impact on conservation programs, since small populations must often be preserved in zoos, botanical gardens, and small reserves. Adequate knowledge of relatedness within these populations can be used to avoid inbreeding and consequent loss of vigor.

Significant advances have also been made in the theoretical description and understanding of evolutionary ecology and behavior over the past two decades. Many complex traits of organisms are not amenable to quantitative genetic analysis, and the connection between most complex phenotypes and molecular genetics cannot yet be made. Nevertheless, progress has been made in understanding the selective basis of many complex traits; that is, why they have evolved.

The Union of Behavioral, Population, and Genetic Ecology

Ecology in the Next Decade Will Forge New Links with Evolution and Behavior

The synthesis of ecology, behavior, and evolution will contribute to the solution of some of society's most important problems—problems that are both complex and diverse. The problems range from understanding the biological basis of human aggressive behavior to the evolutionary ecology of disease, and to the ecological causes and consequences of species extinction. For the first six decades of this century, the disciplines of ecology, behavior, and evolution were pursued largely in isolation from one another, but recently they have converged remarkably. The past 25 years have seen dramatic progress empirically, experimentally, and theoretically in the new fields of behavioral and evolutionary ecology.

An Important Area of Research in Evolutionary Ecology Is the Study of the Evolution of Different Kinds of Life Histories and Their Associated Behaviors

Life-history theory attempts to explain the evolution of growth and reproduction schedules over the life-span as a function of imposed mortality schedules. It also attempts to explain the evolution of the mortality schedules themselves. In part, the schedules are determined by external factors, but they also interact with evolutionarily determined allocations of resources and time at different stages in the life history.

Behavior belongs in such theories because it plays a central role in the reproductive biology and survival of organisms and also in growth—insofar as limiting resources must be found, sequestered from competing organisms, and collected or consumed. Over the next decade, we can expect to see the development of more synthetic theories of life history that make more explicit use of behavioral mechanisms and concepts. Promising signs of such developments include sex-allocation theory, which seeks to understand the evolution of breeding systems—in particular when to expect bisexual species, hermaphroditic species, or sex-changing species that begin life as one sex and later change to the other. The following examples illustrate how behavior and life-history theory can be connected.

Some reef-dwelling fishes of the group known as wrasses are female when they are young fish; as they grow larger they often become males. Male wrasses are large territorial fish that exclude other males and control a harem of females. In this system, small males have no opportunity to mate in the presence of a large male. However, small fish can mate and reproduce if they are female—as members of a large male's harem. If a fish is large, it is advantageous to become a male, because a male can produce many more offspring with a harem than it could as a single female member of another male's harem. Whether or not a fish becomes a male is a function of the social environment. If the current male disappears, one or more of the larger females begins to exhibit male-like aggressive behavior, and one of these females will emerge as the dominant fish and change into a male. The age and size at which the sex change occurs also depends on the social environment—on the relative size and number of rival females in the population at the time the male role becomes vacant. Models have been developed that successfully predict when the switch should occur as a function of wrasse population structure and social environment. Thus, the wrasses provide an example in which behavioral biology and life-history theory have been successfully integrated.

Studies of host-parasite relationships have also provided insights. Parasitic wasps, for example, adjust the sex ratios of their progeny according to the size of the host (a fly pupa in which they lay their eggs) and the number of females that have already parasitized the host. It turns out that the female wasp cannot only assess the size of the host, but can tell whether other females have laid their eggs in it by the time she gets there. Theoretical studies of the wasps, based on the idea that natural selection should lead the female to maximize the number of her grandchildren, make predictions that agree remarkably well with what the wasps actually do—another example of the successful integration of behavior, life-history theory, and genetics.

Another promising avenue of interdisciplinary research in population biology and behavior is represented by the strong new interest in how behaviors of individuals in interacting groups combine and shape group behavior. In individuals, behaviors unfold in a linear temporal sequence, but in groups many behaviors can be expressed simultaneously. In a kin group, for example, polymorphic behavior patterns can evolve in which behavioral specialization leads to adaptive, cooperative behavior among related individuals. In true social insects, many cases of behavioral polymorphisms have been identified in which closely related workers perform different tasks in the colony (an example of this is the development of worker and soldier ants from a single brood of sisters). However, much research needs to be done to understand the origin and mechanisms of behavioral polymorphisms, including their genetic bases.

Game Theory and Optimization Theory Have Much to Contribute to Evolutionary Ecology

An approach that has been developing for several years, the use of techniques from the theory of games and optimization and control, will continue to grow in power and sophistication. We still have much to learn about how selection operates in nature and why particular traits are adaptive. The prospect of understanding complex ecological, morphological, and behavioral adptations in molecular terms is still a distant goal. Phenotypic approaches, within an evolutionary framework, that suppress some (unknown) genetic detail have become increasingly sophisticated, realistic, and general in the past few years. Such approaches are essential for the study of many traits and may have such practical benefit as the understanding of the evolutionary ecology of disease.

Ecosystem and Community Ecology

Ecosystems and Communities Are Dynamic and Constantly in Flux

Community ecology involves the description of species and population assemblages, an understanding of the similarities and differences in their responses to the environment, and an elucidation of their interdependence on one another. The central fascinating questions of community ecology concern why there are so many kinds of plants and animals, why they are distributed as they are, and how they interact. In other words, to what extent does community structure represent only the sum of the properties of the component species and history, and to what extent does it represent the inevitable result of interactions between the species?

The degree to which communities actually are integrated networks, rather than simply individualistic assemblages of species, has been the subject of debate and investigation for many years. Is succession—the more or less regular development and elaboration of communities from simple to complex—an orderly progression of species, each paving the way for the next, or do the observed patterns simply reflect the individual life histories? Are communities assembled according to specific sets of rules? Are the structures of food webs determined by dynamic interactions among species, or are they determined primarily by energetic constraints? If species interactions are important, does that interdependency leave its mark on the evolutionary record, so that coevolved complexes of species represent higher order evolutionary units, or do any observed patterns simply reflect coadaptation to common environmental features? These represent some of the most hotly debated and exciting topics in ecology today. The "either-or" style of the debate is probably obscuring the answers and delaying progress; ecologists probably need to reframe many of these questions in order to make further progress.

Ecosystem Studies Are Concerned with Flows of Energy and Materials Among Groups of Organisms and Between Biological Communities and the Abiotic Environment

Historically, ecosystem studies have proceeded from simple descriptions of the amounts of energy or nutrients in an area through measurement of rates of flow to an analysis of the regulation of these flows. Energy and element transfers are substantially regulated by organisms as well as by the chemistry and physics of abiotic systems, while the rate and nature of the flows themselves feed back to affect both organisms and the environment. Studies of the regulation of ecosystem processes therefore depend on population, community, and physiological ecology, and such studies must integrate across levels of biological organization.

The central principles of ecosystem ecology are relatively easily summarized, although the details are complex. First, energy is dissipated in its transfer through biological systems, whereas the elements that make up organisms are conserved. The dissipation of energy drives the circulation of elements, thereby maintaining productive ecosystems. Second, biological processes tend towards defined ratios of the elements they utilize, because life itself is based on the use of a defined set of elements in particular ratios. Alterations in the availability of elements arising from their use to build structural materials (such as carbon in wood, calcium and phosphate in mammals) have profound consequences for the functioning of ecosystems. Anthropogenic alterations in the ratios of elements circulating in nature (as with the mobilization of sulfur during the combustion of fossil fuels) also may have large-scale consequences. Finally, interactions among trophic levels in ecosystems can affect the resources available to all trophic levels by inhibiting or hastening the cycling of essential elements.

In practice, the study of the regulation of ecosystem processes is extraordinarily complex because of the the large number of kinds of organisms involved and the diversity of their interactions, which are constantly changing. Despite these difficulties, substantial progress has been made in analyzing the regulation of ecosystem processes in terms of these principles.

The development of the watershed-ecosystem approach in the 1960s, based on a comprehensive study of the Hubbard Brook drainage in New England, had significant effects on the way ecosystem research has been conducted subsequently. It defined a bounded system in which inputs and outputs of materials and

MICROBIAL ECOLOGY

Microbial ecology was regarded as a subdiscipline of soil science, limnology, or oceanography until it developed rapidly during the past two decades into a prominent discipline in its own right. It comprises all the subdisciplines of general microbiology, from physiology through biochemistry to genetics, and it deals with bacteria, fungi, protozoa, certain microalgae and viruses, their interactions with the environment, with each other, and with higher organisms. By virtue of their high metabolic diversity and their ubiquity, microorganisms and their specific activities play a fundamental role in the existence and stability of ecosystems and in the biogeochemical cycles of inorganic and organic matter. It is characteristic of basic research in microbial ecology that its results lead, more often than in any other field of biology, to new developments in the applied biosciences, bioengineering, or biotechnology.

Recent headlines were made by the discovery of new forms of life observed at deep-sea hydrothermal vents. Microbial ecologists were able to show that complex ecosystems of dense animal populations in permanent darkness were primarily supported by microbial chemosynthesis rather than photosynthesis. Using geothermally provided oxidizable sulfur compounds, so-called chemoautotrophic bacteria replace the function of plants in these ecosystems. Much of this primary productivity takes place in hitherto unknown forms of symbioses between certain invertebrates and chemosynthetic bacteria. More recently, submarine nongeothermal springs have been found with similarly dense animal populations that seem to live entirely on methane, again in symbiosis with bacteria. These surprising findings have fundamentally altered our understanding of the nature of life on earth.

Applied aspects of this work may include using waste hydrogen sulfide, immediately addressing the acid rain problem, and reassessing single-cell protein production from waste methane. Furthermore, microbiological studies of submarine geothermal vents have discovered new forms of bacteria that are able to grow at temperatures between 80° and 110°C and have initiated a biochemical-biophysical search for the upper temperature limit of life.

energy could be measured and upon which experiments could be imposed; it also provided a context and set of constraints for studies of ecological processes within watersheds.

The development of realistic models for air and water circulation offers a similar opportunity for a much broader array of ecological systems. These models can be used to calculate the past and future trajectories of air and water masses over periods of days (for atmospheric models) to years or more (for oceans). When coupled with chemical analysis of air or water masses, they allow the calculation of balances of energy and materials up to the scale of a continent. This capability is essential to the development of a science of the biosphere; without it, our measurements will be carried out on a much finer scale than the conclusions that we wish to draw from them.

Approaches Based on Understanding the Functioning of Natural Ecosystems Have Proved Most Useful in the Analysis of Renewable Resources, Such as Water, Forests, and Soils

Declining production in temperate-zone agriculture has often been reversed by massive applications of fertilizer, by the use of genetically improved crops, and by chemical and physical measures that are sometime extraordinarily intense. Large-scale alternative practices that place more emphasis on recycling of nutrients than on massive inputs and outputs have proved feasible in many sites and should be developed further as the basic information on which they depend becomes available.

A critical lack of comparative data hampers our ability to predict the impact of different land-use practices on the long-term productivity of the managed site, the ecology of the surrounding lands, and regional climatic patterns. Large-scale experiments should be set up to investigate the flux of materials (including limiting nutrients, pesticides, carbon, and water) in different tillage systems, crop rotations, and grazing regimes. These experiments will produce information that can be applied to management decisions of importance for both temperate and tropical agriculture and forestry.

Agricultural systems in turn provide ecologists with opportunities to explore fundamental questions that are difficult to investigate in nature. For example, large-scale experimental manipulations are possible because crops are adapted to management with standard farm equipment. Similarly, the unraveling of complex processes in agroecosystems is facilitated by the rich background of details that is already available on the genetics, physiology, biotic stresses, and nutrient requirements of crops. Finally, agricultural systems lack the complexity and integration that develop within natural systems during the course of evolution. By skillfully exploiting situations in which these "missing interactions" occur, ecologists can gain a deeper understanding of the forces that organize mature systems and cause the instabilities that we associate with landscapes modified by humans.

Large-Scale Development Projects Often Generate Controversy Because of Their Potential Adverse Impacts, but They Also Represent a Tremendous Underused Scientific Resource

Many large projects that affect the environment cannot be prevented, even if there is concern that they will have adverse impacts. One reason is that we often do not know what will happen. And one reason for our ignorance is that we do not take the opportunity such projects provide to treat them as scientific experiments. Once it is known that the project will be done, it is relatively inexpensive to design it so that information can be obtained. Scientific controls and monitoring programs can be built into the design of the project; at the least, the information gained could be used in mitigating the project's adverse effects.

Many projects represent ecological experiments on a grand scale—a scale that would never be supported by any scientific funding agency. The cutting of canals between bodies of water with widely different biota, the changing of topography, the large-scale applications of pesticides, the building of roads, the fragmentation of forests, agriculture, and the building of dams can potentially produce enormous ecological effects of great theoretical as well as practical interest. Regardless of their possible environmental effects, many such projects could be taken advantage of to gain scientific understanding; the scientific resource they represent certainly should not be wasted.

In addition to studying planned large-scale projects, ecologists can learn a great deal from investigating catastrophes such as volcanic eruptions and forest fires. For example, the 1988 forest fires in Yellowstone National Park, where 800,000 acres were burned, provide an excellent opportunity to investigate many important ecological processes. Studies on nutrient flow and sediment dynamics, succession, and interactions between species can now be investigated on a larger scale than normally found in standard ecological studies. The effects of natural disasters, such as hurricanes, earthquakes, and disease epidemics, would be better understood if ecologists had the resources to study them as experimental systems.

HUMAN-CAUSED ENVIRONMENTAL CHANGES, THE PROBLEMS THEY CAUSE, AND SOME SOLUTIONS

Humans Have Been Changing the Natural Environment Since Prehistoric Times

As human populations and technological ability have increased, human-caused environmental changes have increased also, both in their variety and in their potential to threaten life on earth. The most direct short-term threat seems to be posed by the destruction of natural ecosystems by development, especially in the tropics. Since this destruction is leading to the loss of many species of plants and animals, it is irreversible.

Another potentially serious threat, but acting on a much longer time scale, is that of human-caused global change. As we continue to inject chemically and physically active gases into the atmosphere (for example, chlorofluorocarbons, "greenhouse" gases), we increase the likelihood that significant changes in climate will result. Although the climate has been changing for as long as we know anything about it, some of the changes that might result from human-caused

MOUNT SAINT HELENS

On May 18, 1980, Mount Saint Helens volcano in the state of Washington erupted explosively, spewing forth enormous quantities of volcanic debris. A landslide of 3 cubic kilometers preceded the blast. Plants and animals were obliterated or damaged over an area of 1,500 square kilometers in a 90° arc to the north of the volcano, and the mountain itself lost almost 400 meters of its previous height. The damage to living forms was caused by directed volcanic blast, avalanches of debris, mudflows, flows of hot pumice (pyroclastic flows), and falls of ash. This natural experiment provided opportunities to study a variety of ecological processes on a large scale—a scale impossible to achieve through deliberate experimentation. The scientific possibilities were apparent to Congress when it created the 44,000-hectare Mount Saint Helens National Volcanic Monument in 1982.

The major ecological issue studied around Mount Saint Helens has been succession. The way communities of organisms succeed one another has been a subject of study and debate for 100 years or more. Usually, even if an area is cleared, succession is strongly affected by the seeding of dominant species nearby. The area cleared by the Mount Saint Helens eruption was so large that processes such as sprouting and growth of seedlings could occur without being swamped by nearby dominant species.

The terrain, even after the eruption, is characterized by great variation. Some areas are deeply buried in ash and mud, some are only thinly covered (by less than 1 meter of ash), some are scoured to bare rock; others have been eroded at different rates. Some areas were covered by snow at the time of the eruption. The extent of the affected area and the variation within it have allowed comparisons to be made and have led to insights into the importance of different mechanisms under different circumstances. The studies indicate that succession occurs through a mix of mechanisms and that no simple model is able to explain all cases.

In addition to providing information about natural succession, Mount Saint Helens National Monument has provided insights into artificial succession, or rehabilitation ecology. The devastated area was divided into regions. In some, natural processes were allowed to occur unimpeded. Others received various degrees of management for such purposes as recreation, forestry, and fishing. This division allowed a unique opportunity to experimentally assess the effects of diverse management approaches to ecosystem rehabilitation. Comparisons have made it clear that rehabilitation efforts also affected ecosystem processes involving organisms higher on the food chain. Active planting of shrubs and trees speeded revegetation of some devastated areas. But the removal of woody debris from streams for flood protection had negative effects on stream recovery, and planting trees for forestry and clearing debris for recreational purposes and to provide access reduced scientific opportunities.

> Although the eruption was tragic in many ways, the national monument has been of great scientific and educational value; it represents an excellent example of scientists' taking opportunistic advantage of a natural catastrophe to perform enlightening studies.

atmospheric changes could happen faster than natural ones or have more serious consequences.

Both types of threat provide many research opportunities in ecology—indeed, they mandate such research. They represent ecological experiments at the grandest scale. In this section, we consider first the potential loss of species through habitat destruction. Later, we consider some global changes that might be caused by pollution in its broadest sense.

Species Loss and Conservation Ecology

Issues in Conservation Cannot Be Addressed Successfully Without Also Addressing Economic and Cultural Issues

In coming decades, conservation will become the increasing concern of both natural and social sciences. Solving problems of conservation, especially in underdeveloped countries, will require (1) the development of viable economic strategies for the rural poor, (2) more public education about conservation and birth control, (3) economic incentives from the developed countries for adopting effective conservation and resource management programs, and (4) the financial, institutional, and legal means for upholding conservation law. During the decades ahead, many interdisciplinary efforts will be mounted to develop solutions to the biological, social, and economic problems specific to the major threatened ecosystems of the world.

The Loss of Biological Diversity Throughout the World Should Be a Matter of Extreme Concern

Because the tremendous diversity of form and function among organisms illuminates all fields of biology, knowledge of it is important both theoretically and practically. Yet, our knowledge of that diversity, even at the most rudimentary level, is strictly limited. Thus, of the estimated minimum of 5 million species of plants, animals, and microorganisms in the world, only about 1.4 million have been named and classified. At least two-thirds of the total occur in the tropics, where our knowledge is poor and where the majority of species are uncharacterized.

During the past decade or two, biologists have become convinced that a large proportion of species contributing to global biological diversity will probably disappear long before we have a chance to understand and appreciate them. The problem results from the explosive growth of a record human population, which has doubled to its present size of more than 5 billion people since 1950 and promises to double again within a few decades; extensive poverty, especially in the tropics and subtropics; and a lack of knowledge about appropriate, sustainable systems in forestry and agriculture that are suitable for the tropics. A majority of the human population resides in the tropics and subtropics and has access to only about a tenth of the world's wealth and the commodities that support human life. Their relation to the environment is often highly destructive of that environment. Throughout the world, pollution of various kinds, much of it associated with industry, likewise threatens biological diversity. The widespread use of pesticides, herbicides, and other chemicals in agriculture also threatens thousands of species with extinction.

These activities are resulting in the rapid destruction of natural communities throughout the world. The tropical lowland rain forests, richest and least well known of all the earth's biological communities, are now limited to about half of their former extent; they will be completely destroyed or reduced to small, mainly degraded patches by the middle of the next century. It can reasonably be concluded that human beings constitute an ecological force without historical precedent; yet our numbers will double again in about 40 years, given current rates of growth.

Except for conspicuous organisms such as birds and mammals, the actual effect of large-scale habitat destruction on species extinction is difficult to estimate because such a small fraction of the earth's biological diversity has been described scientifically. Of those 5 to 30 million species of plants, animals, and microorganisms in the world, a minimum of 3 million are in the tropics, no more than 500,000 of which have even been named. Tropical forests will be largely destroyed or degraded over the next 20 or 30 years, which suggests that at least a quarter of the total biological diversity is likely to be lost.

The conservation crisis is real and worsening; there are many questions, few answers, and little time. Nevertheless, some progress has been made and there is some reason for optimism if we take action in a timely manner. A new synthetic discipline, conservation biology, is already making important contributions to our ability to preserve as much as possible of the world's remaining biological diversity. We now outline some of the major biological problems to be addressed and promising avenues of attack.

Conducting Inventories of Threatened Ecosystems Is a Pressing Need

Since the majority of the world's terrestrial species are restricted to the tropics and subtropics, much of this attention must be devoted to the systematics and biogeography of the organisms found there. The application of comprehensive sampling techniques has led to estimates of a total insect diversity in the tropics of between 10 and 30 million species, incredible numbers that command our attention. No more than 300,000 species of tropical insects have been named so far. Some groups of vertebrates and the plants are reasonably well known, but most of the details of their distributions, life histories, and possible utility to humans remain to be discovered.

Comprehensive inventories of the areas that are being destroyed most rapidly, such as the forests of tropical and subtropical Asia, Madagascar, Africa outside of the central Zaire Basin, Central and northern South America, and most of the rest of the tropics are urgently required as a basis for conservation action and because of their scientific importance. To meet the need for biological inventories, we need to train more students in systematic biology and biogeography and especially to enhance cooperative research by involving students and scientists from developing countries, where only about 6 percent of the world's scientists live. The national and regional institutions of these countries must also be strengthened as part of the global effort to understand and conserve biological diversity.

Recent Studies Using Quantitative Samplers Show That Biodiversity at Single Sites in the Ocean Is as High as in the Richest Tropical Environment on Land

Because of its unfamiliarity, the deep-sea environment is not usually thought of as an important reservoir of biological diversity. However, despite low temperatures, darkness, great pressure, and a limited food supply, it is not apparently a hostile environment for the rich diversity of life that has evolved there. Samples taken in the 1960s provided the first glimpse of the magnitude of deep-sea species richness. About 200 samples from a 200-kilometer strip of sea floor between 1,500 and 2,500 meters deep contained about 800 species of macrofaunal animals. Since many of the species have only been collected once, and most only a few times, extrapolation of this diversity estimate to larger areas is difficult. Therefore, our knowledge of the evolution and functioning of deep-sea ecosystems is inadequate for making decisions concerning development of mineral resources and use of the ocean for waste disposal. The fauna of any given oceanic site is so poorly known that we would not be able to determine whether or not changes were occurring after a disturbance.

The biota of the deep-sea floor play a significant role in global geochemical cycles. Rates of microbial activity in deep-sea sediments are low, and many of the animals grow slowly and live a long time. The susceptibility of this fauna to slight shifts in the chemical composition of particles settling to the bottom, or of deep water that originates from near the surface at high latitudes, has not been studied. Considerable effort will be required in the coming decades to determine whether widespread mortality or shifts in species composition of this fauna are likely. Study of community structure and its natural variation should proceed in conjunction with experiments that illuminate the population biology, feeding behavior, and interactions of the animals with sediments. Interpretation of the fossil record and geochemical cycling require measurement of rates of turnover and burial within the sediments and flux of materials across the sediment-water interface. These processes cannot adequately be studied by using a few chemical and physical measurements since they depend on the little-known activities of individual species.

The study of deep-sea populations will advance as a result of in situ experiments at stations on the bottom. These studies are best done in the context of long-term measurements of the biology of the entire water column. Continuous studies of the water column above the bottom station are needed to understand the impact on bottom communities of surface events such as settling blooms of phytoplankton and gelatinous zooplankton.

Deep-sea hydrothermal vent environments require special consideration. The total numbers of species found at vents is not especially high, but extremely interesting and unusual microorganisms and animals have coevolved to take advantage of geothermal energy. More than 20 new families and 50 new genera of animals have been discovered in the few years that vents have been studied. Hydrothermal circulation is driven by hot magma associated with the formation of the earth's crust along a mid-ocean ridge. Hydrothermal fluid pours out along the ridge at temperatures in excess of 350°C. Many metabolic types of bacteria live on chemically reduced compounds, in some cases at temperatures of more than 100°C. At low-temperature areas along the continental margin, seeps of fluids rich in methane and sulfide support communities similar to those at hydrothermal vents.

The study of interactions of organisms with the poorly understood geochemical and geophysical processes of the mid-ocean ridge has required cooperation among biologists, chemists, physicists, and geologists. The full range of techniques, from the most advanced molecular techniques for studying microorganisms to the most sophisticated in situ instrumentation and large-scale survey systems, have already been applied to the study of vent biology. The interdisciplinary RIDGE initiative sponsored by the National Science Foundation should advance our understanding of ridge processes. A similarly broad approach could be applied to the study of organisms in other parts of the deep sea. For example, microtopography, variation in patterns of turbulence close to the bottom, and sediment resuspension are important determinants of feeding behavior and patterns of recruitment.

The Choice and Design of Reserves Must Be Based on Biologically Sound Criteria if They Are to Protect as Many Species of Plants, Animals, and Microorganisms as Possible

Organisms survive poorly in small patches of vegetation. Their populations are small and subject to both inbreeding and consequent loss of vigor as well as to loss by chance alone. In addition, the environmental conditions around the margins of small areas of vegetation differ greatly from those characteristic of large, undisturbed stands of the same vegetation. For all of these reasons, a great loss of species of organisms of all groups is expected as much of the remaining vegetation is destroyed throughout the world.

Choices must be made, and the political and economic setting of the reserves must also be taken into consideration. If the economies of the countries in which they are established—largely the developing countries of the tropics and subtropics—are too weak to support the reserves and to provide at the same time an adequate standard of living for their people, there needs to be compensation for the reserves at several levels. None of these critical steps will be sufficient, however, if the biological criteria are not properly established first.

The principal biological questions for establishing nature reserves are these: (1) What are the major biogeographic regions within a given area? (2) What is the habitat diversity within the area, and how does species richness depend on it? (3) What major factors are likely to lead to extinction of the principal species? Critical factors responsible for the maintenance of species in the ecosystem need to be evaluated, and the interactions among them need to be elucidated if the reserves are to be managed properly. The sizes of species populations adequate to avoid the results of inbreeding and to ensure survival need to be established.

The Loss of Genetic Diversity That Can Occur When Populations Become Too Small Is an Important Concern of Conservation Biology

In zoo populations and domestic livestock, the effects of inbreeding depression on growth, development, survival, and reproduction are well known. However, we know little about the actual sizes of breeding populations of plants and animals in the wild and only a modest amount about genetic variation in natural populations. In animals and plants, the available evidence suggests that individuals that are heterozygous (having two different allelic forms of genes) for more genes often exhibit enhanced survival, disease resistance, growth rates, and developmental stability. The loss of heterozygosity occurs because of inbreeding and genetic drift when populations become small. Conservation of genetic diversity is also of importance to the maintenance of evolutionary potential; species that have lost most of their genetic variability may be in serious trouble. The cheetah, for example, has almost no genetic variation at the major histocompatibility locus associated with disease resistance; it is therefore seriously threatened by diseases, including feline infectious peritonitis. For our major crops and domestic animals, genetic diversity is a major source of disease resistance.

Research is needed to determine the effective breeding population sizes of animals and plants in nature. We know how genetic variation is distributed and maintained in only a handful of species. Research is needed into methods for preserving genetic diversity not only in wild strains of crop plants and domestic animals, but also in organisms as varied as threatened species of primates and commercially important tropical timber trees. In plants, seed banks and tissue culture centers are promising aids to preservation, and analogous techniques are available for animals also.

Research in Ecosystem Restoration and Regeneration Will Yield Many Observations Useful in the Development of Ecological Understanding

Over the next several decades, conservation biology will focus increasing attention on the restoration of degraded or destroyed ecosystems. This effort is important for two reasons. First, restored ecosystems are likely to be superior to zoos and arboretums for the maintenance of viable populations of endangered species. Second, regenerated ecosystems will reduce the pressures to exploit nature reserves when other wildlands have disappeared. One of the highest priorities for conservation biology, for example, is reforestation in tropical countries with serious deforestation problems. It is difficult to protect, let alone to justify, nature reserves in third-world tropical countries when people have neither timber nor firewood. However, reforestation in the tropics is not a simple matter either technically or socioeconomically.

Solving these problems will require a major increase in research on ecosystem restoration, particularly in ecosystems of critical worldwide importance. In the case of tropical reforestation, for example, there must be comprehensive investigations of the properties of various trees that might be used for reforestation and the factors that control the establishment of these species in degraded lands. Research in ecosystem restoration will yield a great many observations useful in the development of ecological theory as a whole.

Species Invasions can Result From Environmental Change and Can Alter Ecosystems and Even Exterminate Species

The consequences of species invasion are of current interest because of their relevance to biological control and because of the analogies that have been made between this phenomenon and the fundamentally different one concerning the deliberate release of genetically altered organisms. New disease-causing organisms are being introduced more widely than ever before—some deliberately (as for biological control)—and understanding the properties of such introductions is becoming increasingly important. Where ecological information has been lacking or has received inadequate attention, even deliberate introductions have sometimes led to major ecological problems. A major challenge to ecology, therefore, is the understanding of the characteristics of species and environments contributing to invasiveness, and to likely consequences.

Global Climate Change

Cumulative Environmental Pollution Can Have Surprising Effects and Can Lead to Global Change

Equally important problems are associated with the cumulative effects of environmental pollution. Not only are environmental perturbations often repeated, but also their combined effects may be more substantial than those of the individual events taken separately; sometimes they are qualitatively different. Consequently, traditional methods of assessing the significance of particular actions on a project-by-project basis may fail to predict and, therefore, fail to help us manage these cumulative effects—comprehensive efforts may be necessary.

As we burn fossil fuel, manufacture plastics and other synthetic products, generate electricity, or package consumer goods, we perturb the environment both physically and chemically. The effects of these pollutants are cumulative and usually cross jurisdictional boundaries. The burning of fossil fuel in Ohio can affect lakes in Quebec; the building of dams in Idaho and in Egypt can affect fisheries in Oregon and in the Mediterranean; the use of fertilizers in Maryland and Virginia can affect fisheries as far distant as Nova Scotia.

The problems caused by pollutants released into water and air, or widely into the terrestrial environment, are political as well as scientific. Although it may be economically advantageous for one country not to limit the emission of sulfur dioxide from industrial plants, it may be highly disadvantageous to nations that lie in the path of the pollution. A river may be a convenient dumping ground for chemical wastes, removing them from the area where they are produced; but those wastes may cause considerable economic loss downstream. At a scientific level, we do not yet understand ecosystems' capacities for recovery, for detoxification, or for resisting various kinds of pollution, and the interactions of environmental pollutants are not fully understood, either. Many problems associated with chemical and physical stresses have not yet even been clearly defined. Nor do we yet understand how best to manage environmental problems that cut across jurisdictional boundaries, although the international agreement on protecting ozone in the stratosphere, developed under the auspices of the United Nations Environmental Program and signed in 1987, appears to offer a good model.

The concentration of carbon dioxide in the atmosphere, for example, changes globally as a function of time scale. Annual cycles are controlled by seasonal changes in net carbon uptake and release by biota, whereas a sustained historical increase is caused by the cumulative effects of increasing fossil fuel combustion and deforestation. Even though most fossil fuel is burned in the northern hemisphere, the increase in atmospheric carbon dioxide is global. Large variations are associated with glacial advances and deglaciation. These well-documented changes will undoubtedly stimulate additional efforts to study their effects on climate and on biological processes.

Altered concentrations of carbon dioxide are the best known but far from the only major anthropogenic change in global geochemistry. Other trace gases such as nitrous oxide and methane are increasing globally, acid precipitation has increased regionally, and upper stratospheric ozone levels have decreased even while ozone is increasing in the troposphere locally. Together, spectrally active ("greenhouse") gases other than carbon dioxide are now thought to have an impact equal to that of carbon dioxide on the earth's heat budget. However, natural climatic variability makes it difficult to know just how large the actual climatic effects are.

Ozone and acid deposition are affecting forests and aquatic ecosystems over much of North America and Europe; such changes may become global as the human population increases and more industrial development occurs in tropical regions. Once again, however, it is difficult in many (but not all) cases to separate the effects of natural environmental fluctuations from those of ozone and acid deposition.

TECHNOLOGICAL AND METHODOLOGICAL ADVANCES

Remote Sensing

Evaluation of Global Change Requires the Ability to Document It and to Study Patterns at Fine Scales by Remote Techniques

Remote sensing is not a new technique in ecology—interpretation of aerial photographs has been a valuable part of ecological studies for at least 50 years. The scope and quality of remote sensing has changed dramatically in the last 10 years, however, and we have every reason to believe that its ecological applications are far from reaching their potential.

Many techniques are now available or under development. The best-known of these include the coastal-zone color scanner, which has revolutionized the study of marine primary production because it permits measurements of chlorophyll concentrations over wide regions; and the advanced very high resolution radiometer, used to measure light absorption by leaves on a daily, seasonal, or annual basis worldwide. The radiometer "sees" a pixel 1 kilometer or 6 km square; it is therefore ideal for continental-scale studies. Other sensors, such as the U.S. thematic mapper or the European SPOT, sample much smaller pixels (30 m square for the mapper, 15 m square for SPOT); they are more useful for regional studies.

Other remote-sensing techniques are less well developed but perhaps potentially even more useful to ecologists. Synthetic-aperture radar can "see" both the top of a forest canopy and (under some conditions) below the soil surface to bedrock as well as the ground surface itself, and it can do so at night or through clouds. Laser profilers can measure canopy height and the presence of treefall gaps in intact rain forests; fluorescence measurements may further provide information on chemical composition or physiological status. Airborne sensors with very high spatial and spectral resolution may be especially useful to ecologists; the airborne imaging spectrometer, with 10-nanometer spectral resolution and 10 m square pixel size, is now being used in geobotanical and ecosystem studies that require more determinations of the chemistry of forest canopies. Many of these technologies are still being developed; when fully deployed, they should greatly enhance our capabilities in these critical areas.

Analytic Chemistry

The widespread use of high-pressure liquid chromatography and gas chromatography-mass spectrometry in environmental laboratories has made possible the measurement of chemical substances at concentrations much lower than could be measured a decade ago. Ecologists concerned with air chemistry or trace pollutants can reasonably expect to detect parts per trillion. Improvements in automation and the capacity to analyze multiple samples have helped fulfill the need for adequate replication of samples.

Nondestructive analyses can allow measurements to be made that reflect the spatial organization of chemical constituents and their concentrations under conditions in which they are active physiologically. Nuclear magnetic resonance is useful to ecologists studying ecophysiology and water chemistry; electron microprobes have provided detailed information on soil chemistry in the vicinity of plant roots; and microelectrodes measure dissolved oxygen and other chemicals within living cells or in soil.

Remote chemical measurements represent a special case of nondestructive analysis. Tunable diode lasers can be used to measure the distribution of ozone and aerosols at a range of kilometers or in the neighborhood of an individual leaf. Laser fluorescence can be used to measure the photosynthetic potential of individual leaves. These and other laser-based technologies will be applied more widely to ecological studies in the future, with consequent gains in insight about natural phenomena.

Tools for Studying Paleoecology

For systems at equilibrium, the time dimension is relatively unimportant; consequently ecologists sometimes ignore history in deciphering patterns that can be seen in contemporary biotic systems. Paleoecologists have brought a time dimension to the study of ecology; as the interest in equilibrial systems wanes, ecologists are becoming more receptive to including time among the factors that they routinely take into account. The development of adequate quantitative techniques has allowed critical comparisons to be made between living and fossil communities, providing increased insight into both areas of study. In addition, the ability to make absolute determinations of past time by means of radioisotope determinations has been invaluable. The accurate reconstruction of paleoclimates millions of years into the past has become possible, and the first results are biologically exciting.

Stable Isotopes

The ratios of stable isotopes of particular chemical elements are now determined readily by the use of mass spectrometers. Already such studies have advanced our understanding of physiological processes and fluxes through ecological systems. Studies of plant and animal physiological ecology, food webs, historical ecology, marine ecology, and biogeochemical cycling have all benefited from this approach.

Virtually all elements have at least two stable isotopes, with one of them being far more common than the others. For example, the two stable isotopes of hydrogen—^1H and ^2H (deuterium)—occur at abundances of 99.9844 and 0.0156 percent, respectively. Similarly, there are two stable forms of carbon—^{12}C (98.89 percent) and ^{13}C (1.11 percent). Because of differences in physical properties and in enzyme-based discrimination for or against one of the alternative forms in these systems, natural differences in the stable isotopic composition of biotic and abiotic compounds occur in ecologically relevant processes, including metabolic activities and transfer rates between organisms at different trophic levels.

In plant physiological ecology, the use of stable isotopes provides a reliable means of scaling up from instantaneous metabolic rates to longer term estimates of physiological activity. Thus carbon isotope ratios provide information on water-use efficiencies, hydrogen isotope ratios on water sources, and nitrogen isotope ratios on nitrogen-fixation rates.

Carbon-isotope ratios in plants can be used to distinguish among different photosynthetic pathways. These studies have allowed an extensive evaluation of how plants function in different ecological situations. The ratio of carbon isotopes in a particular plant reflects both enzymatic and diffusion considerations. The initial photosynthetic reaction by the enzyme ribulose 1,5 bis-phosphate carboxylase discriminates against ^{13}C, and ^{13}C diffusion is slower through stomata, the specialized openings on leaves. As the stomata open, thus allowing greater diffusion of carbon dioxide into the leaf for photosynthesis, water loss (transpiration) increases. Consequently, water-use efficiency (the ratio of photosynthesis to transpiration) is strongly correlated with the carbon isotope ratio in tissues.

Such studies are likewise important in animal ecology. Free-ranging animals from rodents to penguins can be injected with doubly labeled water (water enriched both with ^2H and ^{18}O), and then released to resume normal activities in the field. The deuterium (^2H) leaves the animal only when the animal loses water, but the oxygen is lost both in the water and through respiration as carbon dioxide; the oxygen in carbon dioxide and water comes into equilibrium through the enzyme carbonic anhydrase. The rate of loss of the two labeled isotopes thus constitutes an integrated measure both of water loss and of metabolic rates for that animal—important features in understanding how an animal functions.

The old adage "you are what you eat" holds for stable isotopes, which are therefore valuable for studies of food webs. Many single and multiple mixtures of stable isotopes are used to study plant-herbivore interactions as well as relations between animals and other organisms further up the food chain. In a recent study, for example, stable isotopes of carbon, nitrogen, and sulfur were used to trace the flow of organic material within a salt marsh ecosystem and to identify the origins of the detrital substances that were accumulated by the mussels downstream.

Organisms carry some of their history in the isotopic composition of the structures that they form over time (tree rings in plants, scales in fishes, and shell layers in mollusks). The study of these structures can often yield accurate information about past environmental conditions and about the diets of particular kinds of animals in the past.

In whales, for example, the baleen plates (plankton-filtering structures) are formed continuously and reflect the different isotopic compositions of plankton communities in the areas visited by the whales during the course of their migration. Investigators have used this information to trace whale migrations, an ingenious application of isotope ratios to an ecological problem. By adding ^{14}C dating techniques, these investigators were even able to determine the length of time that the whales spent in different areas.

Biotechnology

Modern molecular techniques promise to revolutionize the study of microbial ecology by making it possible to follow the fate of particular genetically engineered bacteria and other microorganisms in the environment. These techniques will likewise enable us to determine the rate of recombination in populations of bacteria much more accurately than we have been able to do previously and will enhance studies of ecological genetics, including those of symbiotic interactions. Our ability to produce precisely engineered bacteria will make it possible to study the adaptive significance of single-gene mutations in nature and under experimental conditions. In principle, such modifications are likewise possible in eukaryotic organisms, and they will eventually allow ecological experiments to be carried out with greater precision in such organisms also.

Models in Ecology

Ecological phenomena consist of processes that take place at different rates. Understanding and predicting such phenomena are facilitated by mathematical methods, which illuminate the relative importance and quantitative characteristics of these processes.

Two philosophies of modeling stand in opposition. In the first, one seeks highly detailed descriptive models, intended for implementation on large computers. These models contain much of the fine structure of ecosystems, but present a number of difficulties: Their specificity hinders their portability to other systems; their large number of parameters presents statistical difficulties in estimation; and their reliability as predictive tools is questionable because of the many ways errors can arise and propagate.

At the other extreme, overly simplified models may be general and portable, but submerge much of what is relevant to the mechanisms underlying their dynamics. They tend to be phenomenological and holistic and to ignore many important factors. Both types of models can be useful in guiding research, but dangers lie in believing the output of either type of model without adequate field testing.

For understanding and managing the environment, a compromise is necessary. No single model suffices, and one needs a combination of models at different levels of detail, much as one might use a nested set of maps to drive to a new location: The broad-scale map, ignoring much detail, is necessary for getting one's bearings and reaching the vicinity of the destination; a more detailed map, limited in scale and objectives, allows one to find one's way past the bridges and old barns that dot the landscape and to reach the final goal.

A case in point is the use of fate and transport models to evaluate the distribution of chemicals in the environment. These models come in two forms: generic and site-specific. Generic models incorporate the basic mechanisms of diffusion, advection, and reaction; parameters are assigned phenomenologically and over broad scales. For near-field effects (for air pollutants near smokestacks or for chemicals in particular estuaries), detailed descriptions of local geometries and topographies become important. In such cases, one must turn to the computer for implementation. Models of both forms are essential for a comprehensive appreciation of the phenomena involved.

The roots of mathematical ecology can be traced to the demographic studies of Graunt and others as early as the seventeenth century. The well-known arguments of Malthus, which indicated that a population growing without bound would soon outstrip the capacity of the environment to support it, were based on detailed analyses of births and deaths in human populations. In turn, these led to the first important mathematical efforts in ecology: namely, those of Verhulst and others to describe the dependence of population growth rate on population size and to infer the consequences of such relations. The most important extensions of these single-species models in the classical literature were to systems of interacting populations, especially the famous differential equation derived independently by Lotka and Volterra. These equations predict the outcomes of situations in which two or more species compete for the same limited resources.

The classical tradition has been carried forth to the present, especially regarding the development of an elegant theory of interspecific interactions and evolutionary relationships. Recent work has extended the theory to an examination of the structure of ecological food webs and the factors governing their organization and has taken the subject into diverse mathematical disciplines such as graph theory and dynamical systems theory.

Increasingly, however, mathematical approaches have been recognized as being valuable not only for theoretical investigations, but also for finding solutions to some of the applied problems that society must confront. Thus, for example, mathematical descriptions of population dispersal, which are among the oldest models in theoretical ecology, are now being used to quantify the movements of agricultural pest species and the rates of advance of exotic species invading new habitats. Further, the theory of island biogeography, which predicts the relation between area and number of species, and more extensive mathematical models of spatially distributed populations are being applied to the design of reserves, in the planning of parks, and in regional landscape ecology.

Optimization and control theory, which have undergone substantial mathematical development in recent years, are being applied as integral parts of programs for the management of renewable resources, especially in fisheries, forestry, and agriculture. Such approaches combine biology and economics through mathematical models; this combination will become an increasing imperative for us as we face energy shortages and resource depletion. Inherent limitations to predictability are apparent in any rigorous mathematical analysis and have made essential the development of adaptive management strategies, which couple short-term prediction with continuously adjusted management rules.

Epidemiology has had a firm mathematical basis since the turn of the century. The impact of epidemiological models has been limited, although the problems that we face in combating the spread of diseases in humans and other animals and in plants are of overwhelming importance. Recent years have seen a dramatic increase in the mathematical modeling of epidemics and an increasing recognition of the need to view such problems in their proper ecological context—as host-parasite interactions. It seems likely that epidemiology will be a most important area of growth in mathematical ecology over the next quarter-century. Current work uses mathematical models to help to understand the factors underlying disease outbreaks and to develop methods for control, such as vaccination strategies.

Finally, the need for environmental protection in the face of threats from such competing stresses as toxic substances, acid precipitation, and the generation of power has led to the development of increasingly sophisticated models that address the stress-related responses of community and ecosystem characteristics; for example, succession, productivity, and nutrient cycling. Such models owe much to their classical origins, but typically differ substantially in form. They recognize the importance of explicitly considering environmental factors, the physical characteristics of the environment, and nonbiotic system components, and they focus attention on holistic measures of system response. Examples

NONEQUILIBRIAL SYSTEMS IN ECOLOGY

The tradition of considering equilibrial systems in theoretical ecology is changing. Early mathematical models of ecological systems dealt with systems at or near equilibrium. Thus, the logistic equation relates population growth to "ecological resistance," or the population's size and environmental carrying capacity. It allows the calculation of the numbers of organisms an environment can support at equilibrium, or the growth rate of the population as it approaches that equilibrium. Similarly, models of community structure have traditionally attempted to estimate the number of species that could occupy an environment at some equilibrial diversity. Although these and other equilibrial models have been useful for guiding thinking and focusing research—indeed, many of them were designed for that purpose—they have too often been interpreted as descriptions of the natural world.

But both the biological and the nonbiological world are constantly changing. Succession occurs in communities; populations of plants, animals, and microorganisms evolve; and ecological efficiencies of organisms and communities change.

Nonbiological changes occur in both space and time. Experiments have shown how the spatial complexity of the environment (for example, presence or absence of refuges or of different habitats) can affect population growth or the coexistence of two or more species. And the degree to which the environment varies over time is still being discovered. This exciting process of discovery is aided on the one hand by satellites and computers that can detect subtle physical changes over short periods and on the other hand by analysis of historical records, both physical (such as ice cores) and written (such as the historical position of glaciers, weather records, and historical records of agriculture). These methods are revealing the complex and changing patterns of variability in atmospheric, terrestrial, and oceanographic climate, which in turn drive changes in biological systems.

In addition to these physical changes are the newly discovered changes that sometimes occur in the mathematics of population biology. About 10 years ago, biologists were made aware that some seemingly simple equations that describe population growth start to behave chaotically for some values of their principal parameters. The study of the chaotic behavior of familiar equations has become a minor growth industry in mathematical sciences, but it has roots in physical reality: In a variety of disciplines, including biology, real systems exhibit chaotic behavior. For all these reasons, increasing attention is being paid to nonequilibrial systems in ecology. Although harder to treat theoretically, they appear to be common.

include large-scale models being developed to examine the likely effects of power plants on estuarine communities or of air pollution on regional patterns of forest productivity.

A major difference between such models and those discussed earlier is that they are in general too large to be analyzed fully; one must rely heavily on computers to simulate possible outcomes. In such applications, however, mathematical analyses are as essential as ever since one must find ways to simplify, to guide simulations, and to derive understanding. Clearly, our applied needs will continue to increase and to represent new and vital challenges for mathematical ecology.

In ecological modeling, the observed degree of variability changes as a function of the spatial and temporal scales of observation as one moves to finer and finer scales. Thus the concept of equilibrium is inseparable from that of scale. The insights from any investigation are therefore contingent on the choice of scale, and there is no single correct scale of observation.

In many efforts to model particular ecological situations, irrelevant details are introduced on the mistaken premise that somehow more detail assures greater truth. In fact, there can be no one "correct" level of aggregation for a given study. If the taxonomic species, for example, is used as the unit of classification, the differences among the individuals within it are automatically ignored. In ecology, functional systems of classification are often preferable to taxonomic ones, and a failure to recognize this relationship has led to difficulties.

CONCLUSION

Research in Ecology Is Brought into Focus by Practical Applications and Needs

These are exciting times for ecological science. The accumulation and organization of experiences from well-crafted experiments and from accidents of nature are providing opportunities to derive basic principles and to formulate concepts. The analytical tools of the science have also seen rapid advances, and powerful computers provide us with limitless new opportunities. Finally, the blurring of the distinction between what is pure and what is applied, necessitated to some extent by environmental crises, will enrich and inspire basic research.

Perhaps the greatest challenge facing us will be in understanding how "physical, chemical, and biological processes that regulate the total Earth system, the unique environment it provides for life, the changes that are occurring in that system, and the manner by which these changes are influenced by human actions." These words are taken from the objectives of the International Geosphere-Biosphere Program, which has undertaken a long-term study of these problems. Atmospheric processes affect and are affected by biological processes in the earth's ecosystems, and we must improve our understanding of these interrelations. To do so, we must find ways to study the dynamics of ecosystems simultaneously on different scales—including landscapes, regions, and continents—and analogous large-scale phenomena in the oceans.

The crisis in biodiversity also necessitates holistic ecological approaches to problem-solving on broad scales. That the conservation of species cannot be separated from the preservation of their habitats has given birth to new approaches to ecosystem restoration and rehabilitation. Tropical forests are among the most severely affected, and they have deservedly received tremendous attention. But the problems are generic ones that affect all our ecosystems.

Biotechnology has been the source of a new set of challenges for ecologists, who have had to view it in terms of both its tremendous potential and its risks. Genetic engineering holds the promise of increasing crop yields, of providing nonpolluting alternatives to chemical fertilizers and pesticides, and of breaking down pollutants that already exist in the environment. Because of a basic lack of information about microbial ecology, however, and a lack of familiarity with the methods now being used to alter the properties of organisms, ecologists have moved cautiously in capitalizing on these opportunities despite their obvious value to society. The coming years should see the resolution of these problems and the widespread application of new techniques for scientific advance and human benefit.

10

Advances in Medicine, the Biochemical Process Industry, and Animal Agriculture

The impact of biological research on health care has long been recognized by the medical and pharmaceutical industries. The rapidly expanding fields of cellular and molecular biology continue to generate possibilities for new pharmaceutical products and medical practices that will have major impacts on the prevention, diagnosis, and treatment of human disease.

The application of biochemical and molecular understanding of cellular processes to product development in plant and animal agriculture lags behind that for medicine. Where such applications have been made in animal agriculture, they have generally resulted from the transfer of biomedical breakthroughs made originally in biomedical research. Currently, however, the biological knowledge is being exploited with increasing effectiveness in the development of new plant and animal agricultural products.

ADVANCES IN MEDICINE

The long and successful overlap of all areas of biology makes it impossible to separate their relative contributions to the development of medical practices and pharmaceutical products. Instead it is more fruitful to discuss some key examples in which the fundamental knowledge of biology is likely to lead to advances in medicine.

Molecular Pharmacology and Human Disease

A Primary Objective of Biomedical Research Is the Development of Chemical Agents That Can Selectively Relieve and Abolish Pathological Processes

Traditionally, the discovery of new drugs has been based on the chemical synthesis of large numbers of analogs, which are then analyzed empirically in screening programs. The success rate has been low relative to the effort and expense involved. In contrast to these traditional methods, most drugs can now be discovered by a concerted effort with a limited number of compounds because the investigators proceed with a rational approach based on a clear understanding of the fundamental properties of the systems—that is, the receptor-recognition properties or the identity of key regulating enzymes. Some notable examples have been the development of propranolol, an agent used worldwide in the treatment of cardiovascular diseases; cimetidine, a histamine-2 agonist, which has dramatically altered ulcer therapy; and lovostatin (Mevacor), a selective inhibitor of cholesterol production—a powerful new tool in the attack against atherosclerosis.

Scientific and analytical developments in the past decade have provided powerful tools for the dissection and understanding of many critical biological processes. For perspective, it took 40 years to elucidate the structure of the neurotransmitter norepinephrine after the demonstration in 1921 that adrenergic neurons released an agent that increased the rate and force of cardiac contraction. It took an additional 20 years to understand the sites and mechanism of synthesis, storage, release, and response to norepinephrine. These inquiries are valuable because the chemical manipulation of intrinsic norepinephrine production, release, and action is the primary basis for the treatment of hypertension in millions of patients.

In contrast to the time required to develop norepinephrine, the application of modern analytical chemical techniques (especially thin-layer chromatography, high-pressure liquid chromatography, and gas-chromatography–mass-spectrometry) have resulted in the rapid discovery and identification of trace amounts of extremely labile but potent substances produced in the body from arachidonic acid that are involved in the regulation of cardiac, pulmonary, allergic, inflammatory, and blood disorders. In 1973, the prostaglandin endoperoxides were described; in 1975, thromboxane A_2 (vasoconstrictor and platelet aggregator); in 1976-1977, prostacyclin (vasodilator and inhibitor of platelet aggregation); and in 1978, leukotrienes (chemotactic and anaphylactic substances). Chemical agents that modify the synthesis of such arachidonate metabolites have already been discovered and proven effective in experimental animals and in clinical trials.

The recognition that normal bodily function, cell-cell communication, and disease processes (such as inflammation, myocardial infarction, immune response, and allergy) are mediated by minute amounts of biologically active substances produced in the body and intricately regulated has provided targets vulnerable to chemical manipulation for therapeutic benefit. Such studies require the development of interdisciplinary programs that bring together cell and molecular biologists, analytical and synthetic chemists, whole-animal physiologists and pharmacologists, and clinical scientists, who collectively focus their research on understanding the molecular bases of disease processes and on their therapeutic modification. The efforts are directed at elucidating (1) the characteristics of the intrinsic biochemical pathways that mediate and modulate normal and pathological function, (2) the determinants and mechanisms of receptor recognition and coupling to intracellular pathways, (3) the design, synthesis, testing, and targeting of new chemical entities, and (4) the evaluation of human disease through the development of new diagnostic strategies, and (5) the application of newly discovered potential therapeutic agents. Thus, opportunities presented by modern analytical, cell, and molecular biological techniques will not only elucidate structure, but when coupled to synthetic efforts, can provide large amounts of material for whole-organ and whole-animal pharmacology studies and ultimately to therapeutic applications. Two specific examples of such efforts are illustrated below.

Prostaglandins, Thromboxane, and Leukotrienes Have Great Potential for Therapeutics

For the past decade, arachidonic acid research has emphasized the discovery and elucidation of locally synthesized potent metabolites that influence regional tissue function. Two primary metabolic routes have been characterized extensively. The first of these is the cyclooxygenase pathway, which produces prostaglandins and thromboxane; the second is the lipooxygenase pathway, which produces the leukotrienes. The cyclooxygenase products have been characterized for their role in renal, cardiovascular, and platelet function. The leukotrienes subdivide into two types: (1) those, such as leukotriene (LT)C/D, that are slow-reacting substances of anaphylaxis and that seem to be intimately involved in pulmonary smooth muscle during anaphylaxis and allergy; and (2) LTB_4, which is an extremely potent chemotactic substance most likely involved in inflammation and tissue injury.

The manipulation of arachidonic acid metabolism provides an ideal approach for the development of new therapeutic modalities. The progression of these investigations has created a situation that dictates the metabolic targets that could be usefully modified by pharmacological agents.

A fascinating discovery has arisen from the dietary manipulation of fatty acids in experimental animals. The depletion of arachidonic acid or its precursor from the diet produces essential fatty acid deficiency, which is life-saving in the autoimmune destruction of the kidney in mice genetically affected with glomerulonephritis (lupus). The lethal consequence of this disease arises from macrophage invasion and destruction of the glomerulus. Essential fatty acid deficiency blocks the leukocyte attack on the glomerulus, apparently by blocking local LTB synthesis. Although essential fatty acid deficiency is not a practical treatment for lupus, it provides a special insight on what might be a useful therapeutic approach. Alternatively, substitution of fish oil for arachidonic acid or its precursors in the daily diet could achieve similar results if they were adequate to interfere with macrophage function.

Thus, on the basis of a sound foundation of studies of arachidonic acid metabolism, specific therapeutic target areas can now be approached on a rational basis. Manipulation of the cyclooxygenase pathway could yield beneficial effects in immune response regulation and thrombosis, while manipulation of the lipoxygenase pathway, especially leukotriene synthesis, could provide a major new class of agents for the treatment of asthma, allergy, anaphylaxis, and inflammation (including myocardial infarction and renal disease).

Atriopeptin Is a Cardiac Hormone Intimately Involved in Fluid, Electrolyte, and Blood-Pressure Homeostasis

Atriopeptin, a peptide hormone, also termed atrial natriuretic factor (ANF) is intimately involved in the regulation of renal and cardiovascular homeostasis. Figure 10-1 illustrates the way in which atriopeptin links the heart, kidneys, adrenals, blood vessels, and brain in a complex hormonal system that regulates volume and pressure homeostasis. Basal levels of circulating atriopeptin can be increased by atrial stretch caused by volume expansion, constrictor agents that elevate atrial pressure, immersion in water, atrial tachycardia, and high-salt diets.

Once in the circulation, atriopeptin exerts a number of effects related to renal and cardiovascular functions. Because it suppresses elevated renin in the plasma and relaxes blood vessels, atriopeptin reduces vascular resistance, thereby lowering blood pressure.

Atriopeptin-immunoreactive neurons have been observed in rat brains, especially in a hypothalamic region that has extensive connections with structures that regulate cardiovascular functions. Animals with lesions that include this region show profound alterations in the regulation of fluid balance, and neither renal hypertension nor salt-induced hypertension develops. Thus, atriopeptin may serve as both a central neuromodulator and a peripheral hormone in the regulation of cardiovascular and renal function.

Current evidence suggests that the major actions of the hormone are to promote loss of fluid and electrolytes and to reduce vascular tone. These combined actions would be expected to reduce atrial stretch and suppress the release and processing of atriopeptigen, the 126-amino acid prohormone that is the primary form of atrial peptide stored in atrial myocytes.

Effects on Renal Function. Atriopeptin also alters salt and water metabolism. The infusion of atriopeptin into laboratory animals produces rapid and transient

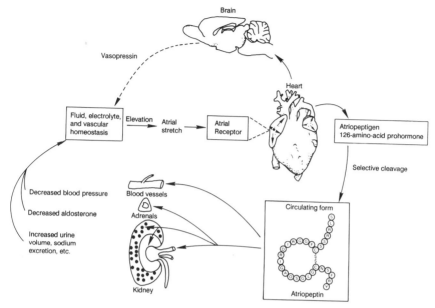

FIGURE 10-1   Schematic diagram of the atriopeptin hormonal system. [Adapted from P. Needleman and J.E. Greenwald, New Engl. J. Med. 314:828 (1986)]

excretion of water, sodium, and potassium. The glomerular filtration rate and the filtration fraction remain elevated throughout the peptide infusion and diminish abruptly when the infusion is stopped. At present, there is no strong evidence of a direct inhibition of sodium transport in the renal tubules.

The atriopeptin-induced increase in glomerular filtration rate is striking, especially since it occurs at doses of peptide that decrease blood pressure and total renal blood flow. Autoradiography techniques have revealed high-affinity receptor sites for atriopeptin localized on the glomeruli and in the papillae of the kidney. Those data, together with the finding that cyclic guanosine monosphosphate (cGMP) increases markedly in isolated glomeruli and medullary collecting-duct suspensions incubated with atriopeptin, suggest that the glomeruli and the medullary collecting duct are the sites that regulate volume and electrolyte excretion. Such action at these sites would be useful in situations in which conventional diuretics, which act through tubular sites, are ineffective.

A potent natriuretic and diuretic agent that is a selective renal vasodilator would have therapeutic potential in pathophysiological states characterized by fluid and electrolyte imbalances. Some promising clinical targets for research on atriopeptin therapy include renal failure, hepatorenal syndrome, and congestive heart failure.

Human Physiology and Pathophysiology. With the development of radioim-munoassays, it has become possible to study the release of human atriopeptin. As in other mammalian species, the concentration of atriopeptin in humans is elevated when intravascular volume and right atrial pressure are increased. Plasma atriopeptin is elevated in adults with chronic renal failure or with congestive heart failure, and a worsening of the condition is directly correlated with the concentration of circulating atriopeptin.

Thus, atriopeptin may have a unique role as a therapeutic agent, especially in critical-care situations in which patients are undergoing intravenous therapy. The demonstration that atriopeptin can increase the glomerular filtration rate in animals with severe renal insufficiency suggests that it has the potential to reduce the frequency of dialysis. The doses of a number of currently used drugs, such as certain antibiotics and chemotherapeutic agents, are limited by their toxicity to kidney tissue. Concurrent administration of atriopeptin with such agents may increase their therapeutic efficacy by inhibiting renal deterioration.

The maintenance of circulatory volume and salt homeostasis, despite changes in diet, fluid intake, posture, physical exertion, and stress, requires the integrated action of several endocrine systems that have contrasting functions. Imbalances of these complex compensatory systems become unmasked in certain disease states. One approach to the therapeutic manipulation of endocrine systems is to administer agents such as atriopeptin to mimic, release, enhance, or antagonize the intrinsic hormone. The discovery of nonpeptide mimics of the atrial peptides or of agents that could stimulate the synthesis and release of endogenous atriopeptin would allow the testing of long-term atriopeptin therapy for hypertension as well as for numerous renal and hepatic diseases.

New Approaches to Understanding Health and Disease: The Lipoproteins

Atherosclerosis Is the Principal Cause of Death in the United States, and Heart Attack (Myocardial Infarction) and Stroke (Cerebral Infarction) Are Its Overt Manifestations

Atherosclerosis, the disease process leading to the life-threatening events of heart attack and stroke, causes the progressive narrowing and eventual blocking of critical regions of the arterial bed. A major feature of the atherosclerotic lesion is the deposition of cholesterol in association with cell proliferation and connective tissue elaboration. Extensive epidemiological and pathological studies have pointed to an important association of lipids and lipoproteins in this disease. Certain species of lipoproteins that accumulate in plasma as a result of aberrant production and catabolism are the source of the cholesterol that accumulates in the artery wall.

Several Major Classes of Lipoproteins Exist

Plasma lipoproteins normally serve important homeostatic functions in biological structure and hormone action as well as in energy production. They are spheroidal particles constituted of a central core of hydrophobic lipids (cholesteryl esters and triglycerides) surrounded by a surface layer of hydrophilic lipid (phospholipids) containing cholesterol and unique proteins called apolipoproteins (Figure 10-2). The major classes of lipoproteins in plasma are the chylomicrons, very-low-density lipoproteins (VLDL), low-density lipoproteins (LDL), intermediate-density lipoproteins (IDL), and high-density lipoproteins (HDL). Each major class contains particles that are structurally and compositionally heterogeneous, are routed into different metabolic channels and may or may not give rise to potentially atherogenic particles. A detailed understanding of this heterogeneity, therefore, is crucial to an understanding of the basic biology of lipoproteins

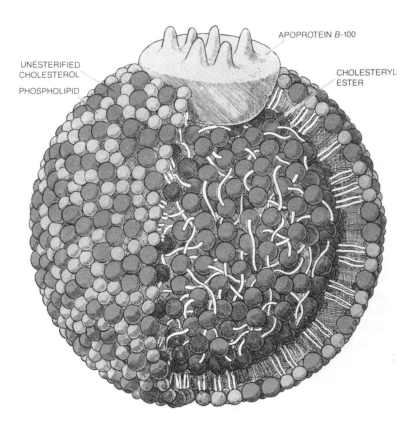

FIGURE 10-2 Low-density lipoprotein. [Reproduced, with permission, from M. S. Brown and J. L. Goldstein, Scientific American 251(5):58-66 (1984)]

and to the establishment of a rational basis for the treatment of lipoprotein abnormalities associated with atherosclerosis.

Cigarette smoking, hypertension, and diabetes increase the chance of developing artherosclerosis. Most individuals who are predisposed to the disease do not show clear-cut defects in lipoprotein metabolism. They may, however, show characteristic features in their plasma lipoprotein profiles, such as high blood concentrations of LDL and low concentrations of HDL. The data suggest that both genetic control and environmental modulation of lipoprotein production, structural and compositional properties, metabolic channeling, and catabolism may produce a net accumulation of lipid, including cholesterol, in arterial cells. The detailed study of these features provides an exciting opportunity for accelerated progress in understanding the function of normal and abnormal lipoproteins.

Research on the Structure and Synthesis of Apolipoproteins Will Lead to Insight into Cardiovascular Disease

Apolipoproteins are determinants of major processes in lipoprotein assembly, secretion, extracellular processing, and catabolic removal, particularly via receptor-mediated pathways. Sequencing of many of the apolipoproteins has been accomplished, DNA clones for some of these have been obtained, and the mapping of apolipoprotein genes has proceeded. Altered forms of some of these genes have been associated with hypertriglyceridemia and with an HDL deficiency state with premature atherosclerosis. The incisive techniques of cell and molecular biology offer exciting prospects for increasing our understanding of apoliprotein gene regulation and expression. It is important to elucidate how dietary, hormonal, and environmental factors regulate apolipoprotein gene expression. In addition, a host of other questions remain on topics such as the assembly of lipoproteins, the role of mutations, and developmental changes in lipoproteins. Studies like these should be linked to studies on heritability. In this connection, research on genetic polymorphism is very important.

Information on Critical Aspects of Lipoprotein Assembly and Secretion Is Lacking

Some of the general features of lipoprotein assembly and secretion are known, but many important components of these processes are poorly understood. For example, signal peptides on apolipoproteins are probably involved in intracellular targeting and secretion. The secreted nascent lipoproteins are exceedingly heterogeneous in particle size and apolipoprotein and lipid composition, and therefore in their metabolic fates. Analysis of the steps in assembly and secretion of the different species of lipoproteins can now be approached by the techniques of molecular and cellular biology.

While the major emphasis in understanding lipoprotein assembly and secretion has been on the VLDL particles, basic information on the determinants of

HDL and LDL assembly and secretion by the liver is equally important but lacking. The processes involved in the assembly and secretion of nascent HDL are still little understood, and they gain importance in view of the role played by HDL in the control of cellular cholesterol. Knowledge of the effects of dietary, environmental, and pharmacological factors on lipoprotein assembly and secretion is rudimentary and in need of detailed molecular characterization.

Insight into What Determines Which Lipoproteins Are Directed into One Metabolic Channel Versus Another One Is Crucial to Our Understanding of the Origins of Atherosclerosis

Once they have been secreted, lipoproteins circulate in the bloodstream, where they interact with enzymes, receptors, and other factors that modulate their catabolic clearance from the blood. As these interactions occur, atherogenic products are sometimes produced. Organisms deficient in components of the extracellular channeling system (lipases, esterifying enzymes, or cofactor lipoproteins) demonstrate extremes of abnormal lipoprotein distributions. For example, the investigation of the cellular biology of the LDL receptor has delineated a finely tuned system for regulating cellular cholesterol that affects plasma LDL levels. Receptor-mediated internalization of LDL and its lysosomal degradation produces free cholesterol within the liver cell which, in excess, can inhibit cholesterol synthesis and the synthesis of the LDL receptor. Intracellular storage of excess cholesterol is facilitated by activation of an esterifying enzyme. Some cholesterol is also secreted from the liver cells in the form of bile acids that enter the intestine and are partially recycled back to the liver.

These control mechanisms have been exploited in pharmacological approaches to the control of plasma LDL levels in humans with hypercholesterolemia. The oral administration of a resin that binds bile salts secreted from the liver into the intestine can remove some cholesterol from the body. This treatment, while increasing the number of LDL receptors, also steps up cholesterol synthesis, thereby reducing the the treatment's effectiveness. Administration of inhibitors of hepatic cholesterol synthesis also increases receptor synthesis, but to a greater degree than the resin. A promising approach for control of LDL levels in hypercholesterolemia is the use of combined drug regimes, which control cholesterol synthesis and reduce LDL in the plasma. The further molecular analysis of the regulation of LDL receptor action and cholesterol metabolism is of high priority. Such studies need to be linked with others at the level of the whole organism, so that dietary and environmental factors in heart disease can be evaluated.

Mechanisms of Atherogenesis Involving Lipoproteins Are Being Determined

In the bloodstream, the diverse classes of lipoproteins and their catabolic intermediates and products make contact with platelets, endothelial cells, macrophages, and smooth muscle cells. Mounting evidence indicates that the interaction of these cells with elevated levels of specific lipoprotein species, such as beta-VLDL (a cholesterol-enriched VLDL remnant) and LDL, directly or indirectly promote changes similar to those observed at various stages in the development of the atherosclerotic lesion.

The frequent observation of lipid-laden macrophage cells in lesions has prompted intensive research on macrophage interaction with lipoproteins. These studies have contributed valuable information on the major features of the macrophage receptors and their function. In addition, a role for the macrophage in converting smooth muscle cells to foam cells (lipid-laden cells) has been suggested by in vitro studies describing incorporation by smooth muscle cells of lipid inclusions derived from the lysis of macrophage foam cells. Detailed information on such systems will greatly enhance our ability to deal effectively with atherosclerosis.

New Opportunities Exist to Prevent and Treat Diseases of Cell Proliferation

Atherosclerosis can serve as an excellent example of a repair process in the artery wall, which is modified in different ways by chronic hypercholesterolemia and by factors generated from cigarette smoking, hypertension, diabetes, or other factors associated with an increased incidence of this disease process. In the United States and western Europe, chronic hypercholesterolemia is the risk factor associated with the highest incidence of atherosclerosis. The study of chronic, dietarily induced hypercholesterolemia in monkeys and pigs has permitted a chronological assessment of the cells involved in the development of the lesions of atherosclerosis. Within a few days after cholesterol concentrations equal to those seen in humans afflicted by the disease have been induced, white blood cells become increasingly sticky and adhere to the surfaces of the endothelial cells lining the artery. The high cholesterol concentrations induce changes both in the circulating white blood cells and in the endothelium. These white blood cells are scavenger cells that take up foreign substances, including some forms of lipid. White blood cells take up lipid in the innermost layer of the artery wall and become foam cells. In so doing, they accumulate and create the first and most common lesion of atherosclerosis, the "fatty streak." In the United States and western Europe, such fatty streaks are found in individuals of all ages, including infants, and represent the precursor lesion, which in many instances goes on to become the advanced, proliferative, occlusive lesion of atherosclerosis, the fibrous plaque that causes myocardial and cerebral infarctions. Thus, experiments with pigs and monkeys were of critical importance in establishing the connection between hyperocholesterolemia and atherosclerotic lesions.

The next change that occurs in chronic hypercholesterolemic monkeys may represent further "injury" to the endothelial cells. These cells seem to separate from one another at particular sites in the artery. After the endothelial cells

POSSIBLE LINK BETWEEN LIPOPROTEINS AND THE CLOTTING SYSTEM

An LDL variant named Lp(a) has been shown to possess an additional apolipoprotein called apo A. This apolipoprotein is attached to the normal LDL apolipoprotein apo B-100 by a disulfide linkage. High plasma concentrations of Lp(a) are strongly associated with accelerated development of atherosclerosis. The manner in which Lp(a) influences atherosclerosis development is still a mystery. However, on the basis of the amino acid sequence of apo A, scientists have recently shown that this apolipoprotein shares considerable sequence homology with plasminogen. Plasminogen is a key component in the clotting system, which binds to fibrin and when activated by tissue plasminogen activator can enzymatically dissolves fibrin clots. It is interesting to speculate on the possible connection between the plasminogen-like structure of apo A and its role in atherosclerosis. There is evidence that Lp(a) is associated with initial atherosclerotic plaques, which contain fibrinogen (fibrin precursor) and fibrin in roughly the same concentration as cholesterol. It has been proposed that Lp(a), mediated by the apo A amino acid sequences that are homologous to plasminogen's fibrin binding sequences, can adhere to the atherosclerotic plaque. Elucidation of possible apo A with components of the clotting system await additional research to determine the exact role of Lp(a) and apo A in the accelerated development of atherosclerosis.

separate from one another, they then retract and expose the underlying white blood cells that have become foam cells. In many instances, another circulating blood cell, the platelet, attaches at these sites of exposure. Platelets seem to release substances that stimulate the localized proliferation of smooth muscle.

We now know that the proliferation of smooth muscle cells at sites where platelets interact can be partly due to the powerful growth factors produced by platelets, endothelial cells, macrophages, and the smooth muscle cells themselves. One of these, the platelet-derived growth factor, is a potent mitogen that stimulates the proliferation of cells that form connective tissue, such as smooth muscle cells in the artery wall. Perhaps of equal importance is the secretion of other growth factors into the local environment by activated macrophages or platelets. Like platelet-derived growth factor, these factors could have a major impact on the local multiplication of smooth muscle cells and macrophages and set the stage for the development of advanced occlusive lesions of atherosclerosis. These lesions could enlarge over a period of months or years and eventually sufficiently occlude the lumen of the artery so that the vascular supply of blood and thus the oxygenation of the tissues are compromised. Such a set of conditions leads to the clinical sequelae commonly associated with heart attacks and strokes.

The approaches of cell biology have permitted us to define the cells that take part in the process and to understand which cells interact, at what points in time, and where. Thus, it has been possible to analyze these cells and determine whether genes are expressed for substances such as growth factors, and, if so, to determine whether these growth factors can be responsible for the observed tissue responses. In addition, cells involved in these tissue responses are being isolated, grown, and purified, making possible their direct investigation in culture. These approaches represent exciting research opportunities both in basic biology and in the medical application of this knowledge.

ARTHRITIS

Since the 1930s, considerable progress has been made in distinguishing such entities as osteoarthritis and rheumatoid arthritis through the use of clinical and biochemical criteria. This progress has led both to an improved understanding of pathogenesis and to the development of new therapies.

The Greatest Success Has Been in Understanding Hyperuricemia and Gout

The major cause of hyperuricemia is decreased excretion of urate, which can occur spontaneously or be induced by drugs, particularly the diuretics used to treat hypertension. Drugs that increase the excretion of urate were introduced nearly 40 years ago and used successfully, but the introduction of the drug allopurinol has revolutionized the therapy of gout and almost completely eliminated chronic gout. Allopurinol, which decreases urinary urate by inhibiting the final reaction in its generation, is effective in lowering serum urate levels. This prevents the onset of gouty arthritis and the formation of uric acid renal stones.

Inflammatory Joint Diseases and Seronegative Polyarthritis Remain Major Problems

Although we know a great deal about its pathogenesis, we still do not understand the cause of the inflammatory joint disease rheumatoid arthritis. Specific antibodies called rheumatoid factors have been found in rheumatoid arthritis patients. These antibodies are specific for the Fc portion of other antibody molecules. The interaction of the rheumatoid factors with other antibodies amplifies inflammatory reactions. The increasing work on rheumatoid factors, with molecular cloning techniques, should provide results far beyond the study of rheumatoid arthritis itself.

In patients with rheumatoid arthritis, considerable attention has been directed toward the study of interactions of T lymphocytes with cells in the joint lining. T lymphocytes interact with other cells by cell contact mediated mechanisms and by the production of cytokines such as interleukin 1, tumor necrosis factor, lymphotoxin, transforming growth factors, platelet-derived growth factor, and certain colony-stimulatory factors. These interactions could result in the proliferation of the joint lining cells that leads to joint fibrosis and in the production of proteinases and lipid metabolites (prostaglandins and leukotrienes) that are responsible for swelling and pain in and around joints. The proteinases and cytokines are being studied in detail, in the hope of identifying new sites for therapeutic intervention.

Factors Influencing Genetic Susceptibility to Rheumatic Diseases Are Being Determined

In 1973, an exciting report indicated a strong association between susceptibility to ankylosing spondylitis and related disorders to the HLA-B27 allele of the class I histocompatibility locus. Since the HLA-B27 gene has now been cloned and sequenced, a new approach to understanding its role in pathogenesis has evolved. This approach involves searching existing protein sequence data bases for homology between HLA-B27 protein and proteins in suspected bacterial pathogens. A sequence has been identified that codes for six consecutive amino acids in *Klebsiella pneumoniae* nitrogenase. Some patients with ankylosing spondylitis have circulating antibodies that specifically recognize this six-amino-acid peptide, which suggests molecular mimicry.

In addition, rheumatoid arthritis has been associated with class II histocompatibility antigens. The use of restriction fragment length polymorphisms (RFLPs) should lead to a more precise genetic analysis and more extensive use of specific probes to define genetic susceptibility.

Lyme Disease Might Be a Good Model for Understanding the Pathogenesis of Rheumatoid Arthritis

Lyme disease was first described in 1975 on the basis of observations of children with what was initially thought to be a form of juvenile polyarthritis. Since then, the disorder has been shown to be caused by the spirochete *Borrelia burgdorferi*. Lyme disease begins with a characteristic skin lesion that is followed by systemic symptoms. Polyarthritis may occur early, but more persistent joint swelling, particularly in the large joints, may take place weeks to months later. The histopathology of the persistent joint swelling is remarkably similar to that of rheumatoid arthritis. The demonstration of small amounts of *Borrelia* antigens in the joint lining suggests that some persistence of infection could be responsible in part for the arthritislike symptoms. Furthermore, the organisms are susceptible to antibiotics; high doses of penicillin not only prevent late complications if given early in the disease, but also produce responses in patients with established arthritis. Thus, when details of the pathogenesis of Lyme disease become known, the information should aid in understanding the pathogenesis of rheumatoid arthritis.

Osteoarthritis Is the Major Joint Problem in Terms of Number of People Affected and Overall Morbidity

The etiology and pathogenesis of osteoarthritis are still undefined. In patients with osteoarthritis, extracellular matrix components degrade through mechanisms that are probably basically similar to those operating in typical inflammatory joint disease. In some individuals, genetic factors may be responsible for abnormalities in matrix proteins (collagens and others). Because of the persistence and unpredictable course of osteoarthritis, it is extremely difficult to evaluate the effects of possible therapeutic agents. For these reasons, the major approach to this disabling disease is joint replacement.

CANCER: CURRENT STATUS AND SOME NEW APPROACHES TO ITS CONTROL

Cancer Has a Number of Causes

Environmental and life-style risk factors—such as the use of tobacco and alcohol, exposure to occupational carcinogens and pollutants, and diet—are major causes of cancer. Only a small number of cancers are thought to be entirely genetic in origin. Environmental and life-style risk factors are believed to contribute to the development of about 90 percent of cancer incidence. For example, a major reduction in tobacco use alone could prevent 30 percent of cancer deaths. In comparison, improvements in cancer diagnosis and therapy have resulted only in a 5-year survival rate of 49 percent. From this, one can see that cancer prevention by avoidance of risk factors has the potential to play a more important role than therapy in the overall reduction of cancer deaths. However, much more study is needed before the full benefits of risk avoidance can be realized.

Even though cancer has a number of causes, many and possibly all cancers seem to have a common component, namely, alteration in the DNA of the transformed cell. That genetic damage is involved in the pathogenesis of cancers is suggested by several lines of evidence including (1) the existence of a subset of cancers influenced by hereditary predisposition, (2) alteration in chromosome structure specifically correlated with certain types of cancers, (3) impairment in mechanisms involved in DNA synthesis or repair of damaged DNA correlated with susceptibility to cancer, (4) the relation between mutagenic potential and carcinogenic activity of various substances, and (5) the discovery that proto-oncogenes, which are normal cellular genes that in an altered form become oncogenes, are linked to tumor development. Cancer is fundamentally a disorder in the regulation of cell proliferation and differentiation, reflecting genetic change that leads to loss of normal constraint. The process of carcinogenesis has many steps; when a genetic abnormality is present it probably represents only one necessary, but not sufficient, step in producing the malignant phenotype. Environmental factors such as viruses, radiation, or chemical carcinogens can also play roles in causing cancer.

The remarkable rate of advance in cellular and molecular biology is leading to an understanding of the causes of cancer and providing more effective approaches to prevention and therapies. This section summarizes certain areas in which this increased knowledge is being applied or is likely to be applied to the problems of human cancers, and it indicates some lines of necessary further investigation.

Cancer Therapy

Advances in Cancer Therapy Are in Surgery, Radiotherapy, and Chemotherapy

Two major premises underlie current treatments of cancer. The first is that the disease, and especially solid tumors, spreads progressively, in discrete steps. It begins locally, permeates local tissues, and then extends to distant sites. For some solid tumors, however, this description is inappropriate. Further, even within tumor types, there may be wide heterogeneity, with some tumors demonstrating early distant metastasis, others showing no proclivity for metastasis, and others behaving according to the premise of progressive spread. A major limitation of present treatment and an opportunity for the future is to develop techniques that distinguish these types of tumors.

The second premise is that tumor cells proliferate more persistently than normal cells. This has given rise to the development of cancer chemotherapeutic agents, which take advantage of putative proliferative differences between normal and malignant cells. Although this has been a useful approach, we have learned that tumors of tissues that are constantly renewing themselves, such as the hematopoietic (pertaining to or affecting the formation of blood cells) system and cells of the epithelial lining, violate this rule. These cell-renewal tissues are usually the most rapidly proliferating cells in the body. Therefore, they become the sites of dose-limiting toxicity when tumors of cell renewal tissues are treated by chemotherapy.

Surgery. Surgery cures more patients with cancer than any other treatment. Examples of its success are in the treatment of colon cancer, cervical cancer, many head and neck tumors, and uterine cancer. Surgery fails when the tumor has already metastasized beyond the bounds of surgical resection or when the volume of the tumor is too great. Efforts for improvement are concerned with developing techniques that meet the same goal for cure, but are less mutilating—because of better reconstruction and graft techniques, improved prosthetic materials, and the use of transplantation. Most importantly, surgical results are improving as surgery is combined with more effective diagnostic techniques as well as techniques such as radiotherapy and chemotherapy.

Radiation Therapy. Radiation therapy is also based on the hypothesis that ablation of tumor can be curative. Its limitations are different from those of surgery. When radiation therapy fails, it usually fails in the center of the tumor. This has given rise to techniques of radiation that administer moderate doses of radiation to larger volumes, but restrict the high doses to the areas with proven involvement. Radiation has the advantage of potentially less functional impairment. Examples of treatment success include the treatment of prostate, cervical, and a variety of head and neck cancers, as well as Hodgkin's disease. New efforts in radiation therapy are concerned with techniques (1) to better localize tumors so that the treatment can be more effectively applied and (2) to deliver radiation more precisely, minimizing the irradiation of normal tissues. Such techniques take advantage of the major advances in electronic and computer technology. Biological approaches are concerned with increasing tumor sensitivity to radiation, and sensitizers of hypoxic cells to radiation are being studied.

Chemotherapy. Cancer chemotherapy has often been based on the second hypothesis, that tumor cells proliferate more rapidly than normal. These therapies have been most successfully applied in those tumors that were associated with rapid proliferation, such as acute leukemia, the lymphomas, and testicular carcinoma. These drugs act at sites shared by malignant and normal cells, thus limiting their usefulness because of their toxicity to normal tissue. Further, these agents may produce alterations for which the cell can develop alternative mechanisms; selection of resistant clones of tumor cells seems to be the most prominent cause of failure, resulting in the development of cells resistant to the action of the chemotherapeutic agent. Mechanisms for this resistance include gene amplification and changes in membrane transport of chemotherapeutic drugs either into or out of the cell. Increasing the drug's dosage and therefore its efficacy has been made possible by the vigorous use of blood products to treat the hematopoietic toxicity. The logical extension of this treatment is the transplantation of bone marrow, which is being attempted with bone marrow obtained either from the patient and stored or from appropriate donors.

A most important type of drug resistance that limits efficacy is pleiotropic: Cells become resistant to a whole host of apparently unrelated agents by one or a few changes. New antitumor cytotoxic agents are being developed that may be less toxic and also more effective against tumor cells that are resistant to available agents. Additional approaches to the treatment of cancer are being based on advances in our understanding of immune mechanisms and of the control of cell proliferation and differentiation.

Combined Use of Therapeutic Modalities Has Evolved Gradually

Surgery, radiation therapy, and chemotherapy have different toxicities and limitations. Surgery is limited by the volume requiring resection in order to remove all subclinical disease. Often surgery fails because of unappreciated microscopic disease beyond the limits of the surgical procedure. Radiation therapy rarely fails at the margins of the treatment for subclinical disease. Rather, limitations in radiotherapeutic treatment efficacy are often caused by a large central tumor: Its toxicity is due to the high dose required to ablate such gross tumor masses. These two forms of therapy can be combined. Surgery removes the gross tumor and radiation therapy is applied to the larger region containing lower densities of cancer cells. Chemotherapy appears to be effective only against small volumes of tumor. It offers the best opportunity for application as a part of a multidisciplinary program where the two regional therapies have eradicated the gross tumor. Examples of combining the three modalities in this fashion are the treatment of Wilm's tumor of childhood, soft tissue sarcomas, breast cancer, a variety of tumors of the head and neck, and osteosarcoma. In some of these, only two of the three modalities are required. All have significantly increased survival and improved the functional and cosmetic results. Similar potential for multidisciplinary treatment is seen in bladder cancer; chemotherapy seems to reduce tumor volumes so that they are amenable to surgical procedures or radiation therapy.

Where chemotherapy has the major role, radiation or surgery may be useful to destroy tumors in sites to which the chemotherapeutic agent cannot penetrate. This appears to be true for the central nervous system, where radiation therapy of the brain has improved the results in the treatment of some patients with leukemia and lymphoma. Surgery has a variety of new uses when combined with radiation and chemotherapy. The use of surgery to evaluate the extent of tumor has proved important in Hodgkin's disease, where removal of the spleen provides vital diagnostic and therapeutic information. A standard hypothesis that may not always be supported is that once the bloodstream has transported cancer through the body, the cancer becomes multiple and therefore resistant to a local therapy. Surgical treatment of limited metastasis in the liver, lung, or even brain has increasingly been attempted with good results. Further improvement may be achieved in combining such surgery with systemic agents.

Newer Approaches to Cancer Control

Prevention of Cancer Involves, in Part, Persuading People to Limit Their Exposure to Environmental Agents That Seem to Be Associated with Increased Risk

The most outstanding example of increased risk for cancer associated with an environmental agent has been the relation between cigarette smoking and lung cancer. On less firm experimental grounds, but with sufficient evidence to be provocative, are suggestions that dietary factors are important causes of human cancers.

Viruses Cause Some Types of Cancer, Especially in Developing Countries

It is estimated that some 10 percent of cancers in the United States and as many as 30 percent in China are associated with virus infections. Among these are carcinoma of the uterine cervix, which has been linked to the human papilloma virus; primary liver cancer (prevalent in Africa and Asia), linked to hepatitis B virus; acute leukemias, linked to human T-cell leukemia viruses; and Burkitt's lymphoma and nasopharyngeal carcinoma, which may be carried by the Epstein-Barr virus. With the identification of such viruses, the possibility of developing immunizing vaccines is being actively pursued, although in each instance this approach is in an early stage of development. Perhaps most advanced is preventive immunization to block transmission of hepatitis B virus. Since the lag period between viral infection and onset of clinical cancer is many years, this program will require some time before its efficacy can be evaluated.

All of the DNA tumor viruses mentioned above (hepatitis B virus, papilloma viruses, and Epstein-Barr viruses) infect differentiated cells that cannot be propagated readily in culture. This inability is a major limitation in studying these viruses. Some cell lines can be infected by human T-cell lymphoma virus 1 (HTLV-1) and human immunodeficiency virus (HIV), but the former virus is cell-associated and therefore infection can be most readily accomplished by cocultivation. A major goal for advancing the study of all of these viruses is to develop efficient and easy modes of infecting cells in culture. For the DNA viruses, this goal will be achieved either serendipitously or by developing more and easier methods of culturing epithelial cells than those available today. It is also most important to identify those viral genes that predispose infected cells to develop into tumors.

Oncogenes. Tumor viruses were first found in chickens more than 75 years ago. Some of the genes coding for the proteins that control growth have been found in a mutated form in certain highly oncogenic retroviruses of animals; the mutated genes are called oncogenes, and the normal form of the genes, proto-oncogenes. Although no virus containing an oncogene has been found in humans, altered proto-oncogenes have been found in several human tumors. The proto-oncogenes are altered in different cases by base-pair mutations, amplifications, or translocations. The presence of these altered proto-oncogenes in human cancer has suggested that they might have a role in the genesis of the tumors. Certainly, the correlation of the location of proto-oncogenes with the break points of chromosomal translocations specific for certain cancers supports this hypothesis. For example, in Burkitt's lymphoma, one of a specific set of translocations involving the c-*myc* proto-oncogene is invariably present (see Chapter 7).

MOLECULAR GENETIC ANALYSIS OF LYMPHOID TUMORS

In addition to providing detailed information about the mechanisms of specificity generation in the immune system, molecular genetic analysis of lymphocytes has provided some important insights into the classification of lymphoid malignancies and, perhaps more importantly, into the mechanisms of disordered growth regulation in these tumors. The determination of the cell of origin and the degree of differentiation of the malignant cell in many leukemias and lymphomas is made possible by establishing whether the immunoglobulin (Ig) genes or the T-cell-receptor genes of that cell have been rearranged and how far the rearrangement process has proceeded. Such information often has considerable value in the choice of therapy and in determining the prognosis of these lymphoid malignancies.

Aberrant gene rearrangements sometimes play an important part in forming human tumors from white blood cells. In the course of detailed examination of the rearrangements of Ig genes in Burkitt's lymphoma cells and in mouse plasmacytomas, a non-Ig gene was found to have become associated with the Ig genes as a result of a reciprocal exchange of material between distinct chromosomes. That non-Ig gene proved to be the cellular oncogene c-*myc*, which is critical for growth regulation. Its changed location is believed to be critical to the process of malignancy in Burkitt's lymphoma and plasmacytomas, possibly because the normal regulation of c-*myc* expression has been disturbed by its being placed in a highly active genetic environment. Other oncogenes are translocated into the region of the Ig and T-cell-receptor genes in other types of B- and T-cell malignancies, strongly indicating that the process of disordered oncogene regulation resulting from genetic translocation in lymphocytes is an important mechanism in malignant transformation. Continued and expanded efforts to determine the mechanisms of oncogene mediation of normal and abnormal growth regulation will be of fundamental value in understanding normal growth regulation and may lead to the development of drugs that can control the growth of malignant cells.

Other genes that have not been identified in detail may also place an individual at increased hereditary risk for cancer. If a person who carries a gene associated with increased risk for a specific cancer could be identified, minimizing exposure to environmental hazards associated with induction of such cancers (for example, skin cancers attributable to ultraviolet light) or applying early detection methods to individuals at risk (for example, familial development of polyps and colon cancers) might help in preventing cancer or treating it more effectively.

New Developmental Therapies Utilizing Biological Response Modifiers, Monoclonal Antibodies, and Other Molecules Offer Prospects for Treatment

Biological Response Modifiers. These modifiers are substances that stimulate or modulate the host's cellular or humoral responses that may play a role in killing cancer cells. Examples include the interferons, interleukins, colony-stimulating factors, and tumor necrosis factors. The genes for most of these factors have been cloned, and the application of recombinant DNA techniques has made these substances available in sufficient quantities to permit evaluation of their clinical usefulness.

A few years ago considerable publicity attended the potential of interferon as antitumor therapy. The availability of abundant amounts of interferon has made adequate investigation possible; it has now become apparent, however, that most cancers do not respond to interferon. One rare malignancy, hairy-cell leukemia, does respond to interferon-α. It is not clear why this leukemia is so sensitive to this interferon.

Interleukin-2 activates killer lymphocytes. Recently interest in it has been rekindled by the demonstration that removing a person's lymphocytes, activating them in vitro with interleukin-2 (LAK cells), and then injecting them back into the patient, along with substantial doses of interleukin-2, results in tumor regression in some patients. The most impressive results have been seen in patients with renal cancers and melanomas. The ultimate role of interleukin-2 in cancer therapy awaits controlled clinical trials along with further studies to increase efficacy and decrease toxicity and the complexity of administration. These studies raise the question whether the human immune cellular system can be manipulated to enhance normal function in dealing with tumor cells.

Tumor Necrosis Factor and Lymphotoxin. These modifiers were initially recognized because of their cytotoxic and antitumor properties. The dramatic effect of tumor necrosis factor on animal tumors aroused considerable interest in its potential clinical application, but initial clinical trials in patients with cancers have not shown similar antitumor effects. Lymphotoxin has limited amino-acid sequence homology to tumor necrosis factor, and, because it has been available in only small quantities, its biological characterization is less advanced than that of tumor necrosis factor. The general consensus is that tumor necrosis factor is part of a network of interactive signals for host immunological and antiinflammatory responses and in maintaining hematopoietic cell development in balance.

Growth Factors. Growth factors regulate the proliferation and differentiation of cells in the body; many of these proteins have been isolated, their genes cloned, and their biological activity defined. Among these growth factors, the hematopoietic colony-stimulating factors are of particular clinical interest. These factors may be clinically useful after bone-marrow transplantation, in cancer chemotherapy to ameliorate toxic effects of drugs in suppressing bone marrow, and in treatment of infectious diseases for which stimulation of hematopoietic cells can be clinically important.

Monoclonal Antibodies (MAbs). Antibodies are proteins secreted into the blood by immunoglobulin-producing cells as part of the body's complex immune response against foreign substances. With the availability of hybridoma technology for the production of individual antibodies in large quantities, it is now possible to evaluate the usefulness of antibodies in the diagnosis and treatment of cancers. Hybridoma technology has made it possible to develop MAbs against many types of tumor cells. Some of them are useful in improving the precision of pathological diagnosis in a number of cancers. Radioactively labeled MAbs have been used as diagnostic tools to identify sites of primary tumors and metastatic deposits within the body; nuclear medicine imaging devices detect the radiation they emit. The factors limiting the usefulness of MAbs in diagnosis are similar to those limiting their application in therapy: (1) inadequate access of the MAb to tumor cell surfaces; (2) heterogeneity of tumors with respect to the amount of surface antigens on individual tumor cells to react with MAbs; (3) alterations in antigen on tumor cell surface causing the cells to be no longer recognized by the MAb; and (4) expression of target molecules for the MAb that are not unique to tumor cells, in that there may be cross-reactivity of an antitumor MAb with portions of molecules expressed on nontarget cells that lead to undesirable toxic side effects.

Rarely does an antibody in isolation kill tumor cells. For example, in order to kill a tumor cell, the MAb can recruit natural effectors such as complement, killer lymphocyte cells, or phagocytes. MAbs can also be used as vectors to deliver to the target tumor cell a lethal agent such as a bacterial or plant toxin, a cytotoxic drug, or a radioactive isotope. MAbs can also potentially suppress a function in tumor cells, which decreases the viability of the cells. Clinical application of MAbs in therapy of malignant disease has had very limited positive results. Aside from one prolonged complete remission of a human B-cell lymphoma, the therapeutic results for the dozens of patients with various cancers that have been treated have been transient and partial at best. Promising areas of current and future research are based on advances in molecular biology that provide a means to tailor antibody fragments for targeting to tumor cells.

Other Molecules. Smaller molecules that cause cancer cells to differentiate have been discovered. Studies with these inducers are based on the concept that cancer frequently blocks differentiation and results in increased proliferation. Growth need not be fast to cause a tumor, only inexorable. The neoplastic state need not destroy the potential of tumor cells to differentiate to an end-state cell that would have lost its malignant potential to multiply. Differentiation of many tumor cells can be induced by a variety of structurally diverse agents, including polar planar compounds, hormones, vitamin derivatives, low doses of certain cytotoxic drugs, physiologically active growth factors, and tumor promoters. Among the several cytodifferentiation agents, the polar planar compound hexamethylene bisacetamide has been most extensively studied with respect to its cellular and molecular effects on transformed cells. It induces differentiation in a variety of cancer cell lines in vitro: leukemia, neuroblastoma, teratocarcinoma, colon, and bladder cancer, among others. Phase 1 and phase 2 clinical trials are attempting to define the toxicity and efficacy of this approach to the treatment of cancer. The possible "reversibility" of the neoplastic state provides new possibilities for therapies as alternatives to treatment based on use of cytotoxic agents.

Substantial Progress Is Being Made in Our Understanding of the Nature of Cancer

It is becoming increasingly possible to identify alterations in gene and gene products in transformed cells and factors that interact with transformed cells. These factors distinguish them from normal cells and cause abnormal cell proliferation and differentiation. New approaches to prevention, earlier diagnosis, and better treatment will contribute to progressive improvement in our ability to control cancer as a major source of human suffering and death.

ADVANCES IN UNDERSTANDING HUMAN GENETIC DISORDERS

Many Genetic Disorders Are Now Being Successfully Investigated

Genetic mapping—the localization of a gene or trait to a specific area on a chromosome—has created a great stir in the popular media and the biomedical literature alike during the 1980s. Diseases of interest include Huntington's disease, cystic fibrosis, polycystic kidney disease, neurofibromatosis, myotonic dystrophy, familial Alzheimer's disease, and Duchenne muscular dystrophy. The excitement is entirely justified. In all of these disorders, the nature of the basic biochemical defect is (or was at the time of mapping) unknown; consequently, no reliable diagnostic method could be designed for prenatal and presymptomatic diagnosis, and no definitive method for carrier detection was available. Furthermore, aside from the current impossibility of gene therapy, treatment strategies directing the fundamental pathogenic processes could not be devised. Knowledge of the location of the chromosomal change responsible for these and many other disorders opens possibilities for diagnosis by genetic linkage analysis or by direct identification of the alteration in the DNA. It also opens prospects of approaching the pathophysiological mechanisms of these disorders through the back door, as it were—a strategy that has come to be termed reverse genetics. It is hoped that elucidation of these mechanisms will suggest ways of interrupting the pathogenetic chain that connects gene to disease phenotype or of compensating for genetic deficiencies.

Elucidation of the Primary Genetic Defect

Two General Methods for Elucidating a Primary Genetic Defect Are the Candidate Gene Approach and Reverse Genetics

What is the protein gene product that is absent or altered in disorder *x*? The candidate gene approach, based on the gene map, often provides a tentative answer (or excludes a possibility). If the map location is known, the gene defect (and thus the abnormal gene product) may be identified through analysis of the segment of chromosome where the mutation lies, recognizing the mutant gene by a difference from the normal, and then characterizing the normal gene in determining the nature and physiological function of its protein product.

A second approach involves reverse genetics. Usually, in unraveling the pathophysiology and primary genetic defect in disorders inherited through classical Mendelian genetics, medical geneticists start with a phenotype that shows a "simple" pattern of inheritance; they identify a biochemical abnormality such as an abnormal metabolite in serum and urine; then they demonstrate the deficiency of an enzyme that could plausibly account for the abnormal metabolite; they clone the normal gene for the enzyme in question; and perhaps ultimately they determine the change in the gene in patients with the disorder.

In many disorders, however, no clue exists to the nature of the defective protein gene product. In these conditions, going directly to the gene (located by mapping studies), determining the structure and function of its normal counterpart, and then working back to the phenotype—reverse genetics—is the approach that has been taken. The approximate location of the gene is determined by genetic mapping methods. When markers are at hand—when defined DNA segments have been cloned—they can be used for locating the disease mutation. In addition to cloned markers, the chromosomal location of genes can be facilitated by cloning large segments of DNA through the use of the newly developed yeast artificial chromosome cloning vectors, followed by new electrophoresis techniques that can separate large DNA fragments. These techniques allow heritability studies of genetic disorders to be readily linked to the techniques of molecular biology.

As an alternative, exploitation of deletions and chromosomal rearrangements can narrow the location of the segment of DNA containing the mutation. With high-resolution cytogenetics (examination of banded prometaphase chromosomes) it has been possible, in X-linked disorders in particular, to find microscopically visible deletions in patients with Mendelian disorders such as Duchenne muscular dystrophy or chronic granulomatous disease. Cloning the normal X-chromosome segment that is absent in the affected male is accomplished by subtractive hybridization: DNA from cells of the male with the deletion is repeatedly hybridized with DNA from cells of a normal male; the DNA that does not hybridize is that which was deleted in the affected male. This strategy was used in cloning the genes for Duchenne muscular dystrophy and chronic granulonatous disease and in identifying the gene for susceptibility to retinoblastoma. However, even after the relevant segment has been cloned (in multiple pieces), identification of the specific gene may still be difficult. Granulonatous disease provides the first example of full utilization of reverse genetics.

Demonstrating that the normal gene cloned by these methods is indeed the one that is mutant in the given disorder can be accomplished through several approaches. Finding the mutation in a protein gene that has been linked to the disease by additional lines of evidence is a good indication that the mutation is responsible for the disease. The demonstration of different types of mutations in the same gene in different patients can also help. Tissue distribution of the corresponding messenger RNA (if it matches the distribution of the clinical phenotype) indicates that the gene is relevant, and evolutionary conservation of the gene in question indicates that it is functionally significant. Correcting the cellular defect with the cloned gene is convincing evidence that the gene is the relevant one.

Gene Diagnosis

Gene Diagnosis Can Take an Indirect Approach by Linkage or a Direct Approach by Demonstrating the Lesions in DNA

Diagnosis by Linkage Analysis. Information on the chromosomal location of a disease-producing mutation can provide a method for diagnosing that disorder by the linkage principle. The approach can be used to diagnose prenatally or before symptoms have appeared, or to detect carriers. The method has shortcomings: To give reliable results, multiple family members must be studied, a linked marker (or preferably more than one) must be segregating into different gametes in the family, and the markers must be closely linked.

Direct Gene Diagnosis. Any genetic disorder exhibits changes in the physical structure of DNA that can be used for diagnosis. By a process that might be called diagnostic biopsy of the human genome, DNA can now be prepared from the circulating white cells in a sample of peripheral blood and appropriately studied for the presence of the lesion characteristic of one or another genetic disorder. A probe of the normal gene is required for direct DNA diagnosis. Genetic modifications in the altered gene can be detected by a variety of laboratory procedures. For example, a specific synthetic oligonucleotide probe can be used when a specific nucleotide substitution is known in a given disorder, such as sickle cell anemia.

GENETICS OF NEUROPSYCHIATRIC DISORDERS

An important application of recent advances in molecular genetics is in understanding the causes of neuropsychiatric disorders such as familial Alzheimer's dementia or bipolar (manic-depressive) disorder. These disorders are familial, but their exact mode of inheritance and underlying protein defect are unknown. Nevertheless, by using hypervariable DNA markers of the human genome it has been possible by linkage analysis to map the susceptibility gene for Alzheimer's disorder and for manic-depressive disorder.

In the case of manic-depressive disorder, in some families linkage was detected by a random search of the entire human genome in a large kindred of the Old Order Amish. The susceptibility locus is near a locus for tyrosine hydroxylase (TH), which is the rate-controlling enzyme in the synthesis of the neurotransmitters dopamine and norepinephrine, themselves implicated in some forms of affective disorder. However, recent experiments suggest that the susceptibility locus to bipolar disorder is not the tyrosine hydroxylase locus itself.

Similarly, the gene defect causing familial Alzheimer's disease was located near the locus for the amyloid beta protein precursor gene, which has been cloned and mapped. Initially it was suggested that duplication of the amyloid protein locus may cause familial and sporadic cases of Alzheimer's disease. However, recent experiments have found no evidence of duplication of this locus in either familial or sporadic cases of Alzheimer's disease. Furthermore, genetic linkage studies have now excluded the amyloid protein gene locus as the site for familial Alzheimer's mutation, which casts doubt on the original conclusion.

These examples show the interactive use of candidate gene and reverse genetic strategies. Mapping a susceptibility locus to a particular region suggests studies of nearby candidate genes. Studies of these genes help to narrow the possible sites for the susceptibility locus. In the case of neuropsychiatric disorders, the genetic regulation of more than 100 neurotransmitters provides specific candidates for investigation because these neurotransmitters are peptides that control the chemical communication within the brain.

Gene Therapy

Replacement or Repair of Genes Is No Longer Science Fiction

Introduction of cloned genes into cultured cells (transfection) by methods of calcium or electrical enhancement or into the developing embryo by microinjection (as in transgenic mice) indicates that somatic cell gene therapy of some genetic disorders may not be far off. For example, it may be possible in vitro to introduce genes into stem cells of a patient's bone marrow and reimplant that marrow. All of these techniques, however, are still in their infancy, requiring much more research to establish effective and safe therapeutic procedures. Scientists are exploring the possibility of gene therapy studies in patients suffering from a rare genetic disease resulting in severe combined immune deficiency caused by defective production of the enzyme adenosine deaminase. However, major obstacles still exist, such as low levels of expression of the gene and possible virus contamination during treatment.

In addition, the chromosomal location of the integrated donor genes is relevant to gene therapy. The experience gained with transgenic mice indicates that integration of the foreign DNA into certain locations in the genome may cause the new gene to fail to be expressed, or worse yet, if it becomes integrated in the wrong place it may cause abnormality. More complete maps of the human genome, now being prepared, will enhance our ability to deal with such problems.

In some disorders in which an abnormal protein leads to abnormality—for example, sickle cell anemia and amyloid polyneuropathy—merely introducing a normal gene may not solve the problem. Ablation of the abnormal gene may be necessary. Research on site-specific (or directed) mutagenesis possibly applicable in this connection is progressing (see Chapter 4 for a discussion on homologous recombination).

NEW APPROACHES FOR CONTROL OF MICROBIAL INFECTION

One of the Great Achievements in American Medicine in the Twentieth Century Is the Reduction in the Incidence of Infectious Disease

Many of the old epidemic diseases have been controlled or, like smallpox, eliminated. Polio and measles are clearly on the road to elimination. The course of this revolution in infectious disease began with dramatic improvements in sanitation, housing, and nutrition. More recently, antibiotics and vaccines have aided in control. Antibiotics have been particularly successful in treating bacterial diseases, although they have been less effective in controlling diseases caused by parasites, viruses, and fungi. Since viruses use many of the host cell's synthetic pathways for replication, it is often difficult to specifically inhibit viral replication without affecting the host.

However, significant breakthroughs have been made in antiviral compound research. For example, the antiviral drug acyclovir is converted into a nucleotide by enzymatic phosphorylation only in cells infected with herpes virus. After acyclovir is converted into the nucleotide, it can then exert its antiviral properties by selectively incorporating into replicating viral DNA, which inhibits viral multiplication. Since phosphorylation is a prerequisite to DNA synthesis, acyclovir is active only in cells infected with herpes virus. This limits its potentially toxic effects on uninfected cells.

Vaccines

The Development of an HIV Vaccine to Prevent Acquired Immune Deficiency Syndrome Is Progressing Slowly

A new disease of humans, acquired immune deficiency syndrome (AIDS) was recognized in 1981. Since that time, more than 30,000 cases have been reported in the United States; it has been estimated that more than 1.5 million people are infected in this country alone. In certain parts of the world, especially in regions of Africa, AIDS is a major and growing problem. Worldwide, 10 million or more persons may be infected. Projections for the future growth of the epidemic indicate continued increases and, despite the uncertainties of these projections, disease and death resulting from AIDS will continue to increase for at least the next decade.

Our understanding of AIDS and its viral cause, HIV, have progressed remarkably because of basic research in molecular and cellular biology, carried out in thousands of laboratories over the past decades. HIV infection has a wide variety of effects on the immune system, resulting in opportunistic infections (infections caused by microbes that rarely cause disease in the healthy host) and cancer (such as Kaposi sarcoma and non-Hodgkins lymphomas). In addition, it has deleterious effects on the nervous system. Although most of the early cases have appeared in homosexual males and intravenous drug abusers, heterosexual transmission has been increasing and is the most common form of spread in Africa.

HIV was isolated in the early 1980s. It is a member of a group of retroviruses that causes many naturally occurring, progressive nonmalignant disorders in animals. Much has been learned about the structure of HIV. It has several genes in addition to the usual retrovirus virion protein genes. The activity of these additional genes probably causes HIV to remain latent in infected cells, and virus infection can thus persist even in the presence of neutralizing antibody in the victim's serum.

The enormous effort currently under way to develop an effective vaccine or drug to control AIDS has thus far produced only drugs with limited effects. Many vaccine candidates are being evaluated, but there is little hope for an effective vaccine in the near future. A major challenge of modern biology is the better understanding of the immunopathogenicity and eventual control of HIV infection. In this regard, an animal model for HIV infection is of utmost importance.

*Conventional Vaccines Have Been Among the Most Successful of
All Biological Products*

Smallpox, for example, has been eradicated. Vaccines against poliomyelitis, measles, rubella, and mumps offer the best means of preventing these diseases. For other viral, bacterial, and parasitic diseases, however, effective vaccines still have not been developed.

Biotechnology has opened two major new avenues for vaccine development. First, biotechnology provides tools for understanding virulence and the nature of the microbial immunogens. Second, biotechnology offers new strategies for engineering vaccines.

Inducing an effective immune response requires precise knowledge of the antigens that stimulate and are recognized by humoral and cellular immunity. A number of new strategies exist for engineering vaccines. When a protein is identified as an important antigen, the gene encoding the immunogenic protein can be inserted into a suitable host (bacteria, yeast, or plant), which can be grown to yield large amounts of the immunogenic protein. The protein engineered and prepared in this manner may be used directly as a subunit vaccine. An alternative to proteins so prepared is the chemical synthesis of immunogenic regions of the protein (synthetic peptide vaccine) based on the DNA sequence of the gene that encodes the immunogen.

In addition to using isolated genes or portions of genes to make immunogenic proteins, isolated genes can be inserted into vaccinia or other DNA viruses. Inoculating the host with such hybrid viruses, can stimulate immunity to the inserted "foreign" proteins in addition to stimulating immunity to its own proteins. For example, the U.S. Food and Drug Administration has approved clinical trials of a HIV-vaccinia vaccine. Hybrid viruses can stimulate immunity against a number of foreign proteins simultaneously. Manipulation of live microorganisms may also reduce virulence while maintaining their capacity to induce immunity. For example, a gene for a toxin can be modified or deleted, causing it to become less virulent while retaining its colonizing abilities. Such a mutant of the cholera vibrio has been shown to immunize human volunteers effectively after oral inoculation with the live, attenuated (less virulent) mutant strain.

Selection of viral mutants can lead to nonvirulent mutants that are immunogenic. It is feasible to reduce the virulence of live microorganisms and acquire live, attenuated vaccine candidates. A special type of attenuated vaccine strain may be generated for segmented genome viruses since new combinations of genes may result in attenuated viruses by reassortment of genome segments.

Finally, it has been possible to mimic the antigenic structure of certain parasites and viruses with anti-idiotypic antibodies. Such "anti-antibodies" have a configuration that resembles the immunogenic protein. Anti-idiotypes prime or actually stimulate an effective immune response. A number of anti-idiotype vaccines stimulate immunity to pathogens such as trypanosomes, rabies virus, and polio virus.

INSERTION OF AN ANTIGEN GENE INTO VIRUS VECTORS

Vaccinia virus, as well as adenoviruses and herpes viruses, have been used as vectors for carrying other genes. The rationale is that part of these viral genomes can be replaced without significantly altering the capacity of the virus to grow in cell culture. Such growth results in a live recombinant virus. In the case of vaccinia virus, these recombinant viruses are less virulent in experimental animals than the vaccinia virus parent, and the recombinants can potentially induce both cellular and humoral immunity. An increasing number of viral, bacterial, and parasitic antigens (including proteins from hepatitis B, influenza, rabies, Epstein-Barr, and human immunodeficiency viruses and from *Plasmodium falciparum*) have been inserted into vaccinia virus in this manner. In many instances, animals inoculated with these recombinant viruses have demonstrated a protective response. One striking advantage of this approach is the possibility of eliciting broad immunity with a single dose. However, the approach still needs to be carefully evaluated in terms of human safety, since the toxicity and possible unanticipated effects remain unknown.

Antiviral Agents

Most Approaches to the Treatment of Viral Infections Relate to the Fact that Viruses Are Intracellular Pathogens

Viruses use the host's normal cell machinery; therefore, it is often difficult to develop effective and nontoxic drugs that are virus specific. However, recent advances in molecular biology, immunology, crystallography, and drug design have initiated a new era of treatment of viral infections. The theoretical possibility of rational drug design based on an understanding of specific steps of viral replication is becoming a reality.

In discussing a rational approach to antiviral chemotherapies, it is useful to divide the viral cycle into a series of steps: early stages (including adsorption, penetration, and uncoating); transcription and replication; and late stages (including assembly, maturation, and release). The development of specific drugs that block each of these stages has been facilitated by many studies that have identified details of the strategies of different viral groups. Selected drugs that affect early events include heparin and other anionic polyelectrolytes, which prevent attachment; oligopeptides, which by mimicking part of the fusion polypeptide of paramyxoviruses inhibit fusion; and amantadine and rimatadine, which inhibit early steps in the growth of type A influenza. Drugs that affect the synthesis of viral nucleic acids include purine and pyrimidine, nucleoside analogs that are active against herpes and other viruses (such as adenine arabinoside, acyclovir, and ribavirin); single-stranded polyribonucleotides, which are thought to inhibit retroviruses by blocking the reverse transcriptase reaction with single-stranded polyribonucleotides; thiosemicarbazones, agents active against pox viruses; and phosphonoacetate and phosphonoformate, which are active against herpes. Drugs active against late events include rifampicin (blocks pox virus assembly) and 2-deoxy-D-glucose, which blocks functions attributed to virus-specific glycoproteins, such as cell fusion, hemaglutination, and hemadsorption, by interfering with the glycosylation of virus-specific polypeptides.

ADVANCES IN THE BIOCHEMICAL PROCESS INDUSTRY

Biological Discovery Is the Driving Force in the Biochemical Process Industry

The biochemical process industry began thousands of years ago with the production of wine, beer, and bread. By the mid-nineteenth century, the technology underlying this industry began to advance in response to a better understanding of microbiology and biochemistry. The discovery that actively growing yeast cells were responsible for the active component in baking and also for the production of alcohol in brewing led to the observation that control of oxygen and sugar supply had greatly improved both processes. These discoveries in the biological sciences led to major process improvements in the evolving biochemical process industry. It was not until the mid-1940s, however, that a true collaboration between the biological scientist and the process engineer began to develop. It has recently become apparent that biological discovery provides the driving force behind process development: The new discoveries in biology during the past decade have become the enabling technologies that underlie a modern and rapidly growing biochemical process industry.

Recent important discoveries include techniques of genetic engineering, cell fusion, nucleic acid and protein sequencing and synthesis, techniques for visualizing protein structure, and an understanding of the molecular basis for enzyme catalysis. With these discoveries, it has become possible to isolate and produce a new set of therapeutic proteins, explore new regimes of biocatalysis, improve diagnostic procedures and plant species, develop better methods for plant protection, develop new biomaterials, and begin to address the effective use of biological systems for hazardous waste disposal and of natural resources. The rapid advancement in biological discovery has created new opportunities and, as a consequence, new problems in industrial biotechnology. The opportunities relate to the possibilities for new products and services in industries devoted to human health care, agriculture, chemicals, and environmental services. The problems relate to the need to translate new discovery in science to technological practice. Problems, such as efficient recovery of therapeutic proteins of high purity or large-scale cultivation of animal cells, were not problems a few years ago; it is only the recent discovery of products requiring these technologies that have forced us to develop advanced manufacturing processes.

Goals in Industrial Biotechnology

The Overall Goal in Industrial Biotechnology Is to Translate New Scientific Discoveries into Processes That Provide Products and Services

Many opportunities for industrial biotechnology are coming from progress in biology. However, for discoveries to be translated into products the techniques for developing biochemical processes must be available and manpower must be adequate.

To fully appreciate and understand the problems and opportunities associated with industrial biotechnology, it is essential to examine the barriers inherent in product development. Reduction or removal of these barriers is essential to maintaining and even accelerating successful translation of science into technology.

Difficulties with the Identification, Isolation, and Characterization of New Products Are the Primary Impediments to Product Development

The primary initial milestones in product development are discovery of new and useful functions followed by the isolation, purification, and characterization of the chemical entity responsible for this function. In many cases, the responsible chemical entity is masked by the complexity of biochemical systems. Many molecules of interest have a regulatory function in stimulating or inhibiting the specific actions of other cells. As a consequence, these molecules are present in very low concentrations in the cell. This makes the identification of their function and the determination of their structure exceedingly difficult. The hundreds of man-years required to isolate and characterize the interferons is a good example of this problem. Even though we have isolated and purified the interferons and characterized their molecular structure, we do not fully understand their mechanism or implications of their function.

Realizing that a product is not truly a product until it can be manufactured in a form suitable for use is an important first principle. The discovery of an exciting new chemical entity with important functions is only the first step in product and process development. For example, recombinant DNA techniques permit us to make a new class of therapeutic drugs composed of pure, highly active proteins. Because these proteins are produced in complex mixtures of similar molecules and require an extremely high degree of purification, developing suitable manufacturing processes has been a major challenge.

Acquisition and Interpretation of Analytical Information Is a Major Problem

A major problem that has plagued efficient process development has been the scarcity of analytical techniques for measuring the product of interest. The biological activity of many new products is difficult to measure, and their structure is often poorly defined; thus, routine measurement of these molecules is difficult. Assays for their biological activities are often cumbersome and imprecise. A powerful analytical technique that has recently evolved is the use of monoclonal antibodies as the basis for molecular recognition and analysis. New sensor technology is needed, including biosensors, instrumentation, and software, that will provide the power needed to support process development more effectively.

Process Synthesis, or Manufacturing, Requires Multiple Steps from Compound Production to Purification

Manufacturing technology for the production of biochemical products requires a multiplicity of unit operations. A typical process for manufacturing a highly pure therapeutic protein is diagnosed in Figure 10-3. The primary unit operation in which the raw materials or nutrient medium is converted to a desired product is carried out in a bioreactor. This reactor may be a fermentor for growing microorganisms or a device for propagation of animal cells; in either case, the cells may have been genetically engineered to overproduce the desired product. Once formed, the product must be separated, concentrated, and purified to its active form and formulated to stabilize, and perhaps enhance, its activity for eventual use as a drug.

A process for manufacturing therapeutic proteins derived from recombinant microorganisms is expensive and complex, not only because of the multiple unit operations that must be carried out to recover highly pure materials, but also because the final product must be made in its native form—that is, with a single three-dimensional structure. Thus, in the case of a protein that has a large number of possible folding configurations, only one is permitted in the final product. This will ensure its efficient use as a therapeutic agent and help prevent the body's immune response from rejecting the molecule as a foreign protein. Both biological and structural integrity must be retained. The ability to achieve this integrity depends on developing a better understanding of how proteins behave during their production and purification and as well as on developing high-resolution analytical techniques that are able to measure structural properties of biological molecules.

Other problems that hinder successful process development include the presence of proteases that modify the product, insufficient reduction in nucleic acid content, and the presence of isomeric forms of a protein product. Glycosylation of proteins is a useful and important posttranslational modification implemented

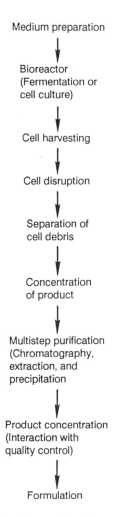

Medium preparation

Bioreactor
(Fermentation or
cell culture)

Cell harvesting

Cell disruption

Separation of
cell debris

Concentration
of product

Multistep purification
(Chromatography,
extraction, and
precipitation

Product concentration
(Interaction with
quality control)

Formulation

FIGURE 10-3 Bioprocessing. [Charles Cooney, Massachusetts Institute of Technology]

by cells, the function of which we need to better understand; we also need to improve the techniques for characterizing, modifying, and recovering properly glycosylated proteins. Additionally, a substantial number of proteins of interest form insoluble aggregates or "inclusion bodies" inside genetically engineered cells, and we need to better understand procedures for recovering these proteins. Such recovery of many proteins could be enhanced by high levels of secretion; however, the molecular basis of secretion is unknown in many microorganisms.

Further advances in understanding the biology and biochemistry of protein secretion will have a major impact on process development.

In recent years, many of the improvements in recovery of new biochemical products have come about through scale-up of known techniques and the use of a wider variety of materials that enhance purification. Improved scale-up has occurred as a consequence of automation of many procedures, especially chromatography, the availability of new chromatographic resins and other adsorbents, the wide use of membrane processes for concentration and desalting, and improvements in mechanical devices such as centrifuges, cell homogenizers, and membrane filters. Improvements are still needed. New techniques, such as affinity chromatography and electrokinetic separations, are still not applicable to large-scale purification because of scale-up cost. There thus remains a need for innovation in both the development and implementation of new recovery procedures, the development of improved materials to enhance unit operations such as extraction and chromatography, and the development of techniques for achieving process integration. As part of this process, computer-aided design software and flow-sheet simulation techniques are likely to become major tools in the hands of the process biochemist and the engineer.

In response to these needs, substantial research has been done to delineate the necessary components of the complex and ill-defined nutrient media that are used for the growth of mammalian cells. From this effort has developed a set of better defined and less expensive cell-culture media that avoid the use of animal sera. These defined low-protein media not only enable efficient and controlled growth of mammalian cells, but also facilitate the recovery of products excreted by these cells. Because these products are often secreted into the medium in very low concentrations (<10 milligrams per liter), the use of media containing high concentrations of protein makes subsequent downstream processing for product recovery difficult. Thus, the use of well-defined, low-protein media greatly enhances efficient product recovery.

Modern Biology Coupled to the Biochemical Processing Industry Has Enormous Potential

Human Growth Hormone. Human growth hormone is a protein produced by the pituitary gland that stimulates the liver to produce somatomedins, which stimulate bones to grow. Children produce large quantities of growth hormone, whereas adults normally produce very little. Pituitary dwarfs represent a population with a defect in the production of the pituitary growth hormone. Children judged deficient in growth hormone used to be injected with the hormone obtained directly from the pituitary glands of cadavers, which they received three times per week. The National Hormone and Pituitary Program processed as many as 50,000 pituitaries per year, supplying the growth hormone derived from them to treat 3,500 individuals. Since the hormone was difficult to obtain, it was always in short supply; the criteria for deciding who to treat were stringent. In 1985 three people who had been treated with the cadaver-derived human growth hormone in the 1960s and 1970s died of Creutzfeldt-Jakob disease, a rare disease that results in dementia and death. The proposed cause of the disease was infectious virus derived from the pituitaries of victims of Creutzfeldt-Jakob disease. At the time of these deaths, biotechnology companies had been working on the application of recombinant DNA technology to produce synthetic human growth hormone.

In this effort, the gene coding for growth hormone was isolated from pituitaries, and genetic expression vectors were developed to produce the hormone by fermentation. This was followed by several years of careful research on the isolation and purification of this rare protein. The successful completion of this task in the fall of 1985 resulted in Food and Drug Administration approval for a genetically engineered version of human growth hormone. Since the growth hormone is manufactured in a bacterium, no human pituitaries are needed and the threat of Creutzfeld-Jakob infections has been eliminated completely.

In addition to solving a major health and social problem, the development of a commercial process for the manufacture of recombinant human growth hormone has made major contributions to the identification of fundamental problems in protein structure and function. Challenging problems—such as determining the factors that influence proper folding of proteins derived from the linear assembly of amino acids, determining the crystalline structure of growth hormone, and understanding its mechanism for effecting the observed physiological responses—are receiving widespread attention. The study of growth hormone will lead to a greater understanding of the biology of other regulatory and mediating proteins, as well as to the design of smaller and more effective molecules. We can anticipate improved treatments for hormone deficiencies, wound healing, osteoporosis, bone fractures, and obesity.

Tissue Plasminogen Activator. Both strokes and heart attacks can occur when a blood clot lodges in a blood vessel, blocking the flow of blood. Tissue plasminogen activator is a naturally occurring enzyme that binds to fibrin clots and is activated to convert plasminogen to plasmin, the latter being capable of dissolving the blood clot. The primary initial applications for this protein are in the treatment of heart attacks, with minor applications for other conditions such as unstable angina, pulmonary embolism, and arterio-venous occlusions. More than a score of companies are currently involved in the research and development of tissue plasminogen activator.

The story differs considerably from that for human growth hormone production. Tissue plasminogen activator is a glycosylated protein; that is, it contains oligosaccharides (sugar complexes) attached to the backbone of the protein molecule. Since bacteria cannot glycosylate proteins, the manufacture of tissue plasminogen activator, or its variants, must be in higher organisms or cells representative of those organisms. At present, major production is achieved through mammalian cell culture technology. This approach has initiated a new cycle of chemical and biological research to develop analytical methods for characterizing and analyzing the sugar attachments to the protein as well as basic biological studies to develop an understanding of the role of these oligosaccharides in the structure and function of tissue plasminogen activator.

Other thrombolytic agents such as urokinase and streptokinase are also being marketed and intensively investigated. Second-generation clot-dissolving molecules are being designed. It is hoped that new products with enhanced specific activities, greater specificity, and fewer side effects (such as nonspecific bleeding) will be generated. The challenge to increase the survival rates of the 900,000 heart attack victims in the United States who reach hospitals each year is great, and the tools available today provide scientists with the means for solving these problems.

Industrial Growth in Biotechnology Is Predicted, but International Competition Will Be Great

Over the past 5 years, biotechnology, defined as the application of biological organisms, systems, and processes to the manufacturing and service industries, has developed dramatically, and commercial products have been made. Although the number of products that have entered the market at this time is limited (for example, insulin, human growth hormone, interferons, tissue plasminogen activator, and hepatitis B vaccine), the momentum to introduce more products continues. Within the next decade we should see the introduction of many new products for medicine, agriculture, and environmental management.

Economic considerations and analyses have replaced some of the euphoric hopes for products derived from biotechnology. This change has resulted in a significant focus toward high-value specialty products and away from large-scale production of general-use products. Combinatorial evaluations of gene engineering, mutagenesis, fermentation optimization, and downstream processing have intensified as economic dictates have become more obvious.

Regulatory issues remain a major question worldwide, particularly with regard to the deliberate environmental release of plants and microbes developed through the use of recombinant DNA technology. As a result, there now exists a regulatory patchwork with little uniformity from country to country. In the United States, regulatory issues are currently under study, and debate continues with regard to what risk-benefit information will be required and who will regulate what.

Worldwide, there is little doubt that biotechnology is seeing major developments in Europe, Japan, and the United States. Other complementary and synergistic technologies that will influence the economic trends and impact of biotechnology must be considered. Examples are more traditional plant breeding, somatic cell culture techniques, and computer science. The Japanese have organized to become a major force in biotechnology. This movement has involved government, industry, and universities and has focused on key target areas such as plant cell culture, hybridomas, bioreactors, and large-scale mammalian culture.

New opportunities to service the growing activities in biotechnology have led to new businesses such as those that provide computer support for molecular biology, build equipment for nucleic acid and peptide synthesis and analysis, and produce mammalian cell cultures. Various government initiatives have led to broader joint ventures and venture capital operations on an international scale, which eventually should have a significant impact on international trade.

ADVANCES IN ANIMAL AGRICULTURE

Challenges Lie Ahead to Increase the Efficiency and Productivity of Domestic Animals Under Widely Disparate Environmental, Economic, and Political Constraints

Of the thousands of animal species that might have been domesticated, human beings have utilized only a handful. These have been selected over the last 10 millennia to occupy increasingly narrow ecological niches. These niches are highly structured management systems designed to increase the efficiency of our domestic species by removing negative environmental effects and maximizing the nutritional status of these animals as they grow to maturity. This intensified approach to domestic animal agriculture has reached its peak in the temperate regions of the world, where the "developed" economies and, generally, a surplus of agricultural products exist. Great discrepancies are found between the agricultural productivity of the temperate regions and those of the tropics and subtropics, where 60 percent of the world's population and half of its domestic animal species reside. For instance, it requires from two to four times as many animals to produce equivalent amounts of meat and milk in the tropics as can be produced by domestic animals in the temperate regions.

Molecular biology has helped provide the ability to address these production problems by unlocking the vast information found within the genome of each animal. Using this information, we can increase productivity of specific tissues, provide new weapons against disease, reduce inefficiencies caused by overproduction of fat in growing animals, increase the efficiency of rumen fermentation, and increase the genetic diversity of the domestic animal gene pool. Since the amount of available information present in the gene pool of the domestic animal population far exceeds our ability to interpret it, it is critical to define those areas most likely to produce real gains in the productivity of domestic animal agriculture.

Growth Regulation

Growth Occurs As a Result of an Increase in Cell Number, Cell Size, or Both

Increasing growth rate has a number of potential benefits to agricultural productivity. Among these are decreased interval to reproductive maturity or market weight, increased feed efficiency, and reduction of maintenance requirements. In addition to increasing the rate at which animals grow it is also possible to consider increasing the growth rates of specific tissues (such as muscle) and decreasing the growth of others (such as adipose tissue).

Regulation of the Somatotropin-Somatomedin System. The growth of whole animals can be accelerated by growth hormone (somatotropin) and insulinlike growth factors (somatomedin). The release of somatotropin into the circulatory system is correlated with a concomitant increase in the circulating concentrations of somatomedin. The release of somatotropin can be effected by the use of a growth-hormone releasing factor, inhibitors of somatostatin, release, or antibodies against somatostatin. Exogenous supplementation with the homologous somatotropin or somatomedins is also an available approach. Largely unexplored is the regulation of the receptors or binding proteins that influence the rate of the degradation and the degree of responsiveness of the target tissues for somatotropin or somatostatin. The approach of designing drugs by studying the receptor-ligand interaction has been successful in the human pharmaceutical area, but it is only recently that a similar approach has been attempted in the domestic animal industry. A recent example is the β-agonists, which stimulate β-adrenergic receptors to partition energy away from adipose tissue and toward muscle. An approach that has been envisioned for the future is to produce peptide mimics of steroids by studying the steroid hormone receptor and building peptides that will activate it. This would be an important breakthrough, since it would alleviate the problem of steroid residues in meat.

To date the use of recombinant homologous somatotropins and chemically and or recombinantly produced growth-hormone releasing factors has been most carefully studied. Since growth hormone does more than produce somatomedins, it is likely that this approach or the use of releasing factors will be the method of choice for the immediate future. However, the use of somatomedins and epidermal growth factors has been proposed for wound healing in valuable animals.

Regulation of Growth of Specific Cell Types. The ability to favor or retard the growth or development of a single cell type has obvious advantages in increasing the productivity of domestic animals. The specificity required to achieve either of these results dictates a greater understanding of the biology of these cell types than may be currently available. However, progress is being made with two cell types: the adipocyte of adipose tissue and the myoblast and myotube of muscle.

Excess storage of lipid is a major cause of reduced feed efficiency in domestic animals and a major health problem in the human population. Reducing excess fat storage in animals addresses both problems simultaneously, since the meat marketed will be of higher quality. Present approaches include the exogenous use of somatotropin and β-agonists during the finishing period to partition nutrients away from adipose tissue and toward muscle synthesis. A more direct approach being tested is the immunization of animals against membrane proteins of adipocytes. This approach has had dramatic effects in laboratory animals, but is largely untested in animal production systems. Much additional work is needed to learn to regulate specific metabolic pathways in adipocytes by regulation of the genome.

Increasing muscle mass in animals may be accomplished by increasing the number of muscle cells or the amount of muscle protein present in cells. Two basic in vitro cellular models are the fetal myoblast cells and myotubes. Two-dimensional mapping of proteins during proliferation and myotube formation has identified specific proteins produced during these processes. However, we do not yet understand the regulation of the synthesis and degradation of these molecules or their specific functions in proliferation and maturation of these muscle cells. Additional basic research is warranted in these areas to provide tools and more sophisticated approaches to increasing the availability of high-quality protein for human consumption.

Reproduction

It Has Been Difficult to Increase the Reproductive Performance of Domestic Animals

A great deal of basic research has been carried out in the past three decades on the regulation of the reproductive process in an attempt to increase reproduction performance. Widespread use of artificial insemination and embryo transfer has increased the rate of genetic gain in certain specifically desired traits, such as milk yield. However, a reliable pregnancy diagnosis is not yet available. In addition, the average litter size in the swine population has not increased in the U.S. herd, although certain breeds in other countries have much higher litter numbers. For example the average litter size of the Chinese hog is 24, versus 10 for the United States.

A major constraint in regulating the reproductive process is the large number of complex cell types that influence reproductive success or failure. In addition, the management systems in place for most domestic animals stress the reproductive process. For instance, increasing milk yield in cattle through genetic selection has had detrimental effects on reproductive performance because of the large metabolic demands to increase milk production. Immediate gains in production could be realized if sexing of semen were accomplished and if pregnancy-specific peptides were identified that could be used to diagnose pregnancy on farms.

In the domestic fowl population, some potential exists for manipulating egg formation to alter the lipid composition (reduce cholesterol), to alter the protein content of eggs, or to express novel proteins that could be harvested. It is also possible to harvest antibodies from eggs of chickens that have been immunized against a foreign protein.

Lactation

Increases in Milk Production Will Occur with the Use of Bovine Somatotropin and Increased Mammary Growth

It is generally believed that use of bovine somatotropin in lactating cattle will increase average milk yields 10 to 15 percent across a 305-day lactation. Bovine somatotropin increases the rate of milk synthesis, but does not increase the amount of secretory tissue present. Increases in the growth of mammary tissue would also increase milk yield since each gram of mammary tissue produces approximately 1.6 grams of milk per day. Although several potential mammary growth factors have been identified, the specific roles these various factors play in regulating tissue growth are not well understood. If specific mammary growth factors could be identified and put to use, it would benefit not only the dairy industry, but also other animal industries as well, since milk production is a major limitating factor on the growth rate of neonatal swine and beef cattle.

Alteration of gene expression in mammary tissue to produce a milk with a higher solid content is a possibility. Milk is 87 percent water, and it is therefore expensive to transport. The major osmotic determinant of milk is lactose, the milk sugar. In marine mammals, the production of lactose is greatly limited, producing a milk that is extremely high in protein and fat. It would be possible to achieve the same effect in cattle by inhibiting the expression of alpha lactalbumin production. This protein, produced only in mammary tissue, is the rate-limiting protein for lactose synthesis. Another possible approach to manipulation of milk synthesis is the production of novel proteins that either improve the quality of milk or are of economic significance.

Infectious Disease

Molecular Biology Will Provide Major Advances in the Diagnosis, Treatment, and Prevention of Some of the Major Diseases That Threaten the Livestock Industry

Disease is a major cause of reduced performance in domestic animals. Some diseases that cause major losses in the livestock industry are foot-and-mouth disease, neonatal enterotoxicosis, trypanosomiasis, mastitis, and respiratory diseases. At present, molecular biology approaches are being used to develop methods of diagnosis and treatment for these diseases.

Some general approaches are the production of animal interferons for treatment of respiratory diseases and the use of monoclonal antibodies for use in diagnostic aids or as treatments against certain bacteria or viruses. Recently, a vaccine against foot-and-mouth disease was developed; it uses a protein-surface antigen encoded by a synthetic gene, which was cloned and expressed at high levels in *Escherichia coli*. Similar approaches might be productive in the search for a weapon against trypanosomes, which, by causing sleeping sickness in cattle, are a major deterrent to increased productivity in the African cattle industry. Use of antibodies against adhesion factors prevents the bacteria that cause neonatal enterotoxicosis from attaching to gut epithelial cells. A similar approach has been successful in preventing the attachment to mammary epithelial cells of certain bacteria that cause mastitis.

Production of Feedstocks

Some Feed Additives, Such As Amino Acids and Vitamins, May Be Efficiently Produced by Recombinant Techniques

The two major amino acids now used as feed additives are methionine and L-lysine. It is doubtful that methionine made by recombinant techniques would cost less than that made by chemical synthesis. However, lysine, which can be absorbed across the gut only in its L-form, can be produced by fermentation that uses coryneform bacteria. It is believed that recombinant techniques will make it possible to produce bacteria that can make 130 g of L-lysine per liter. At these levels of production, the price of lysine will be lower than the equivalent soybean meal source.

Lowering the cost of lysine production may permit the addition to feed of the next two limiting amino acids, L-tryptophan and L-threonine. Success in making recombinant organisms that produce these amino acids in large quantity has been reported.

The production of most vitamins will not be affected greatly by recombinant technology since chemical synthesis will remain financially competitive. However, it has been proposed that vitamin B_{12}, riboflavin, and niacin might be produced competitively by these techniques.

Livestock Improvement

Genetic Engineering and Expanded Breeding Programs Should Result in More Efficient Production of Livestock

The insertion of foreign DNA into animals has received much publicity recently. The challenge is to obtain tissue specificity of gene expression. Most current work in domestic species involves manipulation of the somatotropin system by inserting extra copies of the somatotropin gene into embryos. This approach, which has as its goal improving feed efficiency and meat quality, has its highest chance for success in litter-bearing species such as the pig, where the number of embryos is high. Studies are also being carried out in sheep by using embryo transfer from superovulated animals to increase the odds of success. Although it has been possible to get expression of these extra copies of soma- totropin genes in animals, the results to date have been disappointing. For example, the health and reproduction capabilities of these animals have been impaired, suggesting the need for a tissue-specific approach to gene expression. Ideally, one would like to insert the extra copies of the desired gene into the target tissue to avoid side effects in other tissues. It is likely that this will be a major barrier to the practical use of this approach.

One novel approach to this problem has been to use lactating mammary tissue to produce a foreign protein. This was accomplished by joining the desired gene to the control region of the gene coding for lactoglobulin. This recombinant gene is expressed only in lactating mammary tissue. Therefore, by using an appropriate tissue-specific promoter, investigators can obtain tissue-specific ex- pression.

Another possible example of the same approach would be to insert the gene- regulating expression of the mediator of somatotropin action on mammary tissue. This would permit autostimulation of secreting mammary tissue to increase yield.

As mentioned earlier, the productivity of animals in the temperate regions of the globe far exceeds that of the same species in the tropics and subtropics. Moreover, when the temperate breeds are exported to tropical regions, their productivity falls rapidly, and they often die of diseases to which they have no resistance. An obvious solution is to use local breeds in these regions, provide exogenous hormone treatments to increase their productivity, or to insert genes coding for metabolic enhancers. For example, adaptation to thermal stress is associated with a reduction in circulating somatotropin concentrations. Growth rates and milk yields of cattle in these regions might be enhanced markedly with exogenous supplementation of somatotropin or insertion of extra copies of the growth hormone gene.

The diversity of the domestic animal gene pool is larger than most scientists realize because the predominant breeds of livestock represent only a small portion of the total number of breeds. For example, 90 percent of the dairy cow popula- tion in the United States consists of Holsteins, even though more than 40 different dairy breeds exist throughout the world. No doubt many of these breeds have genetic information of great value to the world cattle industry. In Asia alone, there are five species of nondomesticated ruminants, which are close relatives to cattle although poorly known. Another example is in the swine industry. What genes make it possible for the Chinese sow to produce twice as many piglets in a litter as the average U.S. sows? Southeast Asia is the home of a ruminant pig (*Babirusa*) that might be used to produce a new pig breed, one more efficient at meat production. We should consider not only the relatively few highly utilized domestic animals in our search to increase domestic animal productivity, but also their underutilized relatives.

11

Plant Biology and Agriculture

Plants Constitute the Only Renewable Source of Energy

The enormous quantity of energy from the sun that is captured by the earth every day becomes available for life processes only through the photosynthetic activities of plants, algae, and a few kinds of bacteria. These activities have resulted in the characteristics of the atmosphere that we breathe and have altered our atmosphere's chemical makeup so that it is hospitable to animal life; prolonged over hundreds of millions of years, these activities have given rise to the fossil fuels that power our civilization. At the same time, photosynthesis constitutes the only renewable source of energy that is available to us for the future: a source of energy that is clean and potentially inexhaustible. Since plants directly or indirectly provide for our fuel and fiber needs in addition to being our primary source of food, they are exceedingly important to us from every point of view. Understanding their characteristics is of vital importance for the advance of biological knowledge and for human prosperity as well.

Vegetation plays a major role in maintaining the earth-atmosphere system in a habitable state. Except for the polar icecaps, snow- and ice-covered mountains, and certain of the earth's deserts, all land masses are covered with vegetation. This vegetation contributes to global energy and water budgets through modification of the solar energy, water, carbon dioxide, and nitrogen exchanges at the earth's surface. In short, not the atmosphere, nor the soil, nor any of the other conspicuous features of the earth's surface would exist in their present condition if it were not for the existence of photosynthesis, a process that we now believe to have evolved among the cyanobacteria (blue-green algae) at least 3.5 billion years ago—at least 2 billion years before the origin of any photosynthetic eukaryotes and more than 3 billion years before that of plants.

The solar energy metabolically fixed through photosynthesis constitutes about 0.3 percent of the total solar radiation that reaches the surface of the earth. In addition, a substantial fraction of the solar radiation that reaches the earth is converted into latent heat that leaves the earth's surface through plant transpiration. Some 75 trillion tons of water evaporate each year from the vegetation to the atmosphere. Agricultural vegetation is responsible for about one-third of this water flux and also, because of the coupling of these two processes, for one-third of the total photosynthetic energy fixation. Natural tree ecosystems in the tropical and subtropical zones constitute the major vegetation mediating global water and carbon dioxide exchanges.

The greater part of our food is produced by a few species of annual crop plants, mostly in the temperate and semihumid to semiarid middle latitudes. In food production, accumulation of dry matter is the process of greatest importance. Most of the weight in dry matter comes from the 175 billion tons of carbon dioxide that agricultural plants fix annually through photosynthesis.

In addition to the more obvious activities of plants that occur above ground, extensive activity in modifying the characteristics of the soil is a role of the roots. The amount of water passing through a plant in its transpiration stream is many times the amount required to supply its internal needs. All of this water, together with the inorganic nutrients that plants require for growth, enters plants by way of their roots. Plants accomplish this through physical forces and highly specific transport systems. Interactions between plants and soil microorganisms are of critical importance for certain assimilatory processes: for example, *Rhizobium*, *Frankia*, and certain free-living bacteria for obtaining nitrogen, and mycorrhizal fungi, regularly associated with the roots of approximately 80 percent of all plant species, for phosphorus uptake. In addition, roots produce hormones that are important in directing the characteristics of shoot growth.

PLANTS AND THEIR ENVIRONMENT

Through Evolution, Plants Have Developed Characteristics to Cope with Their Environment

One group of unicellular eukaryotes, the green algae (Chlorophyta), consists of organisms that share a number of biochemical and structural characteristics with plants. The similarities are so great that it is generally agreed that plants were derived from green algae, and specifically from organisms that had many of the features of the multicellular, freshwater alga *Coleochaete*. The cellulose cell walls that are such an important feature of the adaptation of plants on land originated among the green algae, as did the ability to form starch granules within the chloroplasts rather than free in the cytoplasm, and as did certain unique features of cell division that are common to all plants. The ancestors of plants invaded the land at least 430 million years ago, already multicellular and thus protected from the environmental extremes that they were to encounter there. The earliest plants were evidently mycorrhizal, the adaptive features of their symbiosis with fungi assisting them in growing on and eventually molding the features of the raw soils of those ancient times.

With their rigid cellulose cell walls, the bodies of plants are put together as if from a series of bricks. The sorts of cellular movements characteristic of animal embryology are impossible among such organisms, as is the ability to move from place to place in search of more suitable habitats or mates. Consequently, plants have evolved features that suit them to a sessile existence. Their life processes are bathed by a continuous stream of water that moves steadily from their root hairs into their roots, up through specialized conducting cells called xylem through the stems and into their leaves, and then mostly dissipates through the leaves through specialized openings called stomata, which also admit the carbon dioxide that plants require for photosynthesis. A waxy cuticle, similar to the outer covering of many arthropods, evolved among the earliest plants and helps to protect them from drying out.

Rooted in one place, many kinds of plants must tolerate a wide variety of environmental extremes. The consequent selection pressures led to the evolution of plant species that can withstand temperatures ranging from that of liquid nitrogen (-195.8°C) to 90°C and that could grow between temperatures lower than 0°C and higher than 60°C. Some plants are able to grow in solutions as concentrated as saturated salt and to withstand desiccation to the air-dry state.

An additional characteristic of plants not found in animals is the ability to grow endlessly from areas of cell division, or meristems, that occur at the tips of the roots and shoots. New plants can be propagated from such meristems, and, depending on their growth form, may grow through the soil into areas of favorable nutrient status. Each plant can be both embryonic and senescent simultaneously, and the entire history of a plant's development can often be traced in a single organ.

Understanding Plant Characteristics Is Important to Agricultural Development

In nature, the ability of plants to reproduce is of fundamental importance. When plants are grown as crops, it is often their seeds, fruits, or vegetative reproductive structures, such as tubers or fleshy roots, that people desire. The characteristics of crops have been modified by selection and hybridization for at least 11,000 years and are now being modified more precisely by the techniques of genetic engineering. Cultural practices are also important in promoting crop yield. In all types of agriculture, opportunities exist to improve yield in different areas of the world.

The optimal yield of particular strains of plants is usually tested by growing them in nonlimiting conditions—ones in which they do not encounter stress from drought, lack of nutrients, pests, or for any other reason. Such conditions are approximated when crops achieve their highest recorded yields. Record yields may be some three to seven times as high as average yields obtained under more usual conditions in the same year. In the United States, for example corn (maize) had a record yield of 19,300 kilograms per hectare in 1975, but yielded only 4,600 kg/ha on average.

Average agricultural productivity, therefore, falls far short of the genetic potential present in today's crops. Major environmental pressures must affect plants in ways that prevent the expression of their full genetic potential. Thus, improvements in productivity need not rest solely on increases in genetic potential. For this reason, both the identification of the environmental forces and the manipulation of crops to express their genetic potential more fully are important research areas in plant biology.

What are some of the environmental pressures that decrease productivity in plants? Diseases and insects are important contributors: These pests depress U.S. crop yields by an estimated 5.1 and 3.0 percent, respectively, below their genetic potential. In addition, weeds, which compete with the crops, depress yields another 3.5 percent overall, despite the widespread and relatively efficient application of control measures. An additional large depression in yields must be attributed to the only other factor that can be unfavorable, the physicochemical environment. An unfavorable physicochemical environment is found in soils and climates that are ill suited for plants. Adverse physicochemical environments—such as an insufficient supply of water or nutrients—caused yields to be far below their genetic potential. Sometimes the environments in which crops are grown are inherently unfavorable, and sometimes farmers choose not to improve these conditions or cannot afford to do so. At any rate, physiochemical limiting factors are the most important negative influence on U.S. agriculture. In many other parts of the world, crop losses caused by diseases, insects, and weeds are considerably more severe than in the United States, but physiochemical factors usually predominate everywhere in limiting agricultural yields.

The major physicochemical resources for plants are water, soil type, nutrients, carbon dioxide, oxygen, and solar radiation. Of these, water is generally the most limiting. Permanently dry and shallow (drought-prone) soils make up about 45 percent of the total U.S. land area. About 40 percent of insurance payments for crop losses are made to drought-stricken farmers. Cold and wet environments are also important limiting factors, followed by salinity, hail, and wind. To cope with water deficits, farmers have for thousands of years irrigated their fields, which has contributed significantly to higher yields. However, water has become increasingly scarce, and many alternative uses compete with agriculture for it. In addition, water of poor quality causes progressive soil degradation and a consequent loss of overall productivity. Irrigation, therefore, affords only an incomplete solution to the water limitations encountered by plants.

Plant nutrients will probably be less limiting than water in the immediate future because they are more abundant or can be produced in sufficient quantities at an acceptable cost. Supplies of nitrogen, phosphorus, and potassium are likely to be sufficient to support U.S. agriculture for the next 30 to 40 years. However, energy must be used in the manufacture of ammonia (the major source of nitrogen), and this is the largest energy input to the nonirrigated farm. Thus, the cost of energy will be a major constraint on the availability of nitrogen. Similarly, the cost of pesticides will increase with energy costs. Thus, the use of water, certain nutrients, and pesticides will be increasingly restricted, either because resources are limited or for economic reasons. These limitations may be overcome, in part, by the use of plants with lower requirements for these resources, a key aspect of the potential of genetic engineering in combination with other methods for crop improvement.

The Mechanisms by Which Plants Cope with Adverse Environments Have Only Begun to Be Understood in Molecular Detail

Unfavorable environmental conditions depress the rate of photosynthesis, and we are starting to understand the mechanisms by which this process occurs. Such knowledge will be highly applicable for the improvement of crop performance. The expansion of cell walls during plant growth is also affected by unfavorable environmental conditions. These walls contain cellulose as reinforcing microfibrils embedded in a carbohydrate and protein matrix that can flow in a plastic manner. The large pressures inside the cell, which are generated by osmotic forces, can cause plastic deformation of the wall and cell enlargement. The orientation of the cellulose microfibrils determines the direction of growth. Plant cell enlargement is extremely sensitive to certain environmental conditions and is retarded by low temperatures and drought. Such adverse conditions, for example, can cause seeds to fail to germinate or flowers to fail to open. The changes in molecular architecture of the cell wall that take place under such circumstances are largely unknown, as are the roles of water transport and plant hormones. Understanding these factors more completely bears directly on our ability to improve agricultural yield and quality.

The mechanisms of inorganic ion accumulation by plants also constitute a critical area for investigation. Plants differ genetically in their ability to accumulate nutrients—especially nitrogen, phosphorus, and iron—from a given kind of soil, but the molecular bases of these differences are poorly understood. By manipulating these features, performance would be improved.

Plants vary in their ability to withstand freezing temperatures; for example, some plants have developed a way to keep water unfrozen in cells at temperatures as low as -40°C. This ability permits some kinds of trees and shrubs to survive the extreme freezing temperatures that occur seasonally at high latitudes and high altitudes, but we do not understand its structural and molecular underpinnings at all. The reproductive structures of plants are characteristically more susceptible to low temperatures than their stems and leaves, but, again, we do not understand the mechanisms involved.

Many tropical and subtropical plants die when temperatures drop below 12°C, but some, such as cotton, can become acclimated to such chilling conditions. Acclimation is accompanied by a change in the phospholipid composition of the outer membranes of the cells and probably includes similar changes in the energy-transducing membranes of the mitochondria. Energy metabolism seems to play an essential role in the breakdown of cellular functions in cold-sensitive plants. Many of the storage problems of fruits and vegetables can be traced to the breakdown of membranes and the derangement of energy metabolism that occur at these temperatures. The biochemical basis for chilling resistance and acclimation needs to be established much more firmly to form a basis for improving the ability of subtropical plants to resist cool temperatures.

A better understanding of water transport in plants can likewise improve crop performance. The transport of water through the vascular system occurs under great tension (negative hydrostatic pressures), and the continuity of the water pathways is sometimes broken—an abrupt event that seems to be caused by cavitation of water under tension. An embolism that forms in the vascular tissue blocks further transport in that section of the system. Modern methods of electronic analysis indicate that such events occur frequently and are influenced by vascular architecture. Knowledge of how to keep the vascular pathways intact and filled with water is an important need.

Accurate studies of plant biology demand access to controlled environments. Growth chambers and similar facilities permit the efficient evaluation of factors affecting growth of plants throughout their entire life cycles. In addition, tissue culture and seedling culture systems provide convenient ways to study problems of plant growth. Such systems provide opportunities to explore how limiting water affects the growth of roots and shoots and allows the use of biophysical methods, growth regulators, genetic mutants, and molecular genetics to explore some of the reasons for altered development. Tissue culture systems likewise permit experimentation under controlled conditions. They have the additional advantages that metabolites can be supplied in the culture medium and that selection pressures can be created to identify desired genotypes at the cellular level.

Taken together, these research areas illustrate some of the ways a better understanding of plant growth could improve agricultural productivity. In principle, most of the features we have discussed should have a genetic basis. Thus, selection for more efficient water use and nutrient acquisition, as well as for the ability to avoid toxic ions, should help produce plants able to withstand unfavorable environments. The genetic and molecular mechanisms of plant resistance to disease and insect attack are also becoming known. Pest organisms are not only responsible for crop loss in the field, but also for a large amount of loss during storage. In these areas, as in many others, an improved understanding of the ways plants grow and develop will enhance our ability to produce better crops.

PHOTOSYNTHESIS

A Better Understanding of Photosynthesis Is Crucial for Our Future

Plants, like all organisms, depend on the products of photosynthesis for their growth (Figure 11-1). The accumulation of plant biomass is a measure of the plant's total photosynthesis less the respiratory losses that have occurred during its growth. Crop productivity is linked to the seasonal photosynthetic performance of the crop canopies. For this reason, knowledge of the relation between productivity and photosynthesis has largely provided the incentive for the broadly based research effort into this elementary plant process.

Advances in Photosynthesis Research Utilize the Full Range of Modern Biological Approaches from Biophysics to Molecular Biology

Tremendous strides have been made in gathering information about the catalytic components of photosynthesis at the level of atomic structure. Wide-ranging discoveries have created the opportunity to understand photosynthetic mechanisms at a molecular level. The most significant of these breakthroughs has

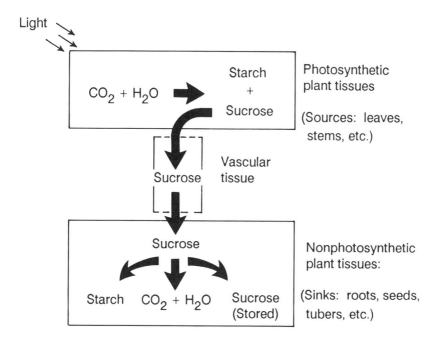

FIGURE 11-1 Diagrammatic illustration of carbon processing in green plants. [B. B. Buchanan, University of California, Berkeley]

been the recent crystallization of the photosynthetic reaction center of the purple bacterium *Rhodopseudomonas viridis* and the determination of its three-dimensional structure by x-ray diffraction analysis (see Figure 3-2). The wealth of existing knowledge concerning the mechanistic features of these complexes, which lends significance to this structural information, calls for corresponding structural work on other catalytic components of photosynthesis. The new structural information contributed immediately to our understanding of the molecular functioning of photosynthetic bacterial reaction centers. The three-dimensional structure along with the biochemical and biophysical information about the various catalytic and redox-active sites (sites of electron transfer) have focused attention on specific regions of the amino acid sequence, which seem to have special significance in light absorption and in charge-transfer processes. In this prokaryotic photosynthetic organism, designed alterations in the genes coding for polypeptides that make up the reaction center are possible and becoming routine. This sort of molecular engineering, coupled with the sophisticated capabilities of molecular spectroscopy and biochemistry, is certain to contribute much to our understanding of photosynthesis.

The crystallization of the *Rhodopseudomonas* reaction center has significance beyond the information obtained from its x-ray structure since it represents a fundamental discovery pertaining to the crystallization of integral membrane proteins. An intensive effort is under way to crystallize other major polypeptide complexes of bacterial and plant photosynthetic membranes.

The development and refinement of numerous other physical techniques are contributing to the revolution in structural information about the catalytic components of photosynthesis. In particular, dynamic information about structural transformations occurring during catalysis, which cannot be obtained from the static picture provided by x-ray analysis, is now becoming available through the use of powerful physical methods. The development of high-resolution nuclear magnetic resonance (NMR) techniques and their application to biology have been particularly successful. For instance, spin-echo NMR techniques allow the selective detection of a small subset of highly mobile, charged amino acid side chains that extend from the protein into the surrounding aqueous environment. The focus of this technique can be narrowed further to those amino acid residues that respond during catalysis; in other words, attention can be focused on the catalytic site as has been done for the chloroplast's coupling-factor enzyme. Even greater detail about the identity and rearrangements of catalytic site groups can come from other NMR techniques, such as double resonance and two-dimensional methods.

NMR is but one example of the array of physical techniques being used to analyze the structural basis of photosynthetic reaction mechanisms. Neutron scattering, electron scattering and electron microscopy, linear and circular dichroism, resonance Raman spectroscopy, Fourier transform infrared spectroscopy, extended x-ray absorption fine structure, and electron paramagnetic resonance spectroscopy are all contributing to the accumulating wealth of information about the molecular structure of the photosynthetic apparatus.

The amino acid sequence of photosynthetic membrane polypeptides has recently been determined from the nucleotide sequence of the corresponding genes. This advance, in turn, has permitted estimation of the two-dimensional folding patterns of these proteins by hydropathy analysis. This information has been taken into account in the most recent models of electron transfer through the complex.

Much Has Been Learned About Regulatory Mechanisms in Photosynthesis, but Much Remains to Be Done

With our current knowledge about the component processes of photosynthesis, it has become possible to investigate specific questions about their interdependence. The most important mechanism in the regulation of chloroplast processes is light activation, a central feature that coordinates the light-driven reactions with the so-called dark reactions of photosynthesis. Light is absorbed by chlorophyll and is converted to regulatory signals that modulate the activity of selected enzymes. Such regulation is essential because enzymes that degrade carbohydrates coexist in chloroplasts with enzymes of carbohydrate synthesis. Some biosynthetic enzymes are activated by light, whereas degradative enzymes are deactivated by light. In this way, the concurrent functioning of pathways that operate in opposing directions (futile cycling) is minimized and the efficiency of temporally disparate metabolic processes is maximized.

A number of soluble enzymes of photosynthetic carbon dioxide assimilation and other biosynthetic pathways show a similar activation response to light. Light regulates specific enzymes through a number of complementary mechanisms that have been identified during the past few years. These include changes in the concentration of certain ions, the concentration of regulatory metabolites, and the oxidation state of thiol groups (-SH) on key regulatory enzymes. Important in such thiol changes is the ferredoxin-thioredoxin system, a system in which thioredoxins—small regulatory proteins—are reduced in the light by the photosynthetic apparatus. The reduced thioredoxins, in turn, reduce and thereby activate selected target enzymes. In this way, the cell can adjust flux through the metabolic pathways associated with oxygenic photosynthesis in accordance with energy and metabolite status.

The catalytic activity of ribulose bisphosphate carboxylase/oxygenase (rubisco) is also modulated by light. Rubisco, which is the most abundant enzyme in the biosphere, performs the carboxylation reaction, the basis for photosynthetic carbohydrate production. Studies on the mechanism of rubisco activation have taken an unexpected turn with the recent discovery of the involvement of a polypeptide dubbed "activase." In certain plants, a newly identified inhibitor, 2'-carboxyarabinitol-1-phosphate, turns off the enzyme at night. The mechanisms controlling the formation of this inhibitor and the mode of regulation by the activase are currently under active investigation.

During the past years, much progress has been made in understanding the regulation of sucrose production in plants. Sucrose is the mobile form of energy that most plants form for transport to photosynthetic sinks; for example, to storage organs such as tubers and seeds that are the source of most of the world's food. During photosynthesis, chloroplasts convert carbon dioxide, water, and phosphate to triose phosphates, which migrate to the cytosol and combine to form sucrose. Photosynthesis requires inorganic phosphate, which is released during sucrose synthesis; therefore, photosynthesis and sucrose synthesis must be closely coordinated. There is evidence that this coordination is provided in part by phosphate. Recently, a second compound specifically serving this function has been identified. Fructose-2,6-bisphosphate coordinates the metabolism of sucrose and starch and, in so doing, links metabolic processes of the chloroplast with those of the cytosol. Recent results suggest that fructose-2,6-bisphosphate may also coordinate cytosolic and amyloplast (a starch containing plastid) metabolism in sink tissues.

Our Current Knowledge of the Biochemistry and Physiology of Photosynthesis Has Made It Possible to Study the Process in Whole Plants or Intact Tissues

Studies on photosynthesis have important applications to the improvement of agricultural productivity. Low temperatures; drought; photoinhibition; the accumulations of herbicides, pesticides, or fertilizers; and pollution are examples of frequently encountered conditions that compromise the efficiency of production in crops. Impaired photosynthesis is a major contributor to these losses, and we need to understand the mechanisms by which it takes place, something we are now in a position to do.

One area poised for major advances is the application of recently developed and adapted physical techniques to investigate component processes of photosynthesis in situ. Techniques such as kinetic absorption spectroscopy, delayed light imaging (Figure 11-2), NMR spectroscopy, flash fluorescence, and photoacoustic spectroscopy are now applied to diagnostic studies of how particular environmental conditions may influence intersystem electron transfer, adenosine triphosphate formation and consumption, enzyme activation, light regulation, photosynthetic reaction center activity, and the transfer of light energy. The usefulness of these techniques depends on an underlying experimental basis for interpreting the often complex results obtained from in situ measurements. This fact points to the need for expanding this information base.

The development of "model" plant systems will unquestionably contribute to the solution of problems in photosynthesis related to agriculture. A notable recent advance has been the development of vigorous photoautotrophic cultured cell lines. Because of the difficulty of growing plants to maturity under heterotrophic

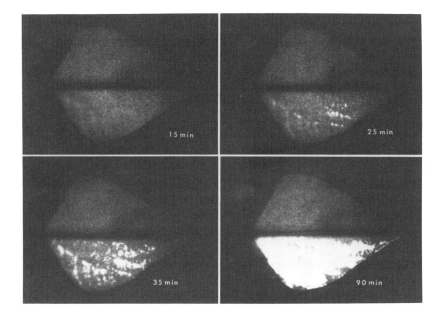

FIGURE 11-2 Delayed light imaging in herbicide-treated bean leaves. Leaves normally emit a tiny portion of the photosynthetically active radiation that they absorb. Defects in certain chloroplast processes increase the amount of light that is emitted. This feature of photosynthesis has been exploited by using the increase in light emission to reveal the photosynthetic performance of different portions of leaves under stress. The four phytoluminographs depict the spatial distribution of delayed light emission in a red kidney bean leaf after the lower half of the leaf blade was sprayed with an herbicide. The herbicide was applied 5 minutes before illumination; the notations refer to the length of time the leaf had been illuminated. The herbicide used inhibits the enzyme glutamine synthetase, and its action results in the accumulation of ammonium ions, which is the most likely metabolic cause of delayed light emission. [Donald R. Ort, University of Illinois]

conditions, the screening and selection of photosynthetic mutants is generally limited to positive selection methods. The advent of these photoautotrophic cell lines and the promise they hold for expanding the use of mutants in photosynthesis research emphasizes the need for substantial effort aimed at developing reliable plant regeneration procedures. It is becoming increasingly evident that cyanobacteria represent a highly useful model system for chloroplasts. The progress that has been made in developing a genetic transformation system for cyanobacteria highlights their potential for the investigation of photosynthetic processes and events generally. Among plants, *Arabidopsis* and *Petunia* are genetically tractable, and they lend themselves particularly well to being modified through genetic engineering. They offer many approaches to long-standing agricultural problems with new research strategies having potential far beyond what could be imagined just a few years ago.

Another area with potential for advances concerns the central enzyme of the carbon reduction cycle, rubisco. Competition by molecular oxygen for the carboxylation substrate at the catalytic site of this enzyme considerably lowers the efficiency of photosynthesis. Since the discovery of natural interspecific variation in the severity of the competition between carbon dioxide and oxygen, the possibility of substantially reducing the enzyme's oxygenase activity, perhaps to negligible levels, has been recognized. Accumulating information about the catalytic site and its reaction mechanism, coupled with the ability to make designed alterations in the gene, is a promising approach toward elucidating the molecular factors that control the discrimination between CO_2 and O_2. The evolution of rubisco in an ancestral anaerobic atmosphere rich in CO_2 produced an enzyme with a flaw that now burdens plants living in CO_2-poor, oxygenic atmospheres. Molecular genetics, guided by an in-depth understanding of the molecular basis for the competition between O_2 and CO_2, may enable scientists to design a more efficient enzyme. Genetic transformation of the plant gene coding for the catalytic subunit of rubisco appears now to have gone beyond cloning in bacteria. Transfer of the chloroplast gene into the nucleus and addition of a chloroplast-targeting sequence to the protein is a major step toward producing plants with an "engineered" rubisco gene, which will yield a more efficient enzyme.

Photosynthesis depends on processes occurring elsewhere in the plant. In particular, the developing portions of the plant and specialized storage organs, which are considered "sinks" for photosynthate, exert a poorly understood control on processes that occur in the chloroplasts. Basic information is lacking about the mechanisms that control the development of photosynthate sinks and that determine the priorities of individual sinks for available photosynthate. Little is known about how the various chemical forms of photosynthate cross cellular and organellar boundaries in either source or sink tissue. In many cases, it is not even known whether specific transporters are involved. In contrast, excellent progress has been made in discovering the mechanistic basis for carbohydrate transport across bacterial membranes. Much of this recent success has been fueled by the application of elegant new techniques of immunology and molecular biology. The approaches and mechanistic principles established by this pioneering work in bacterial carbohydrate transport will have a great impact on research in photosynthate transport and partitioning.

The diverse disciplines of photosynthesis research are beginning to converge in a meaningful and synergistic fashion. As a consequence, the prospects for applying what has been learned about photosynthesis to problems relevant to agriculture and the prospects for seminal discoveries about photosynthesis have never been better.

NITROGEN FIXATION

Nitrogen, Which Is Abundant in the Atmosphere, Is Essential for All Organisms

Even though nitrogen constitutes some 80 percent of our atmosphere, it is relatively difficult for organisms to obtain. Since it is an essential constituent of proteins, nucleic acids, many enzymic cofactors, and other essential metabolites, nitrogen is required by all organisms, which obtain it primarily as a result of the nitrogen-fixing abilities of a very few kinds of bacteria. These bacteria convert nitrogen from its gaseous form (N_2) into ammonia (NH_3), which can be used by other organisms. Because it is so scarce in an appropriate form, nitrogen deficiency is a common limiting factor in the growth of plants, animals, and microorganisms. Biological nitrogen fixation in bacteria, in which nitrogen is converted to ammonia catalytically by the enzyme nitrogenase, has been studied intensively as a biological process of considerable fundamental interest and of potentially substantial energy savings through the use of less nitrogen fertilizer.

Research on nitrogen fixation is carried out at levels of biological organization ranging from ecology to molecular biology. The biochemistry and molecular genetics of nitrogen fixation have been greatly advanced by studies on a model organism, *Klebsiella pneumoniae*, whose relationship to the common colon bacterium *Escherichia coli* allows the application of many sophisticated techniques that have been developed for use with its extensively studied relative. Other bacteria have also been important as experimental material for investigations on how nitrogen fixation functions in various ecological niches and takes place in connection with a number of different biochemical strategies. Among the achievements of the past few years have been the identification of all the genes required for nitrogen fixation (*nif*) in *Klebsiella* and the demonstration of function for several of them. Among these genes are the three coding for the enzyme nitrogenase and several whose products are important for combining nitrogenase with its molybdenum-iron cofactor and for the delivery of electrons to the enzyme. The energetics of nitrogen fixation is an important concern if this process is to find new agricultural uses. It is being explored in several systems, including complex symbiotic associations such as those that involve the nodule-forming bacterium *Rhizobium*, which lives on the roots of the plant family Fabaceae, the legumes.

Several major advances have been made in understanding how nitrogen fixation is regulated. These findings tie in with new understanding of the overall regulation of nitrogen metabolism. Specifically, cells with adequate nitrogen reserves do not fix nitrogen because *nif* genes are not transcribed. Their activation requires the general nitrogen regulatory system Ntr to induce the expression of a nitrogen-fixation-specific activation system (Nif). After activation, the gene product of *nifA* then induces transcription of all the other *nif* genes. It has been found that the Nif regulatory system is evolutionarily related to the Ntr regulatory system, which also controls genes responsible for ammonium assimilation and amino acid catabolism. Furthermore, both *nif* and *ntr* genes have specialized promoters whose nucleotide sequences differ from the promoter sequences of most prokaryotic genes. This information is significant with regard to our concepts concerning the regulation of transcription in bacteria and the regulation of numerous physiological systems.

Another major achievement has been the demonstration that gene expression of the nitrogenase loci (*nifHDK*) in the filamentous cyanobacterium *Anabaena* involves the rearrangement of the DNA itself. In vegetative photosynthetic cells, the *nifHD* and *nifK* genes are separated in the genome. When cells differentiate to become nitrogen-fixing heterocysts, the DNA is rearranged to align *nifHDK* as a continuous operon, as in *Klebsiella*. Our ability to understand the cyanobacterial system has been revolutionized recently by the development of techniques for genetic conjugation in these bacteria, which have made possible experimentation on genes and their expression. Unicellular cyanobacteria that show temporal separation between oxygen-producing photosynthesis and oxygen-sensitive nitrogen fixation are ideal models for the possible compatibility of nitrogen fixation and photosynthesis in plants in the absence of symbiosis.

The most extensively studied association between plants and nitrogen-fixing microorganisms is that of *Rhizobium* and its legume hosts, which include such important crop plants as alfalfa, soybean, peanut, vetch, cowpea, beans, peas, and clover, as well as a number of important tropical timber trees, the winged bean, and the "miracle tree," *Leucaena*, now being used to vegetate large areas in the Asian and Pacific tropics and as a ready source of fuel. The family Fabaceae consists of some 18,000 species of plants; because of their ability to grow in relatively infertile soils, they are often locally prominent in vegetation. Colonies of *Rhizobium* form nodules on the roots of legumes and live within them, where both the presence of the bacteria and the structure and biochemistry of the nodules play key roles in the process of nitrogen fixation. This association is the most important single contributor to the supply of nitrogen on earth that is available for biological reactions.

Substantial advances have been made in understanding the genetics of the *Rhizobium*-legume nitrogen-fixing system during the past decade. The *nif* genes of *Rhizobium meliloti* were identified by their DNA homology to the cloned *nifHDK* of *Klebsiella*. Subsequently, their functionality was proven by a site-directed gene replacement technique, which has since become indispensable for the genetic manipulation of *Rhizobium*, *Agrobacterium*, and other bacteria associated with plants. Other genes important for host recognition, formation of nodules, and efficiency of nitrogen fixation have been identified, cloned, and analyzed. In most species studied so far, these genes lie on large native plasmids. An exception appears to be the genes for symbiosis and nitrogen fixation in the *Bradyrhizobium* (slow-growing *Rhizobium*) strains, which nodulate soybean, peanut, cowpea, and other legumes.

A new view of what happens in the rhizosphere as soil microbes such as *Rhizobium* encounter their legume hosts has emerged from studies of nodulation gene expression. The *nod* genes of *Rhizobium* are required for recognition and invasion of the plant hosts and for nodule formation on their roots. As such, they appear to be the earliest-acting genes in the legume-*Rhizobium* association. These genes are not transcribed by *Rhizobium* cells grown in pure culture; their expression is activated in the presence of host plants, indicating that a signal is sent from the host to the bacteria.

Legumes themselves play an important role in the symbiotic fixation of nitrogen by *Rhizobium* bacteria. Certain proteins, which are produced only in nodules and not in uninfected roots, seem to be essential for nodule function. One of these is leghemoglobin, an oxygen-binding protein that helps to protect the oxygen-sensitive enzyme nitrogenase. Soybean leghemoglobin genes have been cloned, and their synthesis is controlled at the transcriptional level by an unknown signal from the bacterium. The primary amino acid sequence of plant leghemoglobin strikingly resembles that of animal myoglobin. Whether this similarity reflects common descent is not known, but nitrogen-fixation is an ancient process, still carried out under the anaerobic condition in which life first evolved. Other nodule proteins, called nodulins, appear at specific times during infection; their synthesis is also regulated at the transcriptional level. Further investigations of the loci encoding nodulins may help us to understand how nodules form and function and why legumes are appropriate hosts for *Rhizobium*, whereas other plants are not.

Many Questions Concerning the Physiology, Biochemistry, and Molecular Biology of Nitrogen Fixation and Symbiosis Are Still to Be Answered

The active site of nitrogenase and the mode of catalysis need to be elucidated. Why is nitrogen fixation coupled to hydrogen evolution? What is the basis for the oxygen sensitivity of the enzyme? What is the structure of the iron-molybdenum cofactor and what is its relationship to the active site? Understanding the mechanism by which nitrogenase reduces nitrogen may enable us to synthesize catalysts that will perform this process more efficiently. The symbiotic relationship between plants and nitrogen-fixing organisms also raises questions. What determines the host range of symbionts, and why do not all plants form symbioses? What signals pass between bacteria and plants, and how do these signals regulate gene expression in each of them? What metabolic exchanges occur between the symbiotic bacteria and the plant? Why do nitrogen-fixing microorganisms export their fixed nitrogen? What are the energetic costs of symbiotic nitrogen fixation at the cellular, organismal, and ecological levels? More research is also required on the symbiosis between plants and nitrogen-fixing organisms other than *Rhizobia*. Another genus of bacteria, an actinomycete of the genus *Frankia*, commonly forms root nodules within which nitrogen fixation occurs with certain plants other

MESSENGER MOLECULES IN BACTERIAL-PLANT INTERACTIONS

Soil bacteria interact with plants in a variety of ways. Some establish beneficial symbiotic relationships with specific hosts, whereas others invade the plants and cause pathological tumors to form. The successful infection of the plant requires the host to be recognized and genes in both the plant and the microorganism to be activated. Recent discoveries demonstrate that plants release low-molecular-weight organic compounds that activate microbial genes whose products are needed to infect plants.

Bacteria of the genus *Rhizobium* invade the roots of leguminous plants, where they cause root nodules to form. Within these nodules, rhizobia bacteria fix atmospheric nitrogen, which is then used by the plant as an important nutrient. Exudates from alfalfa roots activate a set of genes in *Rhizobium meliloti*, whose expression stimulates the earliest detectable host responses, consisting of root-hair curling and cortical cell divisions. Plant scientists have identified compounds in the exudate of alfalfa roots that induce nodulation genes in *R. meliloti*. These signaling molecules, of which luteolin is the most active, are flavonoids—secondary metabolic products found in virtually all plants. The important flower-coloring pigments known as anthocyanins, which are responsible for most of the reds and blues seen in flowers and leaves, constitute one group of flavonoids. Other flavonoids may have roles in food-choice preference by insects, as blocks to ultraviolet radiation, or possibly in protecting plants from pathogens.

The crown-gall bacterium *Agrobacterium tumefaciens* infects plants through wounds and causes the formation of tumors. Tumorous growth is based on the integration of a specific segment of bacterial DNA into the genome of the plant. For transformation to occur, a set of virulence genes has to be activated in the bacterium. Scientists found that virulence genes are activated by phenolic compounds that are present only in the exudate of wounded, metabolically active cells. Thus, injury to the plant tissue not only provides a portal of entry for the bacterium, but also is necessary for the production of the signaling molecules that activate the infection process.

than legumes, such as *Ceanothus*, *Myrica*, and *Alnus*. We need to understand these interactions better and to determine how they resemble and differ from the better-known one between *Rhizobium* and legumes. Our understanding of the biology of free-living nitrogen-fixing bacteria, some of which are photosynthetic and some not, also contains gaps. How do they solve the problems of protection against oxygen and generation of energy? Can such organisms provide solutions to the limitations in agronomically important nitrogen fixation? Work on nitrogen fixation not only advances our knowledge of a complex process that has great economic impact, but also answers basic questions of biology, such as classical questions concerning the nature of symbiosis. More detailed studies of plant-bacterial associations may point the way to an improved understanding of the functions of plant cells and their components, just as the study of bacterial and animal interactions has revealed fundamental aspects of both bacterial and animal cells.

PLANT GROWTH AND DEVELOPMENT

Plants Have an Open System of Growth in Which the Role of a Very Few Kinds of Plant Hormones Is of Critical Importance

Many developmental processes in plants are regulated by a relatively small number of substances called plant hormones. In addition, environmental cues, such as the duration of the daily light and dark period or the ambient temperature, help synchronize the life cycle of plants with the changing seasons. In at least some instances, environmental effects on plant development are mediated by hormonal factors. An understanding of these regulatory mechanisms is needed if one is to optimize the growth of crop plants.

Plant Hormones

Plant Hormones Have Complex and Often Overlapping Functions

The five known groups of plant hormones are auxins, gibberellins, cytokinins, abscisic acid, and ethylene. These substances often fulfill similar functions. For example, auxins, gibberellins, and cytokinins all induce cell division in different tissues. In addition, auxins and gibberellins both regulate cell elongation, although they probably do so by different mechanisms. Each plant hormone shows a wide spectrum of activities and affects different processes. Ethylene, for example, induces fruit to ripen, flowers to fade, stems of semiaquatic plants to elongate rapidly (for example, in rice growing in deep water), and bromeliads to flower (such as pineapple). The specificity of action of plant hormones is determined by the chemical structure of the compound and by the nature of the target tissue. In some instances, the increased synthesis of a plant hormone initiates a new developmental process. In other cases, the responsiveness of the plant to a given concentration of hormone changes under different conditions of growth.

In recent years, progress has been made in understanding the biosynthesis of plant hormones, most notably that of gibberellins and ethylene. The pathway of gibberellin biosynthesis has been elucidated with a variety of techniques. Enzymological work and the application of radiotracer technology led to the identification of gibberellin intermediates. The availability of so-called growth retardants—compounds that inhibit specific enzymes of gibberellin biosynthesis—has been of great help in isolating gibberellin precursors. Equally important was the use of well-characterized dwarf mutants, which are impaired at different steps of gibberellin biosynthesis. A combination of genetic and biochemical work has brought order into the confusingly large array of different gibberellins; more than 70 kinds are known in plants and in the fungus *Gibberella fujikuroi*. Probably only one of these, however, actively controls shoot elongation in most plants. Other gibberellins are either hormone precursors or inactive metabolites.

Ethylene, the simplest unsaturated hydrocarbon, hardly conforms to our chemical concepts of a hormone; yet it regulates, at extremely low concentrations, a number of key developmental plant processes. Some of these such as fruit ripening are of considerable agronomic importance. Much effort has been invested in elucidating the pathway of ethylene biosynthesis in the hope that control of this process will lead, for example, to extended storage life for perishable agricultural products. Although the enzyme whose activity determines the level of ethylene biosynthesis in most plants is present at vanishingly low levels, even in ripening fruit, it has now been purified, and monoclonal antibodies against it are available.

The genes responsible for auxin (indoleacetic acid) and cytokinin biosynthesis have been isolated from the plant pathogenic bacteria *Agrobacterium tumefaciens* and *Pseudomonas savastanoi*. The cytokinin gene encodes an enzyme that is similar to an enzyme that has been isolated from plants. The bacterial genes for indoleacetic acid synthesis encode two enzymes, a tryptophan monooxygenase and indoleacetamide hydrolase. Auxin biosynthesis in plants is mediated by different enzymes.

Abscisic acid plays an important role in the water relations of plants. When the water supply becomes limiting, a rapid increase in abscisic acid is, at least in part, responsible for the closure of the stomata in the leaves and other green parts of the plant and, as a consequence, for the reduction in the rate of transpiration. The pathway of abscisic acid biosynthesis has not yet been elucidated.

In contrast to the progress made in the elucidation of plant hormone biosynthesis, relatively few advances have occurred in our understanding of the mode of action of these substances. A few exceptions can be mentioned, however. In cereal grains, starch and other reserves are mobilized during germination. This process is initiated by the secretion of gibberellin into the aleurone layer of the seed. There, the hormone induces the synthesis of hydrolytic enzymes, most notably that of α-amylase. From the aleurone cells, hydrolases are secreted into the endosperm, where they break down stored food reserves, for example, starch. The induction of α-amylase by gibberellin is based on enhanced transcription of genes encoding this enzyme.

In the case of the aleurone system, the mechanism of hormone action could be approached successfully because the biochemical response was well defined.

Most other hormonally regulated processes in plants are more complex and, therefore, less tractable. Generally, we know too little about the biochemical reactions underlying particular developmental phenomena, such as growth. It has been known for many years that auxin promotes cell elongation by increasing the plasticity of the cell wall. As a result, water enters the cell, and the hydrostatic pressure extends the wall. A number of polysaccharide and proteinaceous components of the cell wall have been characterized, but their interconnections are only partly understood, and a detailed picture of cell wall architecture is missing. For this reason, it is not clear which bonds have to be broken for the cell wall to loosen and whether this is achieved enzymatically or through the action of protons that are secreted into the cell wall. Molecular biology has permitted scientists to bypass the existing gap of biochemical knowledge and to isolate genes that are activated within minutes as a result of auxin treatment. Study of such hormonally regulated genes may be rewarding, and their hormone-responsive regulatory elements can be identified.

Almost nothing is known about the site of action of plant hormones. Binding proteins have been described for all plant hormones, but no receptor function has been established for any of them. In no instance has it been possible to connect hormone binding and a hormonally regulated biochemical response. Not even in the aleurone system has it been possible to initiate, in vitro, the transcription of hormonally regulated genes.

The Chain of Events from the Initial Interaction of Plant Hormones with Their Receptors to the Manifestation of the Response Must Be Established

Agriculture has benefited greatly from the use of plant hormones and synthetic growth regulators. The first selective herbicides, for example, were synthetic auxins. Growth retardants, which inhibit gibberellin biosynthesis, have been used extensively to stunt the growth of wheat and, thereby, to reduce losses caused by lodging (collapse of the wheat stem from excessive height). Practical applications for plant growth regulators have often been found empirically. The targeted use of plant hormones and synthetic plant growth regulators requires detailed knowledge of their mode of action. In most instances, the response is well characterized at the physiological level. What is urgently needed is the identification of plant hormone receptors and the elucidation of the primary biochemical reactions that underlie the physiological response. Just as mutants blocked in hormone production have helped to establish the pathway of hormone biosynthesis, so can mutants blocked in their response to plant hormones help us to identify hormone receptors and components of the hormonal transduction chain. Isolation of genes whose transcription is regulated by plant hormones will also advance our knowledge of the mechanism of hormone action, especially when the gene products are identified.

Plant enzymes mediating gibberellin and ethylene biosynthesis have been described in recent years. The prospects are good that genes encoding key enzymes in these pathways will be isolated and characterized in the near future. Similar progress has yet to be made in the elucidation of abscisic acid and auxin biosynthesis.

Environment

Environmental Factors Play a Key Role in Plant Development

Since plants are sessile organisms, they must adjust their life cycles to the annual changes in the environment. The timing of such events as seed germination, flowering, the onset of dormancy, and the breaking of dormancy has to be coordinated with the seasons of the year. Plants achieve this coordination by measuring the duration of day and night length and the time over which they are exposed to low temperatures.

When a seedling emerges from the soil and is exposed to light, the growth pattern and the metabolic activities of the plant change completely. The rate of stem elongation is reduced, the leaves unfold, and the photosynthetic apparatus differentiates. These changes are all controlled by phytochrome, the best characterized regulatory photoreceptor. Phytochrome is a protein that occurs in two forms, a red- and a far-red-light absorbing one. Red light of 660 nanometers activates phytochrome by switching it to the far-red-absorbing form. Far-red light (730 nm) converts phytochrome back to its original form and cancels the effect of the initial red illumination.

In many plants, the time of flowering is determined photoperiodically, by the relative length of the daily period of light and dark. This ensures seed production at the proper time of the summer or fall. Photoperiodic induction also prepares perennial plants for the advent of winter. As the nights get longer, buds become dormant, leaves abscise, and the plant acquires cold hardiness. In one instance, the enhanced growth of spinach under long days, the biochemical basis for photoperiodic induction has been elucidated. The greatly increased rate of growth reflects the enhanced activities of two enzymes in the pathway of gibberellin biosynthesis. These photoperiodic processes are under phytochrome control.

Much has been learned in recent years about the phytochrome molecule in terms of its spectral, physicochemical, and immunochemical properties. The gene encoding phytochrome has been cloned and sequenced, and it has been shown that phytochrome controls the expression of its own gene through a feedback mechanism. Phytochrome also regulates the expression of other genes.

In addition to phytochrome, plants contain at least one other pigment that regulates developmental processes, the blue-light photoreceptor. This pigment mediates phototropism and resembles a pigment with analogous functions in fungi. It has not yet been isolated, and the chemical structure of its chromophore has not been determined.

The cold temperatures of winter are often used as an environmental cue for the initiation of developmental processes that take place in spring or early summer. Dormancy in many plant species is broken after exposure to a critical number of cold days; flowering of some plants will occur only if they have experienced a cold period of a certain duration (vernalization). Even though these responses are well characterized at the physiological level, nearly nothing is known about the mechanism of cold perception and the biochemical reactions that underlie the breaking of dormancy or vernalization.

Much Research Is Yet to Be Done on the Effect of Environmental Factors on Plant Development

The perception of nonphotosynthetic light, of cold temperature, and of gravity permits plants to orient themselves in time and space. Much progress has been made recently in research on phytochrome, a pigment of central importance in light perception. Despite this progress, little is known about the transduction of the red-light stimulus perceived by this pigment. What chain of biochemical reactions is set into motion by the activation of the photoreceptor? Because of the central role of phytochrome in the control of many plant processes, research on its mode of action is of prime importance.

Tropic responses to light and gravity probably have a number of reactions in common. How does a plant determine the direction of light and gravity, and how does it orient its growth toward or away from these stimuli? A wealth of knowledge dates back to Darwin on tropic phenomena in plants. However, new approaches are needed if one is to understand, in molecular terms, the mechanisms that govern such responses. Current work with photo- or geotropic mutants of *Arabidopsis thaliana*, a plant with an exceptionally short life cycle, may lead to identification of the blue-light photoreceptor pigment, of gravity sensors, and of biochemical reactions that underlie the tropic response.

The problem of how plants measure temperature, how they determine the duration of the cold period, and how they translate this information into developmental responses requires renewed research efforts. These questions are among the most difficult ones in plant biology because basic concepts, on which testable hypotheses can be built, are largely lacking. Precisely because of this gap in our knowledge, work in this area may be particularly rewarding.

Plant Reproduction

Many Aspects of Plant Reproduction Are Now Amenable to Detailed Analysis

Most crop plants are grown for their seeds and fruits. Understanding the biology of plant reproduction, including flowering, fertilization, and the development of fruits and seeds, is therefore of great economic importance. The production of hybrid plants from inbred parents is an important aspect of reproductive plant biology. Such hybrids often produce substantially higher yields than do the inbred parental lines. Growth of hybrid plants is possible only when self-fertilization is excluded. Reproductive self-incompatibility and cytoplasmic male sterilty are the best known mechanisms to prevent inbreeding in plants. Both are processes of great inherent scientific interest that remain poorly understood at the molecular level despite recent advances.

Genetic Self-Incompatibility Precludes Self-Fertilization in Bisexual Plants

Genetic self-incompatibility, which is widespread among plant species in nature, has been known to plant geneticists for a century. Several mechanisms for self-incompatibility exist; in gametophytic incompatibility, a sperm with a particular haploid S genotype (S is the incompatibility locus) is unable to fertilize an egg having the same allele. Another mechanism, sporophytic incompatibility, is determined by the diploid genotype of a parent plant. The tissues of this plant, including those of its style (part of the flower holding the stigma), will contain two alleles at the S locus, and pollen containing either of these will fail to germinate on the stigma of that plant. How does identity at genetic loci lead to the rejection of a germinating pollen grain? This question can now be approached with new tools, thanks to the identification of the genes responsible for self-incompatibility reactions in tobacco and in mustard.

Cytoplasmic Male Sterility, Which Causes Bisexual Plants to Serve as Female Parents Only, Is Mainly Controlled by Mitochondrial Genes

Cytoplasmic male sterility (CMS) is the basis for the production of hybrids with increased vigor in such important crop plants as corn and sorghum. More than 140 plant species have genes for CMS. Such plants do not produce viable pollen, a trait that is inherited in a non-Mendelian fashion (uniparental inheritance). Substantial evidence now indicates that the CMS trait is encoded by mitochondrial genes in maize, petunia, and sorghum. CMS is probably associated with mitochondrial genes in other plant species as well, although chloroplast genes and viruses cannot be discounted as the cause of CMS in some instances. The CMS trait can be suppressed by nuclear genes known as restorer genes. In the presence of restorer genes, male-sterile cytoplasms are restored to pollen fertility.

In maize, the mitochondrial gene responsible for the *cms*-T type of sterility has been isolated. This gene codes for a polypeptide of molecular mass 13,000 daltons (13 kD), which is located in the inner mitochondrial membrane. The origin of this gene is unusual; it has arisen by a series of recombinational events that have placed its coding sequence behind a mitochondrial promoter. Moreover, this gene is unique in that the gene and its product are not found in other maize cytoplasms or, for that matter, in other plant species. Although the function of the 13-kD protein is unknown, its location in the inner mitochondrial membrane

THE MOLECULAR ANALYSIS OF GENETIC SELF-INCOMPATIBILITY

Genetic self-incompatibility is known in some species of most families of flowering plants and doubtless evolved in the earliest members of the group or their ancestors more than 135 million years ago. In such systems, pollen-tube growth is blocked by incompability mechanisms either in the stigma or in the style. In the mustards, cabbages, and their relatives, the genus *Brassica*, the pollen tubes fail to emerge from the pollen grains or are inhibited at the surface of the stigma if the allele at the *S*, or incompatibility, locus, is the same as one of the two alleles at this locus in the stigma. Researchers have found that inhibition occurs in self-incompatible species of *Brassica* within minutes of the initial contact between the pollen or pollen tube and the papillar cells that line the outer surface of the stigma. Incompatible pollen grains usually fail to germinate or, more rarely, germinate to produce pollen tubes that coil at the surface of the papillar cells and fail to penetrate the surface layer of stigma cells.

The stigma of *Brassica* produces *S*-allele-specific glycoproteins, which, on the basis of several criteria, are believed to be the products of the *S* locus. Nucleic acid sequences derived from the self-incompatibility genes were isolated from a complementary DNA library constructed from stigma messenger RNA. These sequences have been used to study the regulation of the expression of the self-incompatibility genes during flower development. The self-incompatibility sequences are expressed in stigma and anther tissue only during a specific period of the developmental process. The technique of in situ hybridization made it possible to determine that these genes are expressed exclusively in the surface papillar cells, the site of first contact with the pollen (Figure 11-3). The nature of allelic variability at the *S* locus is being analyzed by comparing the nucleic acid sequences derived from different *S* genotypes. In this manner, relatively conserved regions of the *S*-allele-specific glycoproteins, as well as highly variable regions, which may determine allelic specificity, have been identified.

Self-incompatibility has already been used extensively in the production of new hybrid strains of kohl, oilseed rape, and other commercially important species of *Brassica*. The manipulation of genetic self-incompatibility is both agriculturally important and of fundamental biological importance; clearly these systems played a significant role in the evolution of the flowering plants, the dominant photosynthetic organisms on land.

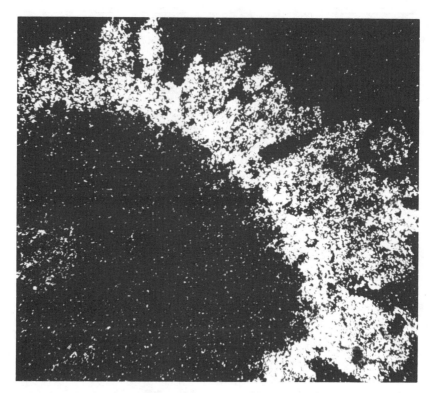

FIGURE 11-3 Expression of the self-incompatibility genes in the papillar cells of the stigma of *Brassica* flowers as shown by in situ hybridization. [June Nasrallah, Cornell University]

suggests that it may impair electron transport or ATP formation. The investigations of *cms*-T have also shed some light on the function of at least one of the nuclear restorer genes. In this case, a restorer gene has been shown to alter the transcription of the gene encoding the 13-kD polypeptide.

Seed-Storage Proteins Are Important in Human Nutrition, but Often Lack Essential Amino Acids

The seeds of certain plants play an important role in human nutrition because of their high content of storage reserves. Some seeds are particularly important because they provide protein as well as calories. However, some of these proteins lack essential amino acids, and people whose diet is based largely on such seeds may experience net amino acid deficiencies. Biochemical studies carried out during the 1960s and 1970s showed that seed-storage proteins are specific to certain stages of embryonic development or to particular embryonic organs and that they are contained within protein storage vacuoles.

The cloning and analysis of the genes for seed-storage proteins revealed that they are encoded by multigene families and that the messenger RNAs for some carry universal signals for sequestering the protein into membrane-bound organelles. Studies of gene expression of storage proteins in transgenic plants have shown that the promoter for embryo-specific gene expression functions across species boundaries and that genes for seed-storage proteins from French bean or soybean are also expressed at the proper developmental time in tobacco seeds. Transcriptional and possibly translational controls for the expression of seed-storage protein genes are influenced both by hormones, such as abscisic acid, and by intrinsic, as yet unidentified developmental signals.

Research on Plant Reproduction Offers Great Potential in Both Applied and Basic Biology

The switch from vegetative to reproductive growth at the shoot apex is the earliest step in flower formation. Physiological experiments have provided strong evidence that photoperiodic induction of flowering is perceived in the leaves and transmitted to the apex by a flowering hormone, often termed florigen. One large gap in our knowledge on the regulation of flowering concerns the nature of this floral stimulus. In many instances, it would be useful to control the time of flowering of crop and horticultural plants. Isolation and chemical identification of the floral stimulus would be a major step toward attaining this goal.

Research on self-incompatibilty in plants is of fundamental as well as applied importance. A major question concerns the biochemical mechanism that operates in the incompatibility reaction. What is the function of the glycoprotein associated with pollen recognition in the stigma or style, and how does it interact with its counterpart in the pollen? Although it is most unlikely that self-incompatibility genes will resemble those of immunoglobin families of animals, it will be intriguing to compare the ways in which the animal and plant kingdoms have generated systems for recognizing self, kin, and foreign cells, permitting common mechanisms to be used in diverse species. Genetic engineering methods could be used to introduce barriers to fertilization in cases in which the production of hybrid progeny may increase yields, and they could also be used to help to remove such barriers when self-pollination would prove advantageous.

Much remains to be learned about CMS. It is not clear how a mitochondrial gene product is involved in pollen development. Several types of CMS and restorer genes have functions that need to be explained. Research on CMS is also relevant to our understanding of susceptibility to certain fungal deseases. A strain of southern corn leaf blight fungus, which destroyed a large part of the corn crop of the United States in 1970, affects only plants that carry the cms-T gene. The basis for the connection between susceptibility and the CMS trait is not yet fully known.

Finally, it is evident that plant mitochondria differ from those of other organisms. Plant mitochondrial genomes are much larger than those of animals;

they are also organized differently and encode additional gene products. Research on mitochondrial functions that are unique to plants offers opportunities for advances in organelle biology.

Continued investigations on seed-storage proteins will help us to understand how external and internal developmental signals regulate gene expression in plants. In addition, such work opens new approaches toward improving the nutritional quality of seeds. It is now possible to correct the amino acid deficiency of seed-storage proteins by altering the gene sequences that encode them. With available transformation systems, such modified protein genes can be replaced into the original plant to complement protein composition there. Since the same controlling sequences for storage-protein gene expression seem to function in distantly related plant species, modified storage-protein genes might also be expressed in the seeds of unrelated species. The consequences of this relation could have considerable economic importance.

PLANT-PATHOGEN INTERACTIONS

Interactions Between Plants and Pathogens Are Biologically Intricate and of Fundamental Scientific and Commercial Interest

Plant pathogens cause serious losses to our major crop plants and have had a substantial impact on society. Even though myriad microbes interact with plants, very few have attained the capacity to cause disease. Susceptibility to invasion by pathogens is the exception rather than the rule in the plant world. This is merely an expression of the highly complex relationship that must be established between host and pathogen in a compatible (susceptible) interaction. Much of the modern research in this area attempts to explain the nature of the signals exchanged between host and potential pathogen and of the genes that control such interactions. Research on these systems has led to exciting new avenues of fundamental inquiry and highly promising results for practical applications.

Crown Gall Is a Disease of Plants That Shares Some of the Properties of Cancer in Animals

Crown gall, a disease of some plants, is caused by a bacterium, *Agrobacterium tumefaciens*, which invades its host through a wound and genetically transforms plant cells into tumorous ones. Bacteria-free tissue from a crown-gall tumor can be cultivated on a synthetic medium and maintained indefinitely in a rapidly proliferating condition. When grafted onto healthy plants, cultured crown-gall tissue produces tumors indistinguishable from those incited by the bacterium. Unlike their untransformed, normal counterparts, tumorous plant cells grown on a synthetic culture medium require no exogenous sources of the growth substances cytokinin and indoleacetic acid. Crown-gall tumor cells have therefore acquired the capacity to produce these growth regulators as a result of their transformation.

Tumorigenicity of the crown-gall bacterium is conferred by genes present on a large plasmid called Ti (Figure 11-4). A fragment of the Ti plasmid, called transfer DNA (T-DNA) is transferred from the pathogen and integrated into a chromosome of the host plant. Genes on the integrated piece of bacterial DNA code for enzymes responsible for the production of the cytokinin isopentenyl adenosine and the auxin indoleacetic acid. Since these genes are expressed only in the plant cell and not in the donor bacterium, their regulatory sequences are designed to function in the eukaryotic environment of the plant cell. Both cytokinins and auxins are natural plant constituents, and their overproduction in the transformed plant cells leads to undifferentiated rapid proliferation characteristic of tumorous growth. T-DNA also contains a gene coding for the synthesis of a novel opine amino acid, such as octopine or nopaline, substances that can serve as nitrogen sources for the bacterium. Their production is also used by scientists to determine whether transformation has occurred. Another region of the Ti

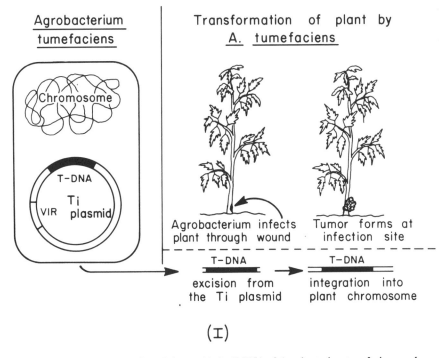

FIGURE 11-4 Transformation of plants with the T-DNA of *Agrobacterium tumefaciens* used as vector. [Tsune Kosuge, University of California, Davis]

plasmid, the *vir* region, contains six genes necessary for the events leading to the integration of the T-DNA into the plant genome. Genes in the *vir* region are activated by messenger molecules that are released by the plant.

In some characteristics, crown gall strikingly resembles a disease on olive and oleander plants caused by the bacterium *Pseudomonas savastanoi*. This bacterium induces tumorous growth by secreting high concentrations of indoleacetic acid and cytokinin into the tissues surrounding the point of infection. The enzymes necessary for the production of these growth regulators in *P. savastanoi* are functionally identical to those encoded by *A. tumefaciens* T-DNA, although the genes in *P. savastanoi* are located on separate plasmids and expressed in the bacterium. No transformation of the plant genome occurs. The nucleotide sequences in the coding regions of the genes for cytokinin and indoleacetic acid synthesis show a high degree of homology with those of the corresponding genes from T-DNA. However, the promoter regions, which control the expression of these genes, are entirely different in *A. tumefaciens* and *P. savastanoi*. This difference in structure was expected since the *Pseudomonas* genes are designed for expression in a bacterial (prokaryotic) cell, whereas the T-DNA genes must function in a plant (eukaryotic) cell. The similarities in the structural genes indicate that the growth-hormone genes in the two tumorigenic systems have a common origin.

The Ti Plasmid Provides a Vehicle for Gene Transfer in Plants

Once it was known that *A. tumefaciens* actually transforms its plant hosts, scientists recognized the potential usefulness of the Ti plasmid as a means of introducing foreign genes into plants. In a number of laboratories, they designed so-called disarmed versions of the Ti plasmid. The genes associated with growth-hormone production were removed from the T-DNA, thereby preventing tumor formation in transformed plants. An antibiotic resistance gene was introduced as a selectable marker in an existing T-DNA gene for opine synthesis, and gene-cloning sites were constructed within the disarmed T-DNA. The genes of the *vir* region necessary for the integration process must be retained on the Ti plasmid or placed into a second helper plasmid. Once the desired gene is spliced into the T-DNA of the vector plasmid, the recombinant plasmid is reintroduced into *A. tumefaciens*. The bacterial cells are incubated with plant protoplasts or with leaf disks to allow transformation to occur. The protoplasts or leaf disks are then freed of the bacterium and placed upon a medium favoring plant regeneration. This transformation procedure has become so standardized that it is now routine in the hands of trained scientists.

The Ti vector system has been used to introduce DNA sequences that cause disease and insect resistance into plants. For example, genes responsible for the production of an insect toxin have been transferred from *Bacillus thuringiensis* into plants. These plants became resistant to certain insects as a result of this transformation. Disease resistance to some viruses has been introduced by genetically engineering tobacco and tomato plants to produce the coat protein of tobacco or alfalfa mosaic virus. These plants show resistance to virulent strains of these viruses. An alternative method to introduce virus resistance into plants makes use of small RNA molecules known to act as "parasites" of some plant viruses. These entities, called satellite RNA, replicate only in plant tissues that are infected with a specific virus, frequently reducing the extent of replication of the virus and ameliorating the symptoms that the virus alone would induce. The satellite RNA becomes enrobed in the coat protein of the virus and thus may be co-transmitted from plant to plant with the virus. The protective effect of the satellite RNA continues even in the subsequently infected plants. Recently DNA copies of cucumber mosaic and tobacco ringspot virus satellite RNA have been introduced into tobacco plants. Plants transformed with either one of the satellite RNAs and then inoculated with the respective virus showed greatly reduced symptoms in comparison with similarly inoculated, untransformed plants.

Virulence Factors of Certain Plant Pathogens Are Natural Herbicides

Many pathogens produce secondary metabolites that are toxic to plants. These chemicals are side-products of amino acid, carbohydrate, nucleotide, and lipid metabolism. One such chemical is tabtoxinine-β-lactam, which is produced by a bacterium, *Pseudomonas syringae* pv. *tabaci*. If secreted into the cells of its host, tobacco, tabtoxinine-β-lactam specifically inhibits the plant's glutamine synthetase, an enzyme essential for the production of precursors for protein and nucleic acid synthesis. The pathogen escapes the inhibitory action of its own toxin by several mechanisms, among them being the production of a glutamine synthetase that is less sensitive to the toxin. This phenomenon, called self-protection, is common among pathogens that produce toxins as a part of their repertoire of pathogenic determinants.

Because toxins produced by plant pathogens kill or injure plant cells, they may be viewed as natural herbicides. The activity of such toxins in selective instances provides a conceptual basis for the use of chemicals for weed control. Indeed, there is an herbicide that imparts its weed-killing effects by inhibiting the glutamine synthetase of plants. An alternative to the use of synthetic chemicals for the control of weeds is seen in fungi that are selectively pathogenic for weedy plants. Since the basis for this host selectivity is the production of a host-selective toxin, this phenomenon is being explored as a way to control weeds.

Recognition and Defense Molecules Function in Pathogen-Plant Interactions

Specific molecules function in the maintenance of many kinds of order in biological systems. Enzymes recognize substrates, cells recognize other cells and pollen compatibility, and incompatibility determines fertility in plants. Recogni-

GENETICALLY ENGINEERED RESISTANCE AGAINST PLANT VIRUSES

The traditional method of protecting crop plants against specific virus diseases has been to search for resistant strains of the crop or its close relatives in the wild and then to introduce the genes responsible for this resistance into cultivated strains with desirable agronomic properties. Typically, the production of disease-resistant plants with favorable agronomic traits requires 6 to 10 years, a relatively long period. Furthermore, suitable resistance genes for many important diseases have not been identified in many cases. Therefore, many crop plants are vulnerable to viral attack, a situation that has negative economic consequences.

Over the past 20 years, plant pathologists have also used the method of cross-protection to generate resistance against viral diseases in crops. This method is based on an observation, made more than 60 years ago, that tobacco plants could be protected against virulent strains of tobacco mosaic virus (TMV) if they were first inoculated with a mild strain of the virus. This approach has its risks since mild viral strains can give rise to virulent ones, which may devastate rather than protect the plant. Also, virus strains with mild effects can develop synergistic interactions with other viruses. Nonetheless, the method has been useful in enhancing crop resistance in some instances.

The phenomenon of cross-protection provided the conceptual basis for genetically engineered resistance against two different types of plant viruses, TMV and alfalfa mosaic virus (AlMV). The *Agrobacterium* transformation system was used to introduce the genes that encode the coat protein of TMV and AlMV into tomato and tobacco plants. Coat protein is normally wrapped around the viral nucleic acid to form the virus particle. Plants regenerated from the transformed cells (transgenic plants) produced TMV and AlMV coat protein, but appeared to be normal in all other respects. When progeny of the transgenic plants were inoculated with virulent strains of TMV or AlMV, the plants either escaped infection or developed a less severe form of the disease than did the nontransformed plants (Figure 11-5). The molecular mechanism responsible for engineered cross-protection has not yet been elucidated. Recent results indicate that fewer infection sites are established in transgenic plants, probably because of some block at an early stage of the infection process. If infection does occur, the rates of viral replication and spread through the plant are reduced. Studies in progress aim at explaining the mechanism of protection, increasing the level of protection, and extending protection to other viruses and other plant species. It is expected that genetically engineered cross-protection will relatively soon become a generally applicable method to introduce viral resistance into plants.

FIGURE 11-5   Genetically engineered cross-protection against tobacco mosaic virus (TMV). (Left) Control tobacco plant (VF36) inocculated with a severe strain of TMV (PV 230). (Right) Transgenic tobacco plant that expresses the TMV coat protein gene (VF36 +CP) also infected with TMV strain PV230. [Roger N. Beachy, Washington University]

tion molecules also mediate the interactions between pathogens and their plant hosts. An example for this is the specific messenger function of small polysaccharide fragments of fungal cell walls, called elicitors, which induce plants to produce chemicals called phytoalexins, which in turn might confer disease resistance to plants because they are toxic to the microorganisms that induce phytoalexin production. A specific molecular configuration is recognized by the plant because any rearrangement in elicitor structure either abolishes or greatly reduces its activity.

The biochemical basis for disease resistance and susceptibility is particularly well investigated in the case of root rot in peas. When attacked by the pathogenic fungus *Nectria haematococca*, peas produce a phytoalexin called pisatin. Some strains of the fungus are sensitive and others tolerant to pisatin. The sensitive strains cause mild disease in peas, from which the plants recover, whereas pisatin-tolerant strains are highly virulent and kill the plants. Tolerant strains respond to pisatin by producing an enzyme, pisatin demethylase, which degrades pisatin to a nontoxic product. Thus, the fungus has developed a way to circumvent a defense mechanism of the plant. Pisatin demethylase is a cytochrome P_{450} monooxygenase, the same type of enzyme that functions in mammalian livers as a detoxifying agent. The gene encoding pisatin demethylase has been isolated from *N. haematococca*. Study of the cloned gene will help us to understand how the fungus recognizes phytoalexins and how it has evolved the capacity to live in their presence.

In solanaceous plants, such as tomato and potato, small cell-wall fragments called oligogalacturonides are released when plant tissue is injured by chewing insects or mechanical rupture. Such oligogalacturonides, or perhaps some other signal molecules, are transported throughout the plant and systemically induce the production of a powerful proteinase inhibitor that interferes with the digestion of proteins. In the initial act of feeding, chewing insects seem to activate a defense system that renders the plant less digestable and may discourage further feeding. The systemic production of proteinase inhibitors in response to injury also occurs in nonsolanaceous plants; it might represent a general plant defense against insect predation.

In some pathogen-plant interactions, a plant's susceptibility or resistance to a particular pathogen is determined by a single gene in the host and another in the pathogen. This gene-for-gene relationship implies a high degree of specificity and recognition between plant and pathogen. Resistance may arise from a specific interaction between gene products of the pathogen and the host plant. Susceptibility would result from a lack of such an interaction. These hypotheses are being tested by isolating genes from bacteria that confer race specificity for their hosts. The products of the genes are being identified, and the structures responsible for specificity will be determined.

With Today's Technology, Fundamental Problems of Plant-Pathogen Interactions Can Be Investigated at the Molecular Level

The intimate relationship that has evolved between plant and pathogen is now the focus of attention by plant scientists. These studies have been greatly enhanced by new techniques in cell culture, chemical analysis, and molecular biology. It should be possible now to obtain answers regarding the responses of plants to challenge by abiotic factors, pathogens, or pests. What is the nature of disease or insect resistance in plants? How are resistant responses induced? What controls the expression of these resistance genes? Why are these genes not expressed in certain host-parasite combinations? Central to much of this research is our current ability to study mechanisms of communication between organisms and between cells. Transmembrane signaling, second messenger activity, and long-distance communication will be areas of active research during the next decade.

Other research areas with exceptional opportunity include the nature of pathogen genes that are essential for causing disease. Is specificity determined by the nature of the plant products that induce the expression of pathogenicity genes or is it determined by regulatory functions that modify the response of pathogens to these products? The recent success in conferring resistance to plants by introducing viral coat-protein genes suggests that, as we understand the mechanisms of cross protection, we will also be able to exploit this phenomenon to control virus disease.

The excitement generated by our ability to transform plants by means of the *Agrobacterium* T-DNA should be tempered by our ignorance as to how this plant pathogen is able to transfer DNA or how this DNA is integrated in the plant chromosome. The search for other pathogens that can serve as sources of vectors to introduce useful genes in our major cereal crops will, in the near future, greatly expand our ability to improve plants by genetic engineering.

Rapidly expanding computer technology has increased our knowledge of how plant pathogens are disseminated. Cooperation among mathematicians, computer experts, and plant pathologists will continue to lead to better prediction of epidemics and, thus, to more rational application of control procedures. There is now renewed interest in the ability of bacteria to colonize leaf surfaces, for example. What triggers the change from epiphytic to parasitic habit in certain bacteria? The recent interest in the possible use of epiphytic, non-ice nucleating bacteria (which do not act as ice-nucleation centers) to prevent frost damage to plants is an example of how answers to some fundamental questions in plant-pathogen interactions can help stimulate the plant biotechnology industry.

GENETIC IMPROVEMENT OF PLANTS

Plant-Breeding Programs Can Now Be Enhanced by Molecular Biology

During this century, plant breeding has led to substantial increases in crop yield through the production of hybrids with increased vigor, the modification of plant chemical composition or morphology, and the genetic transfer of disease resistance, among others. The methods of molecular biology can now be applied to complement those of conventional genetics, especially where barriers of sexual incompatibilty or of sterility preclude the introduction of desirable traits through breeding. Genetic engineering has also made it possible to introduce into plants genes from other organisms, regardless of their genetic relationship. In addition, plants such as *Arabidopsis thaliana* are being used as models for the study of plant molecular genetics, which will provide basic insights into plant biology.

Tissue Culture

Plant Improvement Through Tissue Culture Is Feasible, but Remains Technically Difficult

Plant cell and tissue culture is an important tool for improving plant characteristics. Calli originally derived from plant tissue can be subcultured indefinitely

ARABIDOPSIS—A TOOL FOR PLANT MOLECULAR GENETICS

The use of *Arabidopsis thaliana*, an annual, weedy member of the mustard family (Brassicaceae), for studies in plant genetics was suggested more than 80 years ago by the German botanist Eduard Strasburger. Among the advantages are the small size, short generation time, and copious seed production of this common plant (Figure 11-6). Several plants can be grown per square centimeter, one plant yields thousands of seeds, and its life cycle may be completed within less than 6 weeks. For screening purposes, as many as 10,000 seeds can be germinated in one Petri dish.

Recent studies have revealed additional advantages that make *Arabidopsis* a prime object for studies in plant molecular genetics. *Arabidopsis* has the smallest genome known in plants, with about 70,000 kilobase pairs per haploid chromosome set—a genome about 1/80 of the size of the wheat genome. The relative simplicity of the *Arabidopsis* genome facilitates the cloning of particular genes, which can be used either as probes for isolating corresponding genes from other species or in transformation experiments.

New impetus for intensified research with *Arabidopsis* arose from the realization that desired mutations can be obtained with relative ease. In this process, the seeds are treated with a mutagen and germinated, and the resulting plants are allowed to self-fertilize. The progeny of these plants, called the M_2 generation, are used for screening for mutants. Many mutations with a loss in some metabolic function have been found at a frequency of one in 2,000 M_2 plants. Specific screens yielded well-defined biochemical mutants, which permitted specific metabolic pathways, such as that involved in photorespiration, to be traced. Other mutants that have been isolated include some that are altered in the fatty acid complement of their membranes, others in which starch synthesis is blocked, and still others in which hormone biosynthesis is blocked or hormone sensitivity altered. Such mutants are valuable tools for the study of plant physiology and development.

Recently, *Arabidopsis* was used in the isolation of herbicide-resistance mutants. About one out of 100,000 M_2 seeds sown on an agar medium containing a sulfonylurea herbicide proved to be resistant to this compound. Sulfonylurea herbicides inhibit acetolactate synthase, an enzyme in the biosynthetic pathway of branched amino acids. The mutated gene for acetolactate synthase was cloned from *Arabidopsis* and was used to transform tobacco cells. The regenerated tobacco plants showed stable herbicide resistance. Experiments of this kind underscore the practical and theoretical importance of the *Arabidopsis* system and its potential for future contributions.

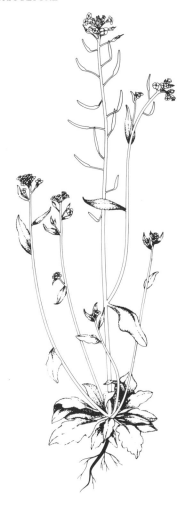

FIGURE 11-6   *Arabidopsis thaliana.* [Chris Somerville, Michigan State University]

on suitable synthetic media and induced to regenerate roots or shoots by altering
the proportion of auxin to cytokinin in the culture medium. The resulting plants
can be grown to maturity. Alternatively, plant cell walls can be digested and calli
regenerated from single protoplasts. The fusion of protoplasts permits the recombination of genetic material, even from unrelated kinds of organisms. Some
plants can also be regenerated by the induction of embryos in cells derived from
tissue that is not normally embryonic. Despite these advances, however, we know
virtually nothing about the principles that allow the regeneration of some kinds of
plants from protoplasts and that seem to preclude such regeneration (at least by

the available techniques) in many others, such as most commercially grown cereals.

A serious problem that limits the utilization of plant tissue cultures for various purposes, including the preservation of desirable strains, is the high frequency of genetic changes in such cultures, a phenomenon called somaclonal variation. Such changes include alterations of chromosome number, chromosomal breakage, genomic rearrangements, and point mutations. Some of these changes may be advantageous, conferring such features as disease resistance, increased sugar yields (sugar cane), tuber uniformity (potatoes), and high levels of fruit solids (tomatoes). In some widely used cultivars of agronomically important plants, in which infertility has precluded the introduction of new traits by breeding, genetic diversity provided by somaclonal variation can be exploited for the selection of desired phenotypes.

We Need to Learn More About the Principles That Underlie Plant Regeneration

How do nutritional and hormonal factors influence the developmental fate of plant cells? What biochemical pathways have to be activated for root and shoot differentiation to occur? How are these pathways regulated? Cultured cells should be particularly well suited for such investigations because their growth conditions can be controlled rigorously. Somaclonal variation raises some basic questions concerning genomic organization and stability in plants. What factors lead to the observed destabilization of the plant genome, and, conversely, what factors maintain stability? What precise changes occur in the genome as a result of culturing? Answers to these questions will have direct consequences for the application of tissue culture technology to plant improvement and for our basic understanding of the mechanisms of plant growth and development.

Plant Cell Transformation

Plant Cell Transformation in Agrobacterium Has Become an Important Research Tool in Plant Molecular Biology

Such fundamental questions as the way gene expression is regulated in plants have been investigated through the use of *Agrobacterium*-induced transformation. Nuclear genes under phytochrome control—for example, the gene encoding the small subunit of the chloroplast enzyme rubisco—have been transferred from one species into another in such a way that the regulation of gene expression by light can be studied against a background of precisely defined gene constructions. In particular, it has been possible to identify the promoter and enhancer sequences that regulate the expression of these genes. These achievements constitute the first steps in clarifying the mechanisms by which environmental signals are transduced in plants.

The T-DNA of *Agrobacterium* has also been used successfully to transfer herbicide resistance into plants. The mode of action of three commonly used herbicides has recently been elucidated. Glyphosate and sulfonylureas inhibit specific enzymes in the biosynthetic pathways of aromatic and branched-chain amino acids, respectively. Triazines block the binding of plastoquinone to an electron transport protein of the photosynthetic apparatus. Mutant plants that are insensitive to sulfonylureas and triazines have now been characterized at the molecular level. In each instance, herbicide resistance was associated with a single nucleotide change in the genes encoding the two target proteins. Resistance to sulfonylureas and triazines has been conferred on susceptible plants by transformation, through the use of the respective genes from herbicide-resistant plants. Plants with increased resistance to glyphosate have also been obtained through genetic engineering. Creating herbicide-resistant plants is especially worthwhile in the case of medium- to low-acreage crops, which do not warrant the development of selective herbicides. The methods of weed control that become possible in such systems decrease production costs and increase yields.

The use of the *Agrobacterium* transformation system is limited by the host range of the bacterium. Although most dicotyledonous plants are susceptible to crown-gall disease, most monocotyledonous plants, including cereals, are not. Therefore, only a few monocotyledonous species have been transformed with *Agrobacterium* until now. However, striking success has been achieved by transforming plant cells directly with DNA. With polyethylene glycol used to perturb the plasma membrane of protoplasts, an antibiotic resistance gene has been introduced into tobacco cells. Plants regenerated from such transformed cells and their sexual offspring express the antibiotic-resistance trait in a stable manner. This same technique was also successful in transforming the cells of a monocotyledonous plant, the ryegrass (*Lolium perenne*).

A technique called electroporation offers another method to transform plants that do not become infected by *Agrobacterium*. Electrical pulses are used to temporarily perforate the cell membrane of protoplasts, permitting DNA to enter the cell. With electroporation, protoplasts of maize have recently been transformed with an antibiotic resistance gene. All that stands in the way of stably transforming many species is our inability to regenerate whole plants from protoplasts or calli. To get around the problem of regeneration from protoplasts or calli, it is possible to shoot DNA-coated microprojectiles directly into intact plant cells. Potential targets include meristems and pollen, which can be cultured in vitro or used in sexual crosses, respectively.

The Techniques of Genetic Engineering Hold Enormous Promise for Agriculture

Potential improvements in crop plants through genetic engineering include increased yield, lowered production costs, improved nutritional qualities, adaptations to unfavorable growing conditions, and new biosynthetic capacities. The feasibility of obtaining a number of traits of these kinds by genetic engineering has already been demonstrated in laboratory trials. Other applications of gene transfer technology, such as the utilization of the plant's biosynthetic machinery for the production of foreign compounds with high commercial value, have not yet been realized. "Custom" crop plants capable of synthesizing specific proteins, valuable oils, or secondary metabolites for medical use could be the bases of new agricultural industries.

To attain these goals, research in the plant sciences must proceed at all levels. Discoveries in plant physiology and biochemistry must keep pace with the rapidly progressing field of plant molecular biology. The metabolic functions of plants must be explored further. Enzymes involved in the synthesis of plant products must be characterized and regulatory mechanisms in plant metabolism elucidated. The mode of action of plant hormones requires intensive study as do the reactions that mediate compatibility and incompatibility between pollen and stigma, symbionts and roots, and pathogens and plants. A thorough knowledge of the processes that could be altered for the improvement of plants provides the basis for the application of recombinant DNA and transformation technologies.

12

Biology Research Infrastructure and Recommendations

In the United States the positive links between basic research and the health of citizens and the economy have always been appreciated. This realization has been justified, as for example in the research findings that have led to improved health care and agricultural efficiency, as well as to the development of a substantial data base of useful biological knowledge. Many components contribute to the overall strength of the field of biology, and these must be considered individually and in combination in order to ensure the health of the field over the next decade. Among these components are training, employment, equipment and facilities, and funding. In addition, the role of large data bases and repositories needs special consideration, as do the relative merits of developing large research centers compared with additional support for individual investigators.[1]

U.S. Scientists Are Finding It Increasingly Difficult to Maintain Their Leading Position in Biological Research

On the basis of the number of publications and citation rates, the United States continues to be the dominant force in biological research. In 1982, American life scientists (clinical, biomedical, biological, and agricultural sciences) published about 87,549 research articles, representing 40 percent of all biology articles published that year. This is a decrease from 42 percent in 1973.[2] An estimate of the quality of U.S. publications can be measured by comparing the number of publications in the top 10 percent of citations (Table 12-1). In 1980, the United States had nearly 13,000 publications in the top 10 percent, while the next most prolific countries—the United Kingdom, Japan, West Germany, and France—had only a small fraction of that (Computer Horizons, Inc., unpublished data, 1987). However, Japan and West Germany are increasing their shares

TABLE 12-1   Quality of Biological Research as Measured by the Number of Publications in the Top 10 Percent of Papers Cited

Year	United States No.	% of Total	United Kingdom No.	% of Total	Japan No.	% of Total	West Germany No.	% of Total	France No.	% of Total
1973	8,795	74	1,965	17	342	3	430	4	318	3
1980	12,869	70	2,641	14	1,081	6	1,058	6	618	3

Source: Computer Horizons, Inc., 1987.

relative to the United States, France, and the United Kingdom. These data pose many interesting questions. For example, is the percent increase demonstrated by Japan a reflection of improved publication quality, increased quantity, or both? Can Japan continue to increase at this rate? To what extent are these numbers merely a function of the level of research investment by governments?

Patent activity in areas related to biology is another measure of our international research position. Two recent reports produced by the U.S. Patent and Trademark Office analyze patent data on genetic engineering and on molecular biology and microbiology. A tabulation of patents of foreign origin granted in genetic engineering shows that patents of U.S. origin have increased from 25 percent of the yearly total (both domestic and foreign origin) in 1973 to 78 percent in 1986.[3] For molecular biology and microbiology, the percentage of patents of U.S. origin has remained steady at about 56 percent of the total since 1973.[4] Since these data were compiled from patents filed in the United States, they may not precisely reflect the international patenting situation. However, in the highly competitive U.S. patent system, the United States is maintaining its leadership role.

The United States has long encouraged and benefited from international cooperation in biological research. As other countries increasingly emerge as valuable sources of quality research, this policy of cooperation should be strengthened.

TRAINING

The Number of Ph.D.s Awarded Each Year in the Biological Sciences in the United States Is Leveling Off

The number of Ph.D.s awarded in the biological sciences increased sharply between 1965 and 1970; the total number in 1985 was 3,766, compared to 3,361 in 1970 [National Research Council (NRC), unpublished data, 1988]. The number of biology Ph.D.s earned in 1985 is more than twice that earned in either chemistry or physics (Figure 12-1A); in this year, an additional 1,982 Ph.D.s were awarded in the agricultural and health sciences.

It currently takes about 6.4 years of registered time to complete a Ph.D. in biology after receiving a bachelor's degree; this is an increase of about a year since 1970 (NRC, unpublished data, 1986). In comparison, it currently takes about 6.5 years to obtain a Ph.D. in physics and about 5.5 years in chemistry (Figure 12-1B).

Support for around 70 percent of the total number of biology students comes from federal and university sources (NRC, unpublished data, 1986). In 1986 there were 36,916 biology graduate students in doctorate-granting institutions (excluding agricultural and health sciences); of these, 8,606 were supported by fellowships and traineeships, 12,059 by research assistantships, 8,609 by teaching assistantships, 5,516 were self-supported, and 2,126 had other types of support.[5]

For postdoctoral support, the government contribution has remained at approximately 60 percent of the total number of postdoctoral positions and the contribution by universities at about 15 percent from 1967 to 1985. However, private foundations have doubled their postdoctoral support, from 8.9 percent between 1967 and 1980 to 17.7 percent between 1981 and 1985. Chemistry and physics Ph.D.s obtain a larger percentage of their support from colleges and universities than do biology Ph.D.s. (NRC, unpublished data, 1986).

As measured by Graduate Record Exam (GRE) test scores between 1964 and 1982, the verbal skills, but not the quantitative skills of students entering graduate school have declined. The GRE Biology Advanced Exam scores have changed little between 1973 and 1986. Students intending to pursue an advanced degree in biology or physical sciences have similar verbal and analytical scores. However, physical science majors score appreciably higher on the quantitative section of the exam.

The proportion of "A" high school students intending to major in biology has dropped from 11.7 percent in 1974 to 6.6 percent in 1983.[2] Concomitantly, the number of biologists being trained at the bachelor's and master's levels has sharply declined in the past decade. In 1983, the number of bachelor's degrees conferred was down 26 percent from 1976, and the number of master's degrees was down 13 percent.[6]

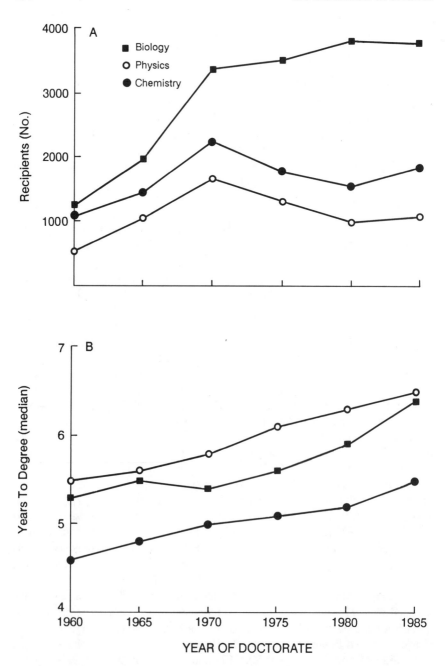

FIGURE 12-1 (A) Number of Ph.D. recipients in natural sciences. (B) Time required to fulfill requirements for the Ph.D.

In an attempt to assess the institutional training quality of universities granting Ph.D.s in the life sciences, the top 25 institutions in total federal obligation to research and development support[7] were compared with the top 25 institutions in numbers of Ph.D.s awarded in the life sciences (NRC, unpublished data, 1986). Of the top 25 research and development institutions, 11 are top producers of life science Ph.D.s.

Funding Support for Training Falls Short of Covering Future Needs

Biological sciences in general and biomedical sciences in particular have enjoyed more support for training at the pre- and postdoctoral levels than most other fields in the natural sciences. This support has been rewarding; it has been one of the important ingredients leading to the spectacular vitality of U.S. biology in recent times. The majority of funds for training have come from the National Institutes of Health (NIH); from the Alcohol, Drug Abuse, and Mental Health Administration; and the National Science Foundation (NSF).

Of the 10,382 NIH training positions awarded in 1986 (5,011 predoctoral and 5,371 postdoctoral), 83 percent were in institutional training grants and the remainder were awarded to individuals. Competitively awarded NIH institutional training grants are a good way to provide training support at quality institutions. Training-grant programs such as this should be encouraged, especially for predoctoral students.

The number of NIH-supported trainees has apparently leveled off after a decrease after 1975 (Table 12-2). Table 12-3 gives the total number of full-time training positions supported by NIH for 1980, 1984, and 1986; it provides a breakdown by the three NIH units most heavily involved in training [National Institute of General Medical Sciences (NIGMS), the National Cancer Institute (NCI), and the National Heart, Lung, and Blood Institute (NHLBI)]. At the predoctoral level, most NIH support has been allocated to interdepartmental training programs. In 1986, NIGMS provided funds for approximately 55 percent of all NIH predoctoral trainees and for about 9 percent of all NIH postdoctoral trainees, the rest of the support being divided among the other units of NIH (William Pittlick, NIH, personal communication, 1987). In total, about one-third of the predoctoral training positions are in molecular and cell biology. The corresponding fraction for postdoctoral fellowships is more difficult to assess.

The decline in training support since 1975 is alarming, especially when the following points are considered. NSF awarded only 760 predoctoral fellowships for 1988, which are distributed over all fields of scientific study. In addition, NSF granted only 40 awards per year for postdoctoral fellowships in plant biology and environmental studies, fields that have not been traditionally covered by NIH. The U.S. Department of Agriculture (USDA) offers 100 postdoctoral fellowships per year in the Agricultural Research Service's Research Associates Program for work with USDA scientists. USDA has also recently initiated a small predoctoral

TABLE 12-2   NIH Support of Graduate Students and Postdoctoral Fellows

Year	(Millions of dollars)	Full-time training positions
1975	154.9	12,272
1980	176.4	10,664
1985	217.5	10,370
1986	212.8	10,382
1987	232.7	11,226[a]
1988[b]	235.6	10,992

[a]Estimated.
[b]Proposed.

training program that provided 57 awards in 1987 (Jane Coulter, USDA, personal communication, 1988). The NIH budget for 1988 allocates funds primarily for continuing NIH training grants; support is not generally available for either competitive renewals or new grants. Moreover, the majority of NIH training grants are now funded below the levels recommended by NIH advisory councils.

> **Despite impressive advances and great opportunities in biology, we are rapidly approaching a crisis in training biological researchers. Current levels of support appear inadequate in the light of the shortages of trained personnel predicted for the late 1990s (see Employment section).**

TABLE 12-3   Training Support (pre- and postdoctoral) for NIH as a Whole and for the Three Largest Sources of Support within NIH

Year	Number of Full-Time Training Positions			
	Total	NIGMS	NCI	NHLBI
1980	10,664	3,765	1,530	1,549
1984	10,514	3,581	1,465	1,688
1986	10,382	3,238	1,394	1,633

NOTE: NCI = National Cancer Institute; NHLBI = National Heart, Lung, and Blood Institute; and NIGMS = National Institute of General Medical Sciences.

Women and Minorities

The Number of Women Receiving Ph.D.s in Biology Has Been Steadily Increasing, but Minorities Are Still Underrepresented

Approximately 1,000 women received Ph.D.s in the life sciences in 1976, versus 2,020 in 1987.[8] In 1987, women received 35 percent of all Ph.D.s awarded in the life sciences.[8] However, in a 1983 survey of academically employed life science Ph.D.s reporting tenure status, only 64 percent of women reporting tenure status had tenured or tenure-track positions, compared with 88 percent of men.[9] Women generally earn less than men in every field of science, although there is some recent evidence that this trend may be changing for women hired within the past decade.

> **The recent employment and educational advances made by women in the life sciences must be fostered and encouraged in order to provide an attractive research and career environment.**

In comparison with the figures for women in biology, the number of minority-group students (U.S. citizens or foreign minority-group students with permanent visas) receiving Ph.D.s in this field has changed little, with minorities receiving between 7 and 8 percent of the degrees in 1975 and 1987.[8,10] The exact percentage of Ph.D.s awarded to minority groups is difficult to determine since the number of Ph.D.s reported with unknown ethnic or visa status is large. The number of foreign students receiving doctorates in the life sciences has remained constant at about 20 percent between 1962 and 1987.[8] In the life sciences, the visa status of foreign students has also remained fairly constant, with about 4 percent of all students having permanent and 16 percent having temporary visas.[8] This is in contrast to engineering, where the number of students with temporary visas has increased from 18 to 41 percent of total students between 1962 and 1987.[8]

> **Every attempt should be made to encourage complete representation of members of minority groups in the biological sciences. This will require in turn that greater attention be paid to their precollege education so that equal training opportunities will exist in college.**

Current Training Needs

Training Mechanisms Need to Optimize Research Opportunities

In addition to a general increase in postdoctoral training, interdisciplinary training is also needed in most areas of biology. Sophisticated research requires not only knowledge of many areas of biology, such as molecular biology, developmental biology, and ecology, but also knowledge of other scientific disciplines, such as chemistry, physics, and engineering. For example, developmental biology requires expertise with several experimental systems. Other areas of research can be studied effectively only at the interface between two or more traditionally separate lines of research: We are now gaining insights into both plant and animal pathogens through pioneering work involving molecular biology and microbial pathogenesis.

Another example is the interface between molecular biology and neuroscience. Here, a synergistic effect between classical neuroscience and molecular biology has created one of the most exciting areas of biology today.

Biology is becoming more chemically and physically oriented, and in many areas training requires an increasing focus on chemical and physical technologies. This is especially important for research in structural biology, which provides an atomic level of analysis for much of modern molecular biological research. Similarly, evolution and diversity, systematics, population biology, and ecosystem studies require an increasingly interdisciplinary approach that includes training in molecular biology, computer sciences, and mathematics.

Structural Molecular Biology Requires Scientists Expertly Trained in the Physical and Biological Sciences

The blend of molecular genetics and structural analysis that will be needed for the next decades of structural biology poses some serious problems for training the next generation of scientists. Today it is relatively rare for an individual to receive intensive training in both physics and biology. In fact, rigorous structural and molecular genetic studies are frequently not even available in the same academic department. One can deal with this problem within existing formats by encouraging individuals to do predoctoral training in one field and postdoctoral in another. However, creating unified programs that blend both disciplines at an early stage in training is worth serious attention as a way of producing the most highly skilled and innovative future investigators. It may be possible to develop curricula that require less mathematics and physics than is common in traditional structural biology, while still providing enough background in these areas to enable a biologically oriented scientist to use structural tools and to appreciate the significance and limitations of structural results. Another approach is to develop joint undergraduate programs between biology and chemistry departments where students can learn about the many opportunities for applying these techniques.

The major advances in developing new nuclear magnetic resonance (NMR) and x-ray diffraction techniques are made by scientists trained in physics. Nuclear spin engineering—the production of sequences of radio frequency pulses—is crucial to the simplification of the complex, overlapping NMR spectra from large biological molecules. New methods to exploit anomalous x-ray scattering to solve the x-ray phase problem require both experimental advances in x-ray detection and the development of new computer algorithms. Thus, the future of structural molecular biology will require interdisciplinary cooperation between molecular biologists and highly trained and specialized x-ray diffraction and NMR specialists. Some of this need may be met by scientists trained in both molecular biology and physics. Some of it, however, requires the participation of those with a rigorous background in traditional physics. Currently, the students and postdoctoral-level scientists trained in the physical sciences needed for the development of structural molecular biology are in short supply.

Ecological and Evolutionary Sciences Require More Scientists Trained in a Greater Variety of Subjects and More Manpower in General

The broad interdisciplinary nature of the modern training required in ecological and evolutionary sciences must be recognized. Not only should evolutionary biologists be conversant with the rapidly developing areas of genetics, molecular biology, cell biology, and biomechanics, but they must also function in the evolutionary framework of thought that emphasizes variation, interaction, history, and the question of why rather than merely how an organism functions as it does. In addition, ecologists routinely rely on statistical models that require an understanding of computer science and mathematics. The perspectives of evolutionary biology, ecology, and systematics could be profitably incorporated into instruction in other areas of biology, with beneficial effects for the entire field. Because the support of postdoctoral personnel by individual research grants is often inadequate, a program of individual or institutional postdoctoral training grants should be developed and targeted toward these areas.

Training at the predoctoral level is also a problem. The individual predoctoral fellowships now available are vitally important for attracting gifted young men and women into these lines of research. In addition, to attract good students it is also important to maintain the promise of the field in terms of availability of future positions and funding for research.

An example of a problem area is systematics. In North America about 4,000 systematists work on 3,900 systematics collections. These numbers are misleading because a large fraction of these specialists, perhaps the majority, are engaged only part time in systematics research. More to the point, few can identify organisms from the tropics, where the great majority of species exist. Probably no more than 1,500 professional systematists in the world are competent to deal with tropical organisms, and their number may be declining because of decreased professional opportunities, reduced funding for research, and assignment of higher priority to other disciplines.

Favorable educational developments should be encouraged by special training-grant programs targeted toward these areas of biology. Moreover, additional attention should be given to the concept of bringing together specialists from relevant fields to investigate ecological and evolutionary problems of mutual and overlapping interest. In recent years, this approach has proven highly effective in some areas of evolution and diversity, both as a research strategy and as a technique for training graduate students.

Training in the Plant Sciences Is Largely Restricted to Land-Grant Universities

Almost nowhere is the need for interdisciplinary training exemplified more dramatically than in the plant sciences. Not only are there traditional disciplinary barriers to research that are experienced by all areas of biology, but also there are institutional barriers. Most departments of biology, for example, have little or no expertise in the plant sciences. Therefore, most research and training activity in plant biology is carried out in land-grant universities in conjunction with colleges of agriculture—thus, land-grant universities employ 80 percent of all plant science faculty. Many outstanding biology departments at private universities have no plant scientists on their faculties. The lack of plant scientists in many biology departments and the consequent lack of exposure of many biology students to the plant sciences seriously limits cross-fertilization by interdisciplinary activity and also limits the influx of talented scientists and students into plant biology research. An interdisciplinary approach to the plant sciences that incorporates them fully into the activities of strong biology departments and research groups is needed to stimulate plant biology research as it has research on microorganisms, *Drosophila*, nematodes, mice, and humans. To some extent, this is happening now, but it must be encouraged and supported.

> **As biological research becomes more sophisticated, the need increases to develop interdisciplinary and flexible training programs for students, postdoctoral fellows, and established scientists.**

EMPLOYMENT

The employment profile for doctoral life scientists is changing. In 1973, approximately 13 percent of all life science Ph.D.s were employed in industry, 67 percent in educational institutions, and 10 percent in government. However, by 1985, employment by industry had risen to 19 percent while employment in educational institutions and government had decreased to 61 and 8 percent, respectively.[5]

It Is Difficult to Forecast the Future Demand for Biology Ph.D.s in Academia, but Some Shortages May Be Experienced in Some Fields

Generally, three sources of Ph.D.s fill academic positions: new Ph.D.s, employees previously supported by grants such as postdoctoral fellows, and

Ph.D.s working for industry. In 1983, slightly fewer than 50 percent of all new hires in academia (3,179 total) were of new Ph.D.s (63 percent of these in a tenure-track position).[11] That year, the new hires represented 6.8 percent of the 46,566 full- and part-time Ph.D.s employed in life-science fields at these institutions; of those, turnover accounted for 4.3 percent and new positions 2.5 percent.[11] It has been estimated that for every 100 job openings in academia, there are about 156 new life science Ph.D.s.[11] In general, this discrepancy between Ph.D.s over academic positions will persist through the 1990s. At about the turn of the century, however, individuals in existing positions will retire in increasing numbers (about 20 percent of the science faculty will reach age 65 in 10 years).[12] At the same time, the undergraduate enrollment rate is projected to increase. Whether or not there will be enough Ph.D.s to fill the projected demand depends largely on the needs of industry during this period. For example, in 1987, 24 percent of new life science Ph.D.s (U.S. citizens and permanent residents) were hired by industry; in 1977 that figure was 17 percent.[8] If industrial hiring trends continue, shortages of faculty members in some subdisciplines of biology may develop.

Industry Has Become a Major Employer of Biologists

In 1985, industry employed 19,200 life-science Ph.D.s, in 1973, 7,100.[5] Industry has always offered an attractive employment option for biologists because of higher salaries and larger research budgets. However, working for industry was generally viewed by many university biologists as intellectually stifling. Such is no longer the case; industrial research and development programs are now contributing to major scientific discoveries. In addition, discoveries in basic biology are leading to practical applications at an increasing rate. As a result, industry is now in direct competition with academia for some of the best biologists in the country.

Biotechnology Research and Development Programs in Industry Are Growing

Biotechnology, as defined by NSF and the Office of Technology Assessment, is a technique that uses living organisms or parts of organisms to make or modify products, to improve plants or animals, or to develop microorganisms for specific uses.

The five main areas of research and development in the U.S. biotechnology industry are health care, plant agriculture, chemicals and food additives, animal agriculture, and energy and the environment. According to an NSF survey, the biotechnology industry spent about $1.1 billion in 1985, which was an increase of 20 percent from 1984. Of the 94 companies responding to the survey, health care accounted for 66 percent of the total spent, whereas plant agriculture—the next highest area in terms of research and development expenditures—accounted for only 13 percent of the total. The biotechnology industry employed approximately 8,000 scientists and engineers in 1986, which is an increase of 12 percent from

1985.[13] Thus, the biotechnology industry is a viable employment option, especially for scientists with expertise in biochemistry, cell biology, microbiology, immunology, molecular genetics, and bioprocess engineering, as well as for technicians trained in instrumentation. There is a growing concern that recruiting scientifically competent people at the bachelor's, master's, and doctoral levels is going to become increasingly more difficult and that this difficulty may impede the growth of the industry. In support of this concern, is the fact noted earlier that the number of students receiving B.S. degrees in the life sciences decreased 26 percent between 1976 and 1983 and the number receiving master's degrees by 13 percent. In contrast, however, the number of Ph.D. degrees awarded remained relatively stable.[6]

> **Shortages of trained technical personnel in biology are now occurring at the bachelor's and master's levels. Attempts should be made to enhance university training programs at these levels, especially in biotechnology-related areas (biochemistry, cell biology, microbiology, immunology, molecular genetics, and bioprocess engineering).**

> **Shortages of Ph.D.s in biotechnology-related areas are anticipated in the late 1990s. Therefore, appropriate educational programs should be initiated and supported immediately.**

At the same time that we recognize the importance of biotechnology industries as important employers of life-science graduates, it is crucial to realize that many fields of biology are not represented, or poorly represented, in industry. Systematics, evolution, and ecology, for example, are all central to our ability to manage the global ecosystem for sustained productivity. If the U.S. scientific enterprise is to continue to be strong, it must continue to find ways to utilize the talents of biologists of all kinds, not simply those who are working in fields that offer immediate rewards in terms of commercial prospects or major grants.

LABORATORY COSTS AND EQUIPMENT NEEDS

More Attention Needs to Be Directed to the Increasing Requirements for Quality Research Equipment and to the Increasing Expense of Laboratory Operation

As biology experimentation becomes more sophisticated, the research equipment and facilities needed often become more advanced and therefore usually more expensive. Currently it takes between $100,000 and $200,000 to adequately equip a laboratory for molecular biology research. For example, a centrifuge can cost about $70,000, a spectrophotometer about $20,000, a liquid scintillation counter about $20,000—and these are only part of the equipment needed to equip a modern biology laboratory. It can cost three to four times that amount to equip a laboratory for x-ray crystallography studies. Once a laboratory is assembled, operation and maintenance costs (excluding salaries) can approach $50,000 per year for a modest research program. Therefore, modern biologists must have a source of research funds adequate to meet the high costs of equipment and research supplies.

The amount of money spent on equipment for the life sciences at colleges and universities increased about 10 percent between 1983 and 1984 and 18 percent between 1985 and 1986.[5] The total amount spent for research equipment at colleges and universities was $318 million in 1986.[5] Results from a national survey,[14, 15] which focused on equipment costing from $10,000 to $1 million, indicated that, in biology, approximately 35 percent of actively used research equipment systems are located in shared facilities.[14] The 1983 national stock of such academic research equipment in the biological sciences and departments of medicine was estimated to have an aggregate original cost of $555 million and a replacement cost (in constant 1982 dollars) of $863 million.[15]

Results of the survey indicated that fewer than 20 percent of biological science department heads characterized the adequacy of their current research instrumentation as "excellent," and nearly 60 percent reported that researchers in their departments cannot conduct critical experiments because of a lack of necessary equipment. In addition, of the equipment in active use in 1983, half of the systems were in some degree of disrepair; 80 percent were not state-of-the-art.[15]

It is clear from this survey that the condition of much of the research equipment used in biology laboratories throughout the country is less than desirable. Even worse, a considerable number of scientists have limited their research programs because they lack crucial equipment.

> **Because of the ever-increasing need for and expense of laboratory equipment, funds to provide for specific pieces of equipment should be available. This is especially true when requested equipment is to be placed in a shared facility.**

The development of instrumentation and general technology should be encouraged nationally to further both basic science and its applications to socially and economically important problems. Specific encouragement by funding agencies should be given to the development of new areas (such as tunneling electron microscopy and fast computation technologies) and techniques that can improve productivity (such as robotics applied to biological and chemical systems).

> **The development of instrumentation to be applied to a variety of biological problems should be accelerated.**

FUNDING

It Is Crucial That Limited Funds Be Spent in Such a Way as to Maximize the Progress of Biological Research

In the United States, the vast majority of biological research support has come from the federal government in the form of research grants, formula funds such as Hatch funds for agricultural research, and intramural research programs. Probably the most important component contributing to the success of U.S. biological research is funding. The U.S. government has made a sustained effort to provide appropriate support for research in the life sciences. From 1970 to 1985, federal funding for life science research increased from $1.44 to $6.37 billion.[16] Although the effects of some of this increase have been offset by inflation, the percentage increase in constant 1972 dollars is still about 72 percent. For 1987, an estimated $2.84 billion was spent on general biology, $263 million on environmental biology, $612 million on agriculture, and $2.18 billion on medical sciences.[16]

In an attempt to evaluate the level of federal support for U.S. biological research (in 1972 dollars), Ph.D. biologists employed in scientific and engineering jobs for 1973 and 1983 were compared with the level of federal support provided during those years. In 1973, approximately $34,000 per Ph.D. was provided, whereas in 1983, approximately $28,000 was provided. (This does not mean that each scientist received that much support, but rather that the federal contribution supported, in part, that number of scientists.) The actual level of support per biologist has decreased over the years as increasing numbers of scientists have been added to the biology work force and even as the average costs of research have increased. Biology is a growing field, and the number of biologists is expanding; it is therefore important that the federal government continue to keep an adequate supply of research money available to reflect both the growth of biology and the increasing costs of doing research.

How research funds are allocated in the future is a matter of prime importance. To what extent should limited funds be used to support research centers versus individual investigators? To what extent should high-priority areas be funded over those of lower priority? To what extent could the funding of individual investigators be made more stable and longer term? Biological research is becoming more complex and therefore more expensive, and long-term commitments are often required if the desired results are to be obtained. This has tended to increase the level of funding needed to support individual research projects. Therefore, along with the overall increase in the number of scientists needed is an increasing need for greater research support. The 72 percent increase in constant dollars for the field as a whole over the past 15 years has not made it possible to attain all the potential results that might have been produced. In addition, there is an increasing demand for research projects that require long-term funding for 5 years or more; funding of this sort has traditionally been rare.

Some granting agencies are now providing long-term support, and it is hoped this trend will continue. If additional long-term support for research becomes available, new and imaginative ways need to be developed to review results of long-term projects.

Funding for new faculty members is also a problem that needs immediate attention. New faculty members are faced with enormous research start-up costs and have considerable pressure to acquire external funding before their research programs have even begun. For many, this is a catch-22 situation; often preliminary data to justify funding are not available, but funding is required if these data are to be obtained. Therefore, most new faculty members are either forced to continue research initiated while in graduate school or as postdoctoral students or to undertake an underfunded research program. Similarly, established scientists find it difficult to obtain funding in areas of research that are not directly connected with their main area of expertise. This greatly limits the growth and diversity of research programs of individual investigators and the level of innovation possible in the system as a whole. The NSF Presidential Young Investigator and the NIH First Awards Programs are good examples of support mechanisms for quality research by new faculty members.

> **Agencies should increase their programs that provide long-term and start-up funding and should look with favor on innovative projects by qualified investigators that propose research in new, creative directions.**

In reviewing the progress of biology over the past 40 years, two things become clear that must be taken into account when considering future funding initiatives. First, throughout this enormously successful process of federal funding, the driving force never came from the granting agencies or from great insights of an individual authority. Second, the role that the funding agencies play is that of the ardent listener, the support role. Inventiveness, creativity, and productivity are rewarded with peer-reviewed infusions of support. The driving force is the individual investigator staking career and reputation on the pursuit of novel insights with the funding agency more or less supporting these individual efforts. Therefore, research opportunities that merit funding should rise from within the research community, from individual investigators, and should not be predetermined. For example, the track record of most prophets of the use of recombinant DNA is dismal. A reading of some of the older literature shows a consistent underestimation of the rate of progress of molecular genetics and a tendency to be blindsided by breakthroughs.

Engineering and the Biological Sciences Need to Develop an Interface

Opportunities exist in instrumentation development and bioprocess engineering, which require engineers and biologists to interact closely. For example, the development of automated procedures, such as those for DNA purification and sequencing, would greatly facilitate the large-scale sequencing efforts that are currently being contemplated. However, limited funds are available at this time. In the NSF biotechnology program, which funds joint proposals of engineers and biologists, many meritorious proposals are not currently being funded because the money is unavailable, yet such projects are probably some of the most significant for the development of science as a whole. In 1987, this program supported 14 proposals at $200,000 per year per proposal (Frederick Heineken, NSF, personal communication, 1988). A new thrust is needed in technology development for the biological sciences so that engineers can work in an interactive environment with biologists. Training programs could provide engineers a working knowledge of the needs of biologists, and vice versa.

Industry Funding and Research Is Playing an Increasing Role in Basic Biological Research

Besides being a key employer of biologists, industry contributes substantially to important basic discoveries and is increasingly involved in funding extramural research projects, both basic and applied, at universities. Industrial support and research collaborations with the academic community provide a mechanism for more rapidly converting basic biological discoveries to the solutions of major industrial and social research and development problems.

Research in the Ecological and Evolutionary Sciences Has Traditionally Been a Relatively Low-Budget Enterprise

Traditionally, the funding of research in these areas involves one or a few investigators working in the laboratory or field, and present funding mechanisms and levels of support reflect this history. The increased need for modern technology in many areas of research has not been matched with increased support, with the result that many research projects are carried out less efficiently or completely than they might be, solely because funds are lacking.

The typical terms of support, which range from 3 to 5 years, are often inconsistent with the long-term nature of some studies in ecology. Investigators may be discouraged from committing their careers to a long-term study when faced with the risk that their funding may be terminated at the end of any 3- or 5-year term of support, even when the project is still incomplete. This is especially true when research funding is tight, as termination of support may not reflect on the quality of the overall research project, but rather the lack of available funds or shifted priorities in the granting agencies.

Some aspects of ecology and related fields are subject to an organization of funding that may be less than optimal. For example, research proposals in population or evolutionary genetics are often reviewed by panelists who, while expert in the genetics of the relevant organism, lack knowledge in the field of population genetics and are therefore unable to review the proposal in its appropriate context. Peer review is thereby ineffective. Because of the composite nature of many of the subdisciplines of ecology and evolution, proposals may be unusually difficult to review adequately. Funding is scarce and provided mainly by a single agency (NSF); it is particularly important that the relevant panels in this area function optimally.

Another problem concerns the neglect of traditional areas for the sake of emerging new ones. Without a revision of such attitudes, we are likely to simply allow a major fraction of the earth's species and their inherent biological diversity to pass into extinction without ever seriously attempting to learn about its existence. The completion of biological surveys of selected groups of organisms for the entire world, and of all groups of organisms for certain areas, is a priority of fundamental importance that can be realized better now than will ever again be possible, owing to the rapid pace of extinction. Such activity may also make possible the preservation of a greater proportion of the organisms than would be possible otherwise. The organization and mechanisms of federal grant support of research in evolutionary biology and diversity should be seriously examined, especially since opportunities are being lost so rapidly.

Plant Sciences Require a Stronger Funding Base to Maximize the Current Interest and Excitement in This Area

The funds available to support basic plant sciences on a competitive basis are very limited. In 1985, competitive grants from federal agencies for basic biological and medical research amounted to $2.2 billion. Of this, only about 5 percent was awarded to the plant sciences. The main competitive grant support for basic plant sciences is provided by NSF (43 percent in 1986), with USDA, NIH, and DOE providing about 19 percent each. It was hoped that the Competitive Research Grants Office of USDA, established in 1978, would assume a major role in funding of basic plant research. This has not been the case. For 1987, the total funding for basic plant sciences was about $25.6 million (excluding forestry). The awards granted by the USDA office are usually for not more than 2 years at an average of $46,200 per year in 1986. This can be compared with an average of $70,000 per year for 3 years for an NSF award.

Parasitology Is an Underfunded Research Area of Great Importance in Tropical Medicine

Parasitology is conventionally limited to the parasitic families of protozoa, helminths (worms) and arthropods, and ectoparasites that include several insect families owing their significance to their role as transmitters of important diseases. For two main reasons, this group of infectious agents remains the premier cause of global ill health. First, human parasitic diseases primarily afflict the less developed nations, especially in the tropics. Second, parasites are technically tedious organisms to work with. The basic conditions for growth and experimental study have often not been defined.

LARGE DATA BASES AND REPOSITORIES

Problems in Information Handling Are Becoming Critical

The exponential increase in biological data has produced information backlogs and overwhelmed data bases. Macromolecular sequencing—the sequencing of the components of proteins and nucleic acids—represents one area in which current information-management systems are inadequate. Therefore, data-base systems and scientific networks need to be established to increase the flow of information among scientists and to allow for the rapid input, manipulation, and update of data. Some additional areas that require attention to information management are ecological modeling, genome mapping, structure-function relationships of macromolecules, and biological inventories.

As more information from diverse sources accumulates, the need to apply a matrix approach or other advanced data-handling methods to maximize the knowledge potential of each item of information will increase. Therefore, it is becoming imperative that biologists receive the necessary education and training in information science. Although some information management issues are being addressed, such as practical aspects of the management of nucleic acid sequence data, other issues, such as information science education for biologists, are not.

> **An assessment of the information-handling requirements for biology should be made. Special emphasis should be given to the training needs of biologists in the information sciences and to the maintenance and enhancement of large-scale data bases.**

Another important source of biological information consists of the resources of systematics collections, mainly held in museums. Comparative biology, evolution, and ecology require such collections, properly curated and preserved and actively studied so that the information in them is readily available. More than a billion samples of plants, animals, and microorganisms are preserved in the museums of the United States, and huge numbers are housed elsewhere. Yet these collections represent only a fraction of the biological diversity on earth, and for the most part even the specimens in them have been inadequately studied. The museums and similar institutions where they are housed are usually inadequately funded. NSF's program on Biology Research Resources, which contributes major funds, provided $5.5 million in 1987 (James Edwards, NSF, personal communication, 1988). When a collection is located in a university, its potential is often neglected. For society as a whole and for the advance of biological knowledge generally, such collections are literally priceless, and ways must be found of funding their needs adquately and making the information in them available. The diverse nature of systematics collections makes it difficult to determine the total funding needed to provide adequate support, but clearly they are currently underfunded.

Collections of living organisms, such as those housed in zoos, botanical gardens, aquaria, seed banks, and tissue culture centers, urgently require increased funding for their proper preservation. The particular strains of organisms used for specific biological experiments should be preserved for future studies; genetic material that has the potential of enhancing either scientific research or the economic potential of our industry likewise merits preservation. National funding for such collections is low (for example, only $1.5 million in 1987 from NSF's program on living organisms genetic stock centers), while the collections themselves are growing rapidly in size and importance. In addition, living organisms from international sources are becoming increasingly difficult to import as a result of strict federal regulations. With as many as a quarter of the earth's species at risk over the next several decades, the selective preservation of species demands a high priority: Our actions now will determine in many instances which kinds of organisms will still exist in the future. Existing centers should be supported, and new efforts should be initiated to gather adequate samples of critical taxa that are approaching extinction. Tissue culture centers, repositories of DNA clones, and other facilities are likewise growing, yet our national effort in preserving them is minimal and must be systematized soon for the common good of science and society.

A unified approach needs to be adopted in organizing and maintaining collections of preserved and living specimens and other biological materials. This will require increased funding and attention.

RESEARCH CENTERS AND THE INDIVIDUAL INVESTIGATOR

Research Centers Should Provide a Valuable Addition to Individual Investigator Research

The past several years have witnessed a move toward the establishment of government-funded, university-based research centers. NSF funds various centers at a total expense that reached $115 million in fiscal year 1987. These centers focus on a wide array of topics, including materials research, engineering, and biology. NSF also launched a presidential initiative on science and technology centers in 1987.[17]

Centers should be established in such a way as to enhance the efforts of individual investigators. In the United States, it has often been the individuality of particular scientists that has made advances possible; some fear exists that the creation of centers might stifle such individual efforts by consuming research funds once used by individual investigators and by institutionalizing an uncreative environment within the center itself. This need not be the case; the research environment in centers is largely a matter of the ways in which the goals of the center are expressed and pursued and the roles that individual investigators play in them. Also, NSF seems committed to keeping their support for centers at 10 percent of the foundation budget. Therefore, it appears that NSF-funded individual investigators will not experience new funding shortages as a result of the creation of NSF centers. It is hoped that centers will allow interactions between scientists representing many different specialties to create something greater than could have been achieved by any number of individuals working in isolation, even if funded generously.

> **Centers can provide a valuable approach to research, but the operation of a center should not interfere with the funding or creativity of the individual investigator.**

NOTES AND REFERENCES

1. Because of differences in methods of presentation among our sources of information, data are presented in this chapter for either the biological sciences or the life sciences. Generally, the life sciences consist of biological, agricultural, and health sciences.

2. National Science Board. 1985. Science Indicators: The 1985 Report. Washington, D.C.: NSF 85-1, National Science Foundation.

3. U.S. Patent and Trademark Office, The Office of Documentation. 1986. Technology Profile Report: Genetic Engineering. Washington, D.C.: U.S. Department of Commerce.

4. U.S. Patent and Trademark Office, The Office of Documentation. 1987. Technology Profile Report: Class 435—Chemistry: Molecular Biology and Microbiology, Washington D.C.: U.S. Department of Commerce.

5. National Science Board. 1987. Science and Engineering Indicators— 1987. NSB 87-1. Washington, D.C.: National Science Foundation.

6. Grant, W. V., and T. D. Snyder. 1986. Digest of Educational Statistics 1985–86. Washington, D.C.: U.S. Department of Education.

7. National Science Foundation. 1987. Federal Support to Universities, Colleges, and Selected Nonprofit Institutions: Fiscal Year 1986. NSF 87-318. Washington, D.C.: National Science Foundation.

8. National Research Council. 1989. Summary Report 1987: Doctorate Recipients from United States Universities. Washington, D.C.: National Academy Press.

9.   National Science Foundation. 1986. Women and Minorities in Science and Engineering. NSF 86-301. Washington, D.C.: National Science Foundation.

10.   National Research Council. 1976. Summary Report 1975: Doctorate Recipients from United States Universities. Washington, D.C.: National Academy Press.

11.   Syverson, P. O., and L. E. Forster. 1984. New Ph.D.s and the Academic Labor Market. Office of Scientific and Engineering Personnel Staff Paper No. 1. Washington, D.C.: National Research Council.

12.   Lozier, G. G., and M. J. Doons. 1987. Is higher education confronting a faculty shortage? Paper presented at the Annual Meeting of the Association for the Study of Higher Education, Baltimore, Md., November 21–24, 1987.

13.   National Science Foundation. 1987. Biotechnology Research and Development Activities in Industry: 1984 and 1985. NSF 87-311. Washington, D.C.: National Science Foundation.

14.   Burgdorf, K., and H. J. Hausman. 1985. Academic Research Equipment in Selected Science/Engineering Fields, 1982-83. Washington, D.C.: National Science Foundation.

15.   Hausman, J., and K. Burgdorf. 1985. Academic Research Equipment and Equipment Needs in the Biological and Medical Sciences. Bethesda, Md.: National Institutes of Health.

16.   National Science Foundation. 1987. Federal Funds for Research and Development: Federal Obligations for Research by Agency and Detailed Field of Science/Engineering: Fiscal Years 1967-1987. Washington, D.C.: National Science Foundation.

17.   Panel on Science and Technology Centers. 1987. Science and Technology Centers: Principles and Guidelines. Washington, D.C.: National Academy of Sciences.

Index

I